The Future of Post-Human Waste

The Future of Post-Human Waste

Towards a New Theory of Uselessness and Usefulness

By

Peter Baofu

CAMBRIDGE SCHOLARS PUBLISHING

The Future of Post-Human Waste: Towards a New Theory of Uselessness and Usefulness, by Peter Baofu

This book first published 2012

Cambridge Scholars Publishing

12 Back Chapman Street, Newcastle upon Tyne, NE6 2XX, UK

British Library Cataloguing in Publication Data
A catalogue record for this book is available from the British Library

ISBN (10): 1-4438-4190-0, ISBN (13): 978-1-4438-4190-0

To Those in the Future World beyond Uselessness and Usefulness

BOOKS ALSO BY PETER BAOFU

Category I: Social Sciences and Related Fields

- 53. *The Future of Post-Human Migration* (2012)
- 52. *The Future of Post-Human Criminality* (2012)
- 51. *The Future of Post-Human Geography* (2011)
- 50. *The Future of Post-Human Gambling* (2011)
- 49. *The Future of Post-Human Education* (2011)
- 48. *The Future of Post-Human War and Peace* (2010)
- 47. *The Future of Post-Human Law* (2010)
- 46. *The Future of Post-Human Organization* (2009)
- 45. *The Future of Post-Human Mass Media* (2009)
- 44. *The Rise of Authoritarian Liberal Democracy* (2007)
- 43. *Beyond the World of Titans, and the Remaking of World Order* (2007)
- 42. *Beyond Capitalism to Post-Capitalism* (2005)
- 41. Volume 1: *Beyond Democracy to Post-Democracy* (2004)
- 40. Volume 2: *Beyond Democracy to Post-Democracy* (2004)
- 39. *The Future of Capitalism and Democracy* (2002)

Category II: Formal Sciences and Related Fields

- 38. *The Future of Post-Human Semantics* (2012)
- 37. *The Future of Post-Human Data Analysis* (2011)
- 36. *The Future of Post-Human Computing* (2011)
- 35. *The Future of Post-Human Formal Science* (2010)
- 34. *The Future of Post-Human Language* (2009)
- 33. *The Future of Post-Human Geometry* (2009)
- 32. *The Future of Post-Human Mathematical Logic* (2008)
- 31. *The Future of Information Architecture* (2008)

Category III: Natural Sciences and Related Fields

- 30. *The Future of Post-Human Transportation* (2012)
- 29. *The Future of Post-Human Architecture* (2012)
- 28. *The Future of Post-Human Chemistry* (2011)
- 27. *The Future of Post-Human Acoustics* (2011)
- 26. *The Future of Post-Human Geology* (2010)
- 25. *Beyond Cosmology to Post-Cosmology* (2010)
- 24. *The Future of Post-Human Personality* (2010)
- 23. *The Future of Post-Human Sexuality* (2010)
- 22. *The Future of Post-Human Urban Planning* (2009)
- 21. *The Future of Post-Human Engineering* (2009)
- 20. *The Future of Post-Human Unconsciousness* (2008)
- 19. *The Future of Complexity* (2007)
- 18. *Beyond Nature and Nurture* (2006)
- 17. *The Future of Post-Human Space-Time* (2006)
- 16. *The Future of Post-Human Consciousness* (2004)

Category IV: The Humanities and Related Fields

- 14. *The Future of Post-Human Culinary Art* (2012)
- 14. *The Future of Post-Human History* (2012)
- 13. *The Future of Post-Human Performing Arts* (2012)
- 12. *The Future of Post-Human Literature* (2011)
- 11. *The Future of Post-Human Humor* (2011)
- 10. *Beyond Ethics to Post-Ethics* (2011)
- 9. *The Future of Post-Human Religion* (2010)
- 8. *The Future of Post-Human Chess* (2010)
- 7. *The Future of Post-Human Martial Arts* (2009)
- 6. *The Future of Post-Human Creative Thinking* (2009)
- 5. *The Future of Post-Human Knowledge* (2008)
- 4. *The Future of Aesthetic Experience* (2007)
- 3. *Beyond Civilization to Post-Civilization* (2006)
- 2. Volume 1: *The Future of Human Civilization* (2000)
- 1. Volume 2: *The Future of Human Civilization* (2000)

CONTENTS

Part One: Introduction

Part Two: Uselessness

Part Three: Usefulness

Part Four: Conclusion

TABLES

Category I. The Theoretical Debate on Waste

Category II: Visions on Nature

Category III: Visions on the Mind

Category IV: Visions on History

Category VII. Visions on Society (Socio-Political)

Category VIII. Visions on Society (Socio-Economic)

Category IX: Visions on Culture

ACKNOWLEDGMENTS

This book, like all other previous books of mine, is written to challenge conventional ideas or, more precisely, in the current context, the ones on waste—and to replace them with an original way of thinking about its future fate (in relation to uselessness and usefulness).

Not surprisingly, this book receives no external funding nor help from any formal organization or institution, because of its political incorrectness—as this is something that I often stressed (and repeated) in all my previous books.

My only reward is those beautiful moments of conceiving something original in the intellectual history of waste.

In any event, I bear the sole responsibility for the ideas presented in this book.

ABBREVIATIONS

ALD = Peter Baofu. 2007. *The Rise of Authoritarian Liberal Democracy: A Preface to a New Theory of Comparative Political Systems*. Cambridge, England: Cambridge Scholars Publishing, Ltd.

BCIV = Peter Baofu. 2006. *Beyond Civilization to Post-Civilization: Conceiving a Better Model of Life Settlement to Supersede Civilization*. NY: Peter Lang Publishing, Inc.

BCPC = Peter Baofu. 2005. *Beyond Capitalism to Post-Capitalism: Conceiving a Better Model of Wealth Acquisition to Supersede Capitalism*. NY: The Edwin Mellen Press.

BCOS = Peter Baofu. 2010. *Beyond Cosmology to Post-Cosmology: A Preface to a New Theory of Different Worlds*. Cambridge, England: Cambridge International Science Publishing, Ltd.

BDPD1 = Peter Baofu. 2004. Volume 1. *Beyond Democracy to Post-Democracy: Conceiving a Better Model of Governance to Supersede Democracy*. NY: The Edwin Mellen Press.

BDPD2 = Peter Baofu. 2004. Volume 2. *Beyond Democracy to Post-Democracy: Conceiving a Better Model of Governance to Supersede Democracy*. NY: The Edwin Mellen Press.

BEPE = Peter Baofu. 2011. *Beyond Ethics to Post-Ethics: A Preface to a New Theory of Morality and Immorality.* Charlotte, NC: Infomration Age Publishing.

BNN = Peter Baofu. 2006. *Beyond Nature and Nurture: Conceivng a Better Way to Understand Genes and Memes.* Cambridge, England: Cambridge Scholars Publishing, Ltd.

BWT = Peter Baofu. 2007. *Beyond the World of Titans, and the Renaking of World Order: A Preface to a New Logic of Empire-Building.* Cambridge, England: Cambridge Scholars Publishing, Ltd.

FAE = Peter Baofu. 2007. *The Future of Aesthetic Experience: Conceiving a Better Way to Understand Beauty, Ugliness and the Rest.* Cambridge, England: Cambridge Scholars Publishing, Ltd.

FC = Peter Baofu. 2007. *The Future of Complexity: Conceiving a Better Way to Understand Order and Chaos.* London, United Kingdom: World Scientific Publishing Co.

FCD = Peter Baofu. 2002. *The Future of Capitalism and Democracy.* MD: The University Press of America.

FHC1 = Peter Baofu. 2000. Volume 1. *The Future of Human Civilization.* NY: The Edwin Mellen Press.

FHC2 = Peter Baofu. 2000. Volume 2. *The Future of Human Civilization.* NY: The Edwin Mellen Press.

FIA = Peter Baofu. 2008. *The Future of Information Architecture: Conceiving a Better Way to Understand Taxonomy, Network, and Intelligence.* Oxford, England: Chandos Publishing (Oxford) Limited.

FPHA = Peter Baofu. 2011. *The Future of Post-Human Acoustics: A Preface to a New Theory of Sound and Silence.* Cambridge, England: Cambridge International Science Publishing, Ltd.

FPHARCH = Peter Baofu. 2012. *The Future of Post-Human Architecture: A Preface to a New Theory of Form and Function.* Cambridge, England: Cambridge International Science Publishing, Ltd.

FPHC = Peter Baofu. 2004. *The Future of Post-Human Consciousness.* NY: The Edwin Mellen Press.

FPHCA = Peter Baofu. 2012. *The Future of Post-Human Culinary Art: Towards a New Theory of Ingredients and Techniques.* England: Cambridge Scholars Publishing, Ltd.

FPHCHEM = Peter Baofu. 2011. *The Future of Post-Human Chemistry: A Preface to a New Theory of Substances and their Changes.* England: Cambridge Scholars Publishing, Ltd.

FPHCHESS = Peter Baofu. 2010. *The Future of Post-Human Chess: A Preface to a New Theory of Tactics and Strategy.* Cambridge, England: Cambridge International Science Publishing, Ltd.

FPHCOM = Peter Baofu. 2011. *The Future of Post-Human Computing: A Preface to a New Theory of Hardware and Software.* Cambridge, England: Cambridge International Science Publishing, Ltd.

FPHCRIM = Peter Baofu. 2012. *The Future of Post-Human Criminality: A Preface to a New Theory of Heroes and Villains.* Cambridge, England: Cambridge International Science Publishing, Ltd.

FPHCT = Peter Baofu. 2009. *The Future of Post-Human Creative Thinking: A Preface to a New Theory of Invention and Innovation.* Cambridge, England: Cambridge Scholars Publishing, Ltd.

FPHDA = Peter Baofu. 2011. *The Future of Post-Human Data Analysis: A Preface to a New Theory of Quantitative and Qualitative Research.* Cambridge, England: Cambridge International Science Publishing, Ltd.

FPHE = Peter Baofu. 2009. *The Future of Post-Human Engineering: A Preface to a New Theory of Technology.* Cambridge, England: Cambridge Scholars Publishing, Ltd.

FPHEDU = Peter Baofu. 2011. *The Future of Post-Human Education: A Preface to a New Theory of Teaching and Learning.* Cambridge, England: Cambridge International Science Publishing, Ltd.

FPHFS = Peter Baofu. 2010. *The Future of Post-Human Formal Science: A Preface to a New Theory of Abstraction and Application.* Cambridge, England: Cambridge Scholars Publishing, Ltd.

FPHG = Peter Baofu. 2009. *The Future of Post-Human Geometry: A Preface to a New Theory of Infinity, Symmetry, and Dimensionality.* Cambridge, England: Cambridge Scholars Publishing, Ltd.

FPHGAM = Peter Baofu. 2011. *The Future of Post-Human Gambling: A Preface to a New Theory of Risk and Caution.* Cambridge, England: Cambridge International Science Publishing, Ltd.

FPHGEOG = Peter Baofu. 2011. *The Future of Post-Human Geography: A Preface to a New Theory of Environments and their Interactions.* Cambridge, England: Cambridge International Science Publishing, Ltd.

FPHGEOL = Peter Baofu. 2010. *The Future of Post-Human Geology: A Preface to a New Theory of Statics and Dynamics.* Cambridge, England: Cambridge International Science Publishing, Ltd.

FPHH = Peter Baofu. 2011. *The Future of Post-Human Humor: A Preface to a New Theory of Joking and Laughing.* Cambridge, England: Cambridge International Science Publishing, Ltd.

FPHHIST = Peter Baofu. 2012. *The Future of Post-Human History: A Preface to a New Theory of Universality and Relativity.* Cambridge, England: Cambridge Scholars Publishing, Ltd.

FPHK = Peter Baofu. 2008. *The Future of Post-Human Knowledge: A Preface to a New Theory of Methodology and Ontology.* Oxford, England: Chandos Publishing (Oxford) Limited.

FPHL = Peter Baofu. 2009. *The Future of Post-Human Language: A Preface to a New Theory of Structure, Context, and Learning.* Cambridge, England: Cambridge Scholars Publishing, Ltd.

FPHLAW = Peter Baofu. 2010. *The Future of Post-Human Law: A Preface to a New Theory of Necessity, Contingency, and Justice.* Cambridge, England: Cambridge Scholars Publishing, Ltd.

FPHLIT = Peter Baofu. 2011. *The Future of Post-Human Literature: A Preface to a New Theory of Fiction and Non-Fiction.* Cambridge, England: Cambridge International Science Publishing, Ltd.

FPHMA = Peter Baofu. 2009. *The Future of Post-Human Martial Arts: A Preface to a New Theory of the Body and Spirit of Warriors.* Cambridge, England: Cambridge Scholars Publishing, Ltd.

FPHMIG = Peter Baofu. 2012. *The Future of Post-Human Migration: A Preface to a New Theory of Sameness, Otherness, and Identity.* Cambridge, England: Cambridge Scholars Publishing, Ltd.

FPHML = Peter Baofu. 2008. *The Future of Post-Human Mathematical Logic: A Preface to a New Theory of Rationality.* Cambridge, England: Cambridge Scholars Publishing, Ltd.

FPHMM = Peter Baofu. 2009. *The Future of Post-Human Mass Media: A Preface to a New Theory of Technology.* Cambridge, England: Cambridge Scholars Publishing, Ltd.

FPHO = Peter Baofu. 2009. *The Future of Post-Human Organization: A Preface to a New Theory of Communication, Decision-Making, and Leadership.*.Cambridge, England: Cambridge Scholars Publishing, Ltd.

FPHP = Peter Baofu. 2010. *The Future of Post-Human Personality: A Preface to a New Theory of Normality and Abnormality.* Cambridge, England: Cambridge International Science Publishing, Ltd.

FPHPA = Peter Baofu. 2012. *The Future of Post-Human Performing Arts: A Preface to a New Theory of the Body and its Presence.* Cambridge, England: Cambridge Scholars Publishing, Ltd.

FPHR = Peter Baofu. 2010. *The Future of Post-Human Religion: A Preface to a New Theory of Spirituality.* Cambridge, England: Cambridge International Science Publishing, Ltd.

FPHS = Peter Baofu. 2010. *The Future of Post-Human Sexuality: A Preface to a New Theory of the Body and Spirit of Love-Makers.* Cambridge, England: Cambridge Scholars Publishing, Ltd.

FPHSEM = Peter Baofu. 2012. *The Future of Post-Human Semantics: A Preface to a New Theory of Internality and Externality.* Cambridge, England: Cambridge Scholars Publishing, Ltd.

FPHST = Peter Baofu. 2006. *The Future of Post-Human Space-Time: Conceivng a Better Way to Understand Space and Time.* New York: Peter Lang Publishing, Inc.

FPHTRAN = Peter Baofu. 2012. *The Future of Post-Human Transportation: A Preface to a New Theory of Networks and Operations.* Cambridge, England: Cambridge Scholars Publishing, Ltd.

FPHU = Peter Baofu. 2008. *The Future of Post-Human Unconsciousness: A Preface to a New Theory of Anomalous Experience.* Cambridge, England: Cambridge Scholars Publishing, Ltd.

FPHUP = Peter Baofu. 2009. *The Future of Post-Human Urban Planning: A Preface to a New Theory of Density, Void, and Sustainability.* Cambridge, England: Cambridge Scholars Publishing, Ltd.

FPHW = Peter Baofu. 2012. *The Future of Post-Human Waste: Towards a New Theory of Uselessness and Usefulness.* Cambridge, England: Cambridge Scholars Publishing, Ltd.

FPHWP = Peter Baofu. 2010. *The Future of Post-Human War and Peace: A Preface to a New Theory of Aggression and Pacificity.* Cambridge, England: Cambridge Scholars Publishing, Ltd.

• Part One •

Introduction

CHAPTER 1
INTRODUCTION—THE EXISTENCE OF WASTE

> Read everything—trash, classics, good and bad, and see how they do it....If it is good, you'll find out. If it's not, throw it out the window.
> —William Faulkner (TE 2012)

A Story of Two Practices

Is waste (or trash) really so useless that, as William Faulkner once wrote, "[r]ead everything—trash, classics, good and bad, and see how they do it....If it is good, you'll find out. If it's not, throw it out the window"? (TE 2012)

Interestingly, this critical view of waste (or trash) can be contrasted with an opposing observation by Isaac Bashevis Singer, who once famously said that "the waste basket is the writer's best friend." (TE 2012a)

Contrary to these opposing views (and other ones as will be discussed in the book), waste (in relation to both uselessness and usefulness) is neither possible (or impossible) nor desirable (or undesirable) to the extent that the respective ideologues (on different sides) would like us to believe.

Of course, this challenge to the opposing views of waste does not imply that waste has no practical value, or that those interdisciplinary fields (related to waste) like epidemiology, global warming, waste management, low-carbon economics, ethical consumerism, resource recovery, freeganism, environmental justice, space debris, and so on are unimportant. Of course, neither of these extreme views is reasonable.

Rather, this book offers an alternative (better) way to understand the future of waste, especially in the dialectic context of uselessness and usefulness—while learning from different approaches in the literature but without favoring any one of them (nor integrating them, since they are not necessarily compatible with each other).

More specifically, this book offers a new theory (that is, *the transfigurative theory of waste*) to go beyond the existing approaches in a novel way.

If successful, this seminal project is to fundamentally change the way that we think about waste in relation to uselessness and usefulness from the combined perspectives of the mind, nature, society, and culture, with enormous implications for the human future and what I originally called its "post-human" fate.

Definition and History of Waste

A good point to start this inquiry is to provide a working definition of the term "waste" and its brief history.

Definition

The term "waste" is also known as "rubbish, trash, refuse, garbage, junk, and litter" and refers to "unwanted or useless materials." (WK 2012)

The European Union, "under the Waste Framework Directive (European Directive 75/442/EC as amended)," defined "waste" as "an object the holder discards, intends to discard or is required to discard." (WK 2012)

And according to the Basel Convention in the United Nations, the term "waste" refers to "materials that are not prime products (that is, products produced for the market) for which the generator has no further use in terms of his/her own purposes of production, transformation or consumption, and of which he/she wants to dispose." (WK 2012; OECD 2003)

However, the term "waste" is not completely the same as "litter," in that "waste is linked to people development," whereas "litter refers to waste disposed of improperly." (WK 2012)

History

With this clarification in mind—historically, say, in antiquity or before, "the amount of waste generated by humans was insignificant due to low population density, coupled with insignificant exploitation of natural resources. Common waste produced during early human history was mainly ashes and human biodegradable waste, and these were released back into the ground locally, with minimum environmental impact." (WK 2012a)

For instance, "the Maya of Central America had dumps, which exploded occasionally and burned. They also recycled. Homemakers brought trash to local dumps, and monthly burnings would occur....Consumption and waste of resources are probably related to supply available more than any other factor." (WK 2012a)

However, things started to change in modern times, especially "with the advent of industrial revolution," when "waste management became a critical issue. This was due to the increase in population and the massive migration of people to industrial towns and cities from rural areas during the 18th century. There was a consequent increase in industrial and domestic wastes posing threat to human health and the environment." (WK 2012a)

That said, "waste has played a tremendous role in history. The Bubonic Plague, cholera and typhoid fever, to mention a few, were diseases that altered the populations of Europe and influenced monarchies. They were perpetuated by filth that harbored rats, and contaminated water supply. It was not uncommon for Europeans to throw their waste and human wastes out of the window which would decompose in the street." (WK 2012a)

A good illustration is "France, specifically Paris," which "seems to have been a leader in poor waste management," because there was the story that, as told by John Saul (1992), "the famed Paris sewer system was created over a long period of time in the second half of the 19th century. The long delays were largely due to the virulent opposition of property owners, who did not want to pay to install sanitary piping to their buildings. The Prefect of Paris, Monsieur Poubelle, succeeded in forcing garbage cans on the property owners in 1887 only after a ferocious public battle. This government interference in the individual's right to throw his garbage in the street—which was, in reality, the property owner's right to leave his tenants no other option—made Poubelle into the 'cryptosocialist' of the hour. In 1900 owners were still fighting against the obligations to put their buildings on the public sewer system and to cooperate in the collection of garbage. By 1910 a little over half of the city's buildings were on the sewer system and only half of the cities in France had any sewers at all." (WK 2012a)

As Saul (1992) continued, "photos of early-twentieth-century Marseilles show great piles of refuse and excrement down the centre of the streets. Cholera outbreaks were common and ravaged the population. In 1954 the last city without, St. Remy de Provence, installed sewers....It was the gradual creation of an effective bureaucracy which brought an end to all this filth and disease, and the public servants did so against the desires of the mass of the middle and upper classes. The free market opposed sanitation. The rich opposed it. The civilized opposed it. Most of the educated opposed it. That is why it took a century to finish what could have been done in ten years." (WK 2012a)

Uselessness and Usefulness in Waste

With the definition and history of waste in mind—this book is to redefine the field in terms of the dialectic relationship between *uselessness* and *usefulness*, as explained below (and summarized in *Table 1.1*).

Uselessness

If the term "waste" refers to "unwanted or useless materials," then it is important to find out what uselessness is.

In the context of waste, "uselessness" can be understood in this book in regard to (a) waste prevention and (b) waste disposal, just to cite two examples, as explained below.

Waste Prevention

The first way to understand uselessness in the context of waste has to do with "waste prevention," which refers to "actions or choices that prevent the generation of waste" and "is used synonymously with the term source reduction"—or alternatively, "waste minimization," which is "the process and the policy of reducing the amount of waste produced by a person or a society." (WK 2012b; CAL 2010)

In the case of "waste minimization," the efforts here are "to minimize resource and energy use during manufacture. For the same commercial output, usually the fewer materials are used, the less waste is produced. Waste minimisation usually requires knowledge of the production process, cradle-to-grave analysis (the tracking of materials from their extraction to their return to earth) and detailed knowledge of the composition of the waste." (WK 2012b)

In general, "the main sources of waste vary from country to country. In the UK, most waste comes from the construction and demolition of buildings, followed by mining and quarrying, industry and commerce. Household waste constitutes a relatively small proportion of all waste. Reasons for the creation of waste sometimes include requirements in the supply chain. For example, a company handling a product may insist that it should be packaged using particular packing because it fits its packaging equipment." (WK 2012b; RCEP 2007)

But this practice (of waste prevention) is limited for two reasons, because, firstly, "waste reduction in one part of the production process may create waste production in another part" and, secondly, "in some cases, it may be best not to minimise the volume of raw materials used to make a product, but instead reduce the volume or toxicity of the waste created at the end of a product's life, or the environmental impact of the product's use." (WK 2012b)

Waste Disposal

And the second way to understand uselessness in the context of waste has to do with "waste disposal" (also known as " waste management"), which refers to "the collection, transport, processing or disposal, managing and monitoring of waste materials. The term usually relates to materials produced by human activity, and the process is generally undertaken to reduce their effect on health, the environment or aesthetics." (WK 2012c)

However, "waste management" should not be confused with a different practice known as "resource recovery which focuses on delaying the rate of consumption of natural resources. All wastes materials, whether they are solid, liquid, gaseous or radioactive fall within the remit of waste management." (WK 2012c)

In general, "waste management practices can differ for developed and developing nations, for urban and rural areas, and for residential and industrial producers. Management for non-hazardous residential and institutional waste in metropolitan areas is usually the responsibility of local government authorities, while management for non-hazardous commercial and industrial waste is usually the responsibility of the generator subject to local, national or international controls." (WK 2012c)

But this practice (of waste disposal) is limited, because it does not address the issues of waste reuse, waste recycling, and waste conversion, for instance.

Usefulness

Because of the limits of "waste prevention" and "waste disposal," another way to study waste has to do with its conversion to usefulness.

In the context of waste, "usefulness" can be understood in this book in regard to (a) waste reuse, (b) waste recycle, and (c) waste-to-energy, just to cite three examples, as explained below.

Waste Reuse

The first way to understand usefulness in the context of waste has to do with "waste reuse," which is "to use an item again" after having been used, so this means that "elements of the discarded item are used again. Initiatives include waste exchange, hand-me-downs, garage sales, quilting, travel mugs, and composting (nutrients)." (WK 2012d & 2012f)

For instance, in the "hand-me-downs" initiative, "a second-hand or used good is…purchased by or otherwise transferred to a second or later end user. A *used* good can also simply mean it is no longer in the same condition as it was when it was first transferred to the current end user. Used goods may be transferred informally between friends and family for free as 'hand-me-downs' or they may be sold for a fraction of their original price at garage sales or in church bazaar fundraisers." (WK 2012e)

But this practice (of waste reuse) is limited, because "second-hand goods may have faults which are not apparent even if examined; purchasing sight unseen, for example, from an Internet auction site, has further unknowns. Goods may cause problems beyond their value; for example, furniture may have not easily seen bedbugs which may cause an infestation which is difficult and expensive to eradicate. Faulty electrical and mechanical goods can be hazardous and dangerous. Goods sold second-hand may have been stolen. In this case in most jurisdictions the goods do not become the legal property of the purchaser; and knowing purchase of stolen goods is a criminal offence." (WK 2012e; MSNBC 2012)

Waste Recycle

The second way to understand usefulness in the context of waste has to do with "waste recycle," which means that "discards are separated into materials that may be incorporated into new products. This is different from Reuse in that energy is used to change the physical properties of the material. Initiatives include Composting, Beverage Container Deposits, and buying products with a high content of post-consumer material." (WK 2012d)

In general, there are two types of waste recycle, as shown below: (WK 2012d)

- "Upcycle"—"converting low-value materials in high-value products (more desirable)"
- "Downcycle"—"converting valuable products into low-value raw materials (less desirable)"

But this practice (of waste recycle) is limited, because "much of the difficulty inherent in recycling comes from the fact that most products are not designed with recycling in mind." (WK 2012g)

Waste-to-Energy

And the third way to understand usefulness in the context of waste has to do with "waste-to-energy (WtE)" or "energy-from-waste (EfW)," which is "the process of creating energy in the form of electricity or heat from the incineration of waste source. WtE is a form of energy recovery. Most WtE processes produce electricity directly through combustion, or produce a combustible fuel commodity, such as methane, methanol, ethanol or synthetic fuels." (WK 2012h)

The most common type of WtE uses the process of "incineration," which is "the combustion of organic material such as waste with energy recovery. All new WtE plants in OECD countries must meet strict emission standards, including those on nitrogen oxides (NO), sulphur dioxide (SO2), heavy metals and dioxins. Hence, modern incineration plants are vastly different from the old types, some of which neither recovered energy nor materials. Modern incinerators reduce the volume of the original waste by 95-96 percent, depending upon composition and degree of recovery of materials such as metals from the ash for recycling." (WK 2012h; EUR 2012; RAMB 2012)

But this practice (of waste-to-energy) is limited, because "concerns regarding the operation of incinerators include fine particulate, heavy metals, trace dioxin and acid gas emissions, even though these emissions are relatively low from modern incinerators. Other concerns include toxic fly ash and incinerator bottom ash (IBA) management. Discussions regarding waste resource ethics include the opinion that incinerators destroy valuable resources and the fear that they may reduce the incentives for recycling and waste minimization activities." (WK 2012h; DREDL 2009)

Diverse Aspects of Waste

This introduction to "uselessness" and "usefulness" in waste thus showss its more complicated nature.

To illustrate this more complicated nature of waste further, consider some diverse aspects of waste below (as summarized in *Table 1.1*).

Waste Types

The first diverse aspect of waste to be illustrated here in regard to its more complicated nature concerns "waste types."

Waste types can come in all shapes and sizes, and hereafter are some good examples: (WK 2012i)

- Ex: "solid wastes"
- Ex: "gaseous wastes"
- Ex: "animal by-products"
- Ex: "biodegradable waste"
- Ex: "biomedical waste"
- Ex: "business waste"
- Ex: "chemical waste"
- Ex: "clinical waste"
- Ex: "commercial waste"
- Ex: "construction and demolition waste (C&D waste)"
- Ex: "domestic waste"
- Ex: "space waste"
- Ex: "electronic waste (e-waste)"
- Ex: "food waste"
- Ex: "hazardous waste"
- Ex: "mineral waste"
- Ex: "industrial waste
- Ex: "metabolic waste"
- Ex: "packaging waste"
- Ex: "radioactive waste (nuclear waste
- Ex: "slaughterhouse waste"
- Ex: "winery wastewater"

Surely, there can be other examples, so the ones above are illustrative, not exhaustive.

Waste Management Topics

And the second diverse aspect of waste to be illustrated here in regard to its more complicated nature concerns "waste management topics."

Like waste types (above)—waste management topics can come in all shapes and sizes, and hereafter are some good examples: (WK 2012j)

- Ex: "anaerobic digestion"
- Ex: "biodrying"
- Ex: "carbon management"
- Ex: "dry cleaning"
- Ex: "low-carbon economy"
- Ex: "gasification"
- Ex: "industrial wastewater treatment"
- Ex: "solid waste treatment technologies"
- Ex: "mechanical biological treatment"
- Ex: "packaging and labelling"
- Ex: "recycling"
- Ex: "sewage collection and disposal"
- Ex: "windrow composting"

Once more, there can be other examples too, so the list above is not exhaustive but only illustrative.

The Theoretical Debate

These more complicated aspects of waste, together with the introduction to its "uselessness" and "usefulness," then lead us to a more important question concerning the theoretical debate on the very nature of waste itself . Indeed, this theoretical debate can also say much about the study of waste as a discipline over time.

Although there can be many different theoretical approaches in the debate—there are five main ones to be summarized hereafter, and they can be called, in the absence of better words, namely, (a) the *mental* argument, (b) the *natural* argument, (c) the *societal* argument, (d) the *cultural* argument, and (e) the *transfigurative* argument (which is my original contribution to the debate), to be elabroated hereafter, respectively.

Lest misunderstanding occurs, three clarifications should be made here, as this is something that I regularly emphasize in my previous books, whenever a theoretical debate is summarized for a topic, *almost verbatim*.

Firstly, there can be other arguments besides the first four here, that is, the "mental," the "natural," the "societal," and the "cultural" arguments (as indicated above). The advantage to select these four here has to do with their illustrative representation of the diverse theories in the literature (which are deemed sufficient for the purpose at hand in this book).

Secondly, the examples (theories) given to illustrate each argument are also supposed to be illustrative, not exhaustive. After all, the arguments will be addressed again in the rest of the book, with other examples (theories) to be introduced.

And thirdly, the four arguments are not necessarily mutually exclusive. For instance, those who advocate the "mental" argument can also consider the "cultural" argument, although they do not focus on the latter.

And the reverse is also true, in that those who make the "cultural" argument can also consider the "mental" argument, although they do not focus on the latter.

In other words, their disagreement is often one in degree, not in kind.

The Mental Argument

With these three clarifications in mind—the first major theoretical approach is called, in the absence of better words, the *mental* argument, which focuses, relatively speaking of course, more on the biological and psychological aspects of waste.

Thus, the word "mental" in the title of the argument is suggestive, because it refers to the focus on the physiological, biological and psychological aspects of waste so understood.

But the mental argument has different issues of concern, and different defenders of the argument offer their own distinctive stands on the issues, depending on their own interests.

For illustration, consider two major versions of the mental argument, namely, (a) biowaste and epidemics, and (b) EPR and the waste-recycling attitude—to be addressed below, in that order.

Biowaste and Epidemics

The first version of the mental argument has to do with biowaste and epidemics in the context of uselessness in the study of waste.

The term "biowaste" refers to "waste consisting of any biologically active materials containing relatively large molecular weight organic compounds. These may for example be plant residues; food wastes from households, catering establishments, and food processors and manufacturers; sewage and animal slurries; abattoir wastes, and animal and fish carcases and residues." (WK 2012k)

One major argument here is that biowaste can cause diseases, because, for example, "the Bubonic Plague, cholera and typhoid fever, to mention a few, were diseases that altered the populations of Europe and influenced monarchies. They were perpetuated by filth that harbored rats, and contaminated water supply. It was not uncommon for Europeans to throw their waste and human wastes out of the window which would decompose in the street." (WK 2012a)

However, for the critics, this study of biowaste and epidemics has some major problems. Consider, for illustration, a few examples below .

Firstly, one main criticism of the study of biowaste and epidemics is that epidemics can produce desirable effects instead.

For example, in the case of the Bubonic plague which caused "the Black Death that swept through Europe in the 14th century,...because the plague killed so many of the working population, wages rose and some historians have seen this as a turning point in European economic development." (WK 2012l; W. Bowsky 1971; A. Bridbury 1983)

And secondly, another main criticism of the study of biowaste and epidemics is that epidemics can be stopped or prevented.

For instance, for the critics, "several classes of antibiotics are effective in treating bubonic plague. These include aminoglycosides such as streptomycin and gentamicin, tetracyclines (especially doxycycline), and the fluoroquinolone ciprofloxacin. Mortality associated with treated cases of bubonic plague is about 1-15%, compared to a mortality rate of 50-90% in untreated cases." (WK 2012l; S. Dufel 2011)

EPR and the Waste-Recycling Attitude

And the second version of the mental argument has to do with EPR and the waste-recycling attitude in the context of usefulness in the study of waste.

The term "EPR" stands for "extended producer responsibility" which is a strategy designed to promote a waste- recycling attitude by integrating "environmental costs associated with goods throughout their life cycles into the market price of the products. The concept was first formally introduced in Sweden by Thomas Lindhqvist in a 1990 report to the Swedish Ministry of the Environment." (WK 2012m; OECD 2001)

In a subsequent report to the Ministry, the following definition was used: "[EPR] is an environmental protection strategy to reach an environmental objective of a decreased total environmental impact of a product, by making the manufacturer of the product responsible for the entire life-cycle of the product and especially for the take-back, recycling and final disposal of the product." (WK 2012m)

However, for the critics, this study of EPR and the waste-recycling attitude has some major problems. Consider, for illustration, a few examples below.

Firstly, one main criticism of the study of EPR and the waste-recycling attitude is that its implementation can be limited.

For instance, "some people have concerns about extended producer responsibility programs for complex electronics that can be difficult to safely recycle, such as Lithium-ion polymer batteries." (WK 2012m)

And secondly, another main criticism of the study of EPR and the waste- recycling attitude is that its application is not effective, even in relevant cases.

For instance, some critics pointed out that "such laws could increase the cost of electronics because producers would be adding recycling costs into the initial price tag." (WK 2012m)

The Natural Argument

The second theoretical approach is called, in the absence of better words, the *natural* argument, which focuses, relatively speaking of course, more on the natural aspects of waste, especially in relation to the state of nature.

Therefore, the word "natural" in the title of the argument is suggestive, because it refers to the focus on the natural aspects of waste so understood.

Like the mental argument—the natural argument has different issues of concern, and different defenders of the argument offer their own distinctive stands on the issues, depending on their own interests.

For illustration, consider two major versions of the natural argument, namely, (a) radioactive waste and treatment and (b) waste reuse and the environment—to be summarized below, in that order.

Radioactive Waste and Treatment

The first version of the natural argument has to do with radioactive waste in the context of uselessness in the study of waste.

The term "radioactive waste" refers to "wastes that contain radioactive material. Radioactive wastes are usually by-products of nuclear power generation and other applications of nuclear fission or nuclear technology, such as research and medicine." (WK 2012n)

Thus, one major argument here is that "radioactive waste is hazardous to most forms of life and the environment" and is to be "regulated by government agencies in order to protect human health and the environment." (WK 2012n)

However, for the critics, this study of radioactive waste and treatment has some major problems. Consider, for illustration, a few examples below.

Firstly, one main criticism of the study of radioactive waste and treatment is that it can be ineffective, more often than previously thought.

For instance, for the critics, problems often occur when "radioactive material was disposed of improperly, shielding during transport was defective, or when it was simply abandoned or even stolen from a waste store. In the Soviet Union, waste stored in Lake Karachay was blown over the area during a dust storm after the lake had partly dried out." (WK 2012n; A. González 1999; GLO 2012)

And secondly, another main criticism of the study of radioactive waste and treatment is that radioactive waste can be useful for some people instead.

For instance, for the critics, there is a lucrative "market for scavenged goods and scrap metal," and therefore, "scavenging of abandoned radioactive material" is not uncommon, even though these scavengers may run the risk of "radiation exposure, mostly in developing nations...The scavengers and those who buy the material are almost always unaware that the material is radioactive and it is selected for its aesthetics or scrap value." (WK 2012n; IAEA 1998)

Waste Reuse and the Environment

And the second version of the natural argument has to do with waste reuse and the environment in the context of usefulness in the study of waste.

The term "reuse" in the context of waste refers to the natural argument that, in order to ensure environmental sustainability, discarded items can be used again, and good examples include "waste exchange, hand-me-downs, garage sales, quilting, travel mugs, and composting (nutrients)." (WK 2012d & 2012f)

For instance, as indicated earlier, in the "hand-me-downs" initiative, "a second-hand or used good is...purchased by or otherwise transferred to a second or later end user. A *used* good can also simply mean it is no longer in the same condition as it was when it was first transferred to the current end user. Used goods may be transferred informally between friends and family for free as 'hand-me-downs' or they may be sold for a fraction of their original price at garage sales or in church bazaar fundraisers." (WK 2012e)

However, for the critics, this study of waste reuse and the environment has some major problems. Consider, for illustration, a few examples below.

Firstly, one main criticism of the study of waste reuse and the environment is that used goods are not of desirable quality.

For instance, for the critics, "second-hand goods may have faults which are not apparent even if examined; purchasing sight unseen, for example, from an Internet auction site, has further unknowns. Goods may cause problems beyond their value; for example, furniture may have not easily seen bedbugs which may cause an infestation which is difficult and expensive to eradicate. Faulty electrical and mechanical goods can be hazardous and dangerous." (WK 2012e; MSNBC 2012)

And secondly, another main criticism of the study of waste reuse and the environment is that it can have criminal consequences.

For instance, for the critics, "goods sold second-hand may have been stolen. In this case in most jurisdictions the goods do not become the legal property of the purchaser; and knowing purchase of stolen goods is a criminal offence." (WK 2012e; MSNBC 2012)

The Societal Argument

The third theoretical approach is called, in the absence of better words, the *societal* argument, which focuses, relatively speaking of course, more on the societal aspects of waste (e.g., organizational, institutional, structural, and technological issues).

In this sense, the word "societal" in the title of the argument is suggestive, because it refers to the societal aspects of waste so understood.

Like the mental and natural arguments—the societal argument has different issues of concern, and different defenders of the argument offer their own distinctive stands on the issues, depending on their own interests.

For illustration, consider two major versions of the societal argument, namely, (a) waste and environmental justice, and (b) waste recycling and technology—to be summarized in what follows, in that order.

Waste and Environmental Justice

The first version of the societal argument has to do with waste and environmental justice in the context of uselessness in the study of waste.

An interesting argument here involves "environmental justice," which refers to "the fair treatment and meaningful involvement of all people regardless of race, color, sex, national origin, or income with respect to the development, implementation and enforcement of environmental laws, regulations, and policies." (WK 2012o; T. Miller 2003; D. Schlosberg 2007)

Unfortunately, for the advocates of environmental justice, injustice is often the outcome, because "many of the environmental burdens...are more often borne by marginalized groups, such as racial minorities, women, and residents of developing nations." (WK 2012)

However, for the critics, this study of waste and environmental justice has some major problems. Consider, for illustration, a few examples below.

Firstly, one main criticism of this study of waste and environmental justice is that what constitutes fairness or its opposite is in the eyes of the beholder.

For instance, for the critics, as the old saying goes, "[w]hat looks fair to the goose seems foul to the gander." (P. Samuelson 2001: 332)

And secondly, another main criticism of this study of waste and environmental justice is that the environmental movement can be elitist in producing the opposite effects.

For instance, for the critics, the environmental movement has been plagued with "three different forms" of elitism, namely, (a) "compositional," in that "environmentalists are from the middle and upper class," (b) "ideological," in that "the reforms benefit the movement's supporters but impose costs on nonparticipants," and (c) "impactful," in that "the reforms have 'regressive social impacts.' They disproportionately benefit environmentalists and harm underrepresented populations." (WK 2012o; D. Morrison 1986)

Waste Recycling and Technology

And the second version of the societal argument has to do with waste recycling and technology in the context of usefulness in the study of waste.

A major argument here is that, with the help of technology, "discards are separated into materials that may be incorporated into new products. This is different from Reuse in that energy is used to change the physical properties of the material." (WK 2012d)

A good example is the invention of the "waste converter," which is "a machine used for the treatment and recycling of solid and liquid refuse material. A converter is a self-contained system capable of performing the following functions: pasteurization of organic waste; sterilization of pathogenic or biohazard waste; grinding and pulverization of refuse into unrecognizable output; trash compaction; dehydration." (WK 2012p)

However, for the critics, this study on waste recycling and technology has some major problems. Consider, for illustration, a few examples below.

Firstly, one main criticism of the study on waste recycling and technology is that its application is limited.

For instance, for the critics, "much of the difficulty inherent in recycling comes from the fact that most products are not designed with recycling in mind." (WK 2012g)

And secondly, another main criticism of the study on waste recycling and technology is that there are competing technologies.

For instance, for the critics, "waste converters do have indirect competition since there are many modes of waste disposal available in the market." (WK 2012p)

The Cultural Argument

And the fourth theoretical approach is called, in the absence of better words, the *cultural* argument, which focuses, relatively speaking of course, more on the cultural aspects of waste (e.g., issues in relation to values and beliefs).

Therefore, the word "cultural" in the title of the argument is suggestive, because it refers to the focus on the cultural aspects of waste so understood.

Like the mental, natural, and societal arguments—the cultural argument has different issues of concern, and different defenders of the argument offer their own distinctive stands on the issues, depending on their own interests.

For illustration, consider two major versions of the cultural argument, namely, (a) waste reduction and simple lifestyle, and (b) waste reclaiming and the ideology of freeganism—to be summarized in what follows, in that order.

Waste Reduction and Simple Lifestyle

The first version of the cultural argument has to do with waste reduction and simple lifestyle in the context of uselessness in the study of waste.

One major argument here is that a simple lifestyle is conducive to waste reduction and refers to "simple living" which "encompasses a number of different voluntary practices to simplify one's lifestyle. These may include reducing one's possessions or increasing self-sufficiency, for example. Simple living may be characterized by individuals being satisfied with what they need rather than want....[A]sceticism generally promotes living simply and refraining from luxury and indulgence....Simple living is distinct from those living in forced poverty, as it is a voluntary lifestyle choice." (WK 2012q; L. Pierce 2000; M. Griffiths 2010)

In this way, "simple living can...be a reaction to materialism and conspicuous consumption" (due to its excessive waste generation), although there are other possible reasons too. (WK 2012q)

However, for the critics, the study of waste reduction and simple lifestyle has some major problems. Consider, for illustration, a few examples below.

Firstly, one main criticism of the study of waste reduction and simple lifestyle is that simple living can be decadent.

For instance, for the critics like Friedrich Nietzsche, asceticism is essentially life-denying.

And secondly, another main criticism of the study of waste reduction and simple lifestyle is that simple living is not possible for everyone.

For instance, for the critics, those who are consumeristic in the contemporary world of capitalism find it impractical to live a life without material comfort and sensual pleasure.

Waste Reclaiming and the Ideology of Freeganism

And the second version of the cultural argument has to do with waste reclaiming and the ideology of freeganism in the context of usefulness in the study of waste.

A refreshing argument here is that waste can be reclaimed by using it in an unconventional way, and this ideology is known as "freeganism," which is "the practice of reclaiming and eating food that has been discarded." (WK 2012r)

A good example of freeganism in action is "dumpster diving," which is "the practice of sifting through commercial or residential trash to find items that have been discarded by their owners, but that may prove useful to the dumpster diver. Dumpster Diving is also viewed as an effective urban foraging technique. Dumpster divers will forage dumpsters for items such as clothing, furniture, food, and other items of the like deemed in good working conditions." (WK 2012r; M. Lewycka 2009; WST 2011)

However, for the critics, the study of waste reclaiming and the ideology of freeganism has some major problems. Consider, for illustration, a few examples below.

Firstly, one main criticism of the study of waste reclaiming and the ideology of freeganism is that dumpster divers are often looked down upon when foraging dumpsters on the street.

For instance, for the critics, dumpster divers often do not receive respect from others and are often viewed as trashy, lazy, and less intelligent people who cannot make it in society.

And secondly, another main criticism of the study of waste reclaiming and the ideology of freeganism is that reclaimed items in dumpsters are not of good quality.

For instance, for the critics, reclaimed items in dumpsters are often used, defective, hazardous, and inferior in quality.

The Transfigurative Argument

And fifthly, unlike the previous four arguments (viz., the mental, the natural, the societal, and the cultural)—the fifth one to be introduced here can be called, in the absence of better words again, the *transfigurative* argument, which learns from all other theoretical approaches (which include the previous four arguments and also other examples as will be introduced in the rest of this book), without, however, favoring any of them, nor trying to integrate them (as they are not necessarily compatible with each other, as the introduction of the theoretical debate in the previous sub-sections has shown).

The Transfigurative Theory of Waste

My *transfigurative* argument can be more precisely called *the transfigurative theory of waste*, since it is to propose a new theory of uselessness and usefulness in the study of waste.

My *transfigurative theory of waste* has four major distinct features.

Firstly, my theory makes good use of all theoretical approaches to the study of waste in relation to uselessness and usefulness, be they about the mental argument, the natural argument, the societal argument, or the cultural argument—especially from the most comprehensive combined perspectives of the mind, nature, society, and culture (as will be dealt with in Chapter Two and Chapter Three).

Secondly, just like many other theories of mine in my previous books, my theory here does not heavily favor any specific theory over others in the literature, nor trying to integrate them (as they are not necessarily compatible with each other).

Thirdly, my theory treats the issue of waste in the distinctive dialectic context of no uselessness without usefulness, and vice versa.

And fourthly, it contains twenty-nine major theses, namely, (a) the first thesis: the formalness-informalness principle, (b) the second thesis: the absoluteness-relativeness principle, (c) the third thesis: the partiality-totality principle, (d) the fourth thesis: the predictability-unpredictability principle, (e) the fifth thesis: the explicability-inexplicability principle, (f) the sixth thesis: the fiction-reality principle, (g) the seventh thesis: the finiteness-transfiniteness principle, (h) the eighth thesis: the preciseness-vagueness principle, (i) the ninth thesis: the simpleness-complicatedness principle, (j) the tenth thesis: the openness-hiddenness principle, (k) the eleventh thesis: the denseness-emptiness principle, (l) the twelfth thesis: the change-constancy principle, (m) the thirteenth thesis: the order-chaos principle, (n) the fourteenth thesis: the slowness-quickness principle, (o) the fifteenth thesis: the expansion-contraction principle, (p) the sixteenth thesis: the theory-praxis principle, (q) the seventeenth thesis: the convention-novelty principle, (r) the eighteenth thesis: the evolution-transformation principle, (s) the nineteenth thesis: the symmetry-asymmetry principle, (t) the twentieth thesis: the softness-hardness principle, (u) the twenty-first thesis: the seriousness-playfulness principle, (v) the twenty-second thesis: the activeness-inactiveness principle, (w) the twenty-third thesis: the selfness-otherness principle, (x) the twenty-fourth thesis: the regression-progression principle, (y) the twenty-fifth thesis: the sameness-difference principle, (z) the twenty-sixth thesis: the stability-reaction principle, (aa) the twenty-seventh thesis: the functionality-nonfunctionality principle, (bb) the twenty-eighth thesis: the intentionality-nonintentionality principle, and (cc) the twenty-ninth thesis: the post-human transfiguration—to be elaborated in the rest of the book, with a summary in the concluding chapter.

Because of the last thesis on "the post-human transfiguration," the word "transfigurative" in the title of my theory are suggestive, because it refers to the striking changes of waste over time—which then require a multifaceted exploration of the dialectic relations between uselessness and usefulness, as well as an understanding of the different views regarding these dialectic relations in the study of waste (together with other debates as will be discussed in the rest of the book)—and in the end it calls for a need to transcend them by future advanced humans and later post-humans in the context of the great transformations of waste in the future world (as will be discussed in the concluding chapter, Chapter Four).

With this clarification in mind—the seminal project here, if successful, will fundamentally change the way that we think about uselessness and usefulness in the study of waste together with other debates as will be discussed in the rest of the book), from the combined perspectives of the mind, nature, society, and culture, with enormous implications for the human future and what I originally called its "post-human" fate.

Theory and Meta-Theory

The *transfigurative* argument, or *the transfigurative theory of waste*, presupposes some kind of meta-theory, in special relation to methodology and ontology—just as all theories, either explicitly or implicitly, require their own versions of meta-theory.

Meta-Theory

As so often clarified in my previous books—theory and meta-theory, broadly speaking, can be further classified into different categories, as shown below:

- On Theory
  - —Ex: in relation to the mind
  - —Ex: in relation to nature
  - —Ex: in relation to society
  - —Ex: in relation to culture
- On Meta-Theory
  - —Ex: in relation to methodology
  - —Ex: in relation to ontology

Each category, in its turn, can then be further sub-divided into different levels of analysis—which will be elaborated in the next two sections (and summarized in *Table 4.43*).

With this reminder in mind, I already proposed (in my previous books) my distinctive contribution to meta-theory, namely, methodology (like "theory of methodological holism," "complex theory of data analysis," and "critical-dialectic theory of formal science," for instance) and ontology (like "theory of existential dialectics" and "contrastive theory of rationality," for instance).

The next two sections are to introduce my meta-theory—especially, existential dialectics and sophisticated methodological holism, respectively—to be repeated, *often verbatim* (with only a few updated revisions), from my previous works, as this is something that I often do in each new book of mine as an introduction to some background information for the convenience of the readers.

In the process, I shall also introduce my numerous new theories (or to be exact, 62 of them in 54 books so far, as of July 2012) in relation to the four categories of the mind, nature, society, and culture in the context of theory (as also summarized in *Table 4.43*).

A Unified Theory of Everything

In the end, my distinctive contribution to both meta-theory and theory serves as a foundation to unify all domains of knowledge (i.e., the social sciences, the formal sciences, the natural sciences, and the humanities) for an overarching understanding of all forms of reality (by way of some ontological principles at the meta-theoretical level and the comprehensive perspectives of the mind, nature, society, and culture at the theoretical level—as will be explained in the next two sections).

This is so, without commiting the sins of reductionism and reverse-reductionism (as often seen in many holistic approaches, with the fad of systems approach as a most recent notorious example, as already critically explained in *The Future of Post-Human Formal Science*, or *FPHFS*), since my distinctive contribution makes good use of different schools of thought without heavily favoring any of them nor trying to integrate them (as they are not necessarily compatible with each other), so as to include (or adjust for) heterogeneity, conflict, subjectivity, and complexity, for example.

In this sense, the word "unified" does not have to mean an integrative approach, in a narrow sense, nor a systems approach, in a broad one, both of which I have critically rebuked (in my previous books), while learning from them.

This overarching project (consisting of numerous theories of mine, as will be summarized in the next section), thus, constitutes my grand unifed theory of everything in the world and beyond.

The Logic of Existential Dialectics

To start, the ontology of existential dialectics can be summarized in this section—*often verbatim* (with only a few updated revisions), from my previous works, as this is something that I regularly do in each new book of mine as an introduction to some background information for the convenience of the readers.

After all, this book, like all others of mine, is in conversation with all previous ones of mine, for the final aim to converge all of them into a larger grand project on the future of intelligent life, both here on earth and there in deep space unto multiverses.

The summary can be organized in relation to (a) the conception of existential dialectics (or its ontics), (b) the syntax of existential dialectics (or its ontomethodology and ontologic), (c) the semantics of existential dialectics (or its ontosemantics), and (d) the pragmatics of existential dialectics (or its ontopragmatics), respectively hereafter—again *often verbatim*.

This means that, for those readers who had read some of my previous books, the summary below is really not much new (*almost verbatim*, with only some updated revisions).

But for those who have never read my previous books, it is a good review of them.

The Conception of Existential Dialectics (or Its Ontics)

This ontology was originally designated as "existential dialectics" in *Beyond Capitalism to Post-Capitalism* (herein abbreviated as *BCPC*), although it was already analyzed in other books of mine like the 2 volumes of *The Future of Human Civilization* (*FHC*), *The Future of Capitalism and Democracy* (*FCD*), *The Future of Post-Human Consciousness* (*FPHC*), and the 2 volumes of *Beyond Democracy to Post-Democracy* (*BDPD*).

Lest any confusion occurs, it is important to stress at the outset that the word 'existential' in "existential dialectics" has nothing to do with Existentialism, which I rebuked in *FHC*, *FCD*, and also *FPHC*. Instead, the word here merely refers to the existence of intelligent life (both primitive and advanced) in a broad sense.

Some Basic Concepts

The conception of existential dialectics (or its ontics) makes use of different concepts (like "sets," "elements," "relations," "operations," "functions," "truth values," "axioms," "postulates," and "principles"—as shown in *Table 4.3*), which are important for the understanding of any logic of ontology.

That clarified—existential dialectics, as a language of ontology, can be analyzed in different ways that a language, as an analogy, is often analyzed, namely, in relation to phonology ("the study of patterns of a language's sounds"), phonetics ("the study of the physical aspects of sounds of human language"), morphonology ("the study of the internal structure of words"), syntax ("the study of how words combine to form grammatical sentences"), semantics ("the study of the meaning of words [lexical semantics] and fixed word combinations [phraseology]"), and pragmatics ("the study of how utterances are used…literally, figuratively, or otherwise…in communicative acts"). (WK 2007)

But since my theory of existential dialectics makes use of the English language for communication here, it does not propose a new way to make the patterns of a language's sounds (as in phonology), to study the physical aspects of sounds (as in phonetics), or even to strive for a new organization of the internal structure of words for the English language (as in morphonology).

For this reason, the language of existential dialectics to be analyzed hereafter is not concerned with the phonological, phonetic, and morphonological aspects of the English language. Instead, the analysis will explore the syntax, semantics, and pragmatics of existential dialectics as a language of ontology to improve our understanding of reality.

The Ontic Dispute

But first, there is an ontic dispute to be addressed, which is related to what I called "the dilemma of ontology" as introduced earlier in *Sec. 1.4* of *The Future of Post-Human Knowledge* (*FPHK*).

A good point of departure concerns what I want to call *the ontic dispute*, on how to understand the very nature of ontology.

Traditionally, scholars in the field used to study the conception of ontology in terms of its entities (e.g., material objects, abstract numbers), properties (e.g., duration, plurality), and relationships (e.g., causation).

This is a good starting point, but it suffers from different problems. Consider three examples of bias below, which are quite common in the literature, and the recent version (of treating concepts and principles in ontology on the basis of "plurality," "dynamism," "duration," "interaction," "life," "consciousness," and "volition") by Reginald Firehammer (2005) only constitutes a latest addition to this old (impoverishing) tradition.

Firstly, the classification of entities, properties, and relationships is too material-centric, in often focusing more on the existence of material entities than on non-material ones.

Secondly, the classification is too anthropo-centric, in often favoring the factor of human agency in depicting reality.

And lastly, the classification is too confused among the three categories relating to each other, in often lumping them together without adequate consideration of their intricated interrelationships (both in relation to each and in relation to all).

A good alternative is precisely what I want to offer in this debate, that is, a new classification of ontological entities, properties, and their relationships in terms of *method*, *structure*, *process*, *agency*, and *outcome* instead. Let me explicate what these words mean below.

In relation to “method”—it concerns how ontology is to be studied. In this sense, the word 'method' here should not be confused with the different usage of the same term 'method' as a technique of investigation in the context of methodology (as already introduced in *Sec. 1.2* of *FPHK*).

In relation to “structure”—it concerns what the general patterns of entities, properties, and relationships are.

In relation to “process”—it concerns how these general patterns change over time.

In relation to “agency”—it concerns how agency affects any change of these general patterns over time (without putting, however, too much emphasis on the role of human agency).

And in relation to “outcome”—it concerns what the outputs of this change over time are.

The Syntax of Existential Dialectics (or Its Ontomethodology and Ontologic)

The syntax of existential dialectics, analogously speaking, refers here to the structure of an ontology which can be studied by way of combining ideas into complex relationships like ontological principles to depict reality in the world—and can be analyzed in seven sub-sections, namely, (a) the selection criteria, (b) further clarifications on classification, (c) ontomethodology, (d) ontologic, (e) the principles as short cuts, (f) the principles as family resemblances, and (g) the dialectic constraints imposed by the principles, respectively hereafter.

The Selection Criteria

To start, not any pair of relationship can be chosen as an ontological principle in existential dialectics; otherwise, there could be billions of them under the sky.

Six illustrations below suffice to clarify this point.

(a) Firstly, an ontological principle cannot *overlap* with other principles, in that it is too closely related with other ones.

Of course, one cannot totally rule out any relationship between two entities, especially in a complex world of everything being linked to everything else—but the comparison here is relative, not absolute.

In this sense, the flexibility-inflexibility pair cannot be an ontological principle, since it has something closely in common with other ones (like the simpleness-complicatedness principle, for instance).

As an illustration, something which is flexible in interpretation is more likely to allow multiple viewpoints and thus makes the overall picture more complicated than otherwise (simple).

(b) Secondly, an ontological principle cannot be *redundant* in relation to other principles, in that it somehow duplicates other principles.

In this sense, for instance, the directness-indirectness pair cannot be an ontological principle, because it somehow duplicates other ontological principles (like the preciseness-vagueness principle, for instance).

As an illustration, something which is direct means to be right to the point and does not need to go to an unnecessarily long loop—so it is less vague, in being clearer about the thing in question.

(c) Thirdly, an ontological principle cannot be *derived* from other ones—as if it is a child-parent relationship.

In this sense, the convenience-inconvenience pair cannot be an ontological principle, because it can be derived from other ones (like the slowness-quickness principle, for instance).

For instance, something which is convenient already implies that it can be obtained quickly (rather than slowly)—all things considered.

(d) Fourthly, an ontological principle cannot be *trivial* (or *parochial*)—as if it is only one component competing with other ones for inclusion in a set.

In this sense, the consistency-inconsistency pair cannot be an ontological principle, because its validity is limited (or parochial), as it can be easily challenged and replaced by opposing major (not minor) theories (e.g., the correspondence theory of truth, the pragmatic theory of truth, the semantic theory of truth, the disclosedness theory of truth, and whatnot), not just in accordance to the consistency (or coherence) theory of truth (for consistency).

Thus, the consistency argument is only one among different opposing arguments in relation to the logic of what constitutes "truth."

(e) Fifthly, an ontological principle cannot be *one-sided*, that is, part of one side (category) of a dialectic pair.

In this sense, the extroversion-introversion pair cannot be an ontological principle, since it is part of the side (category) of "normality" in the normality-abnormaltiy pair (because both extroversion and introversion are two different "normal" personality traits in psychology).

In addition, the normality-abnormality pair is already related to the symmetry-asymmetry principle, the density-emptiness principle, the convention-novelty principle, and the same-difference principle (as will be explained in the section on "the principles as family resemblances" later).

(f) And sixthly, an ontological principle cannot be *misleading*, in that the two apparent opposites are not really opposites at all, upon closer examination.

In this sense, the east-west pair cannot be an ontological principle, since east and west are not really opposites, as there are also north, south, and center to form a set of five opposing elements.

By the same logic, the left-right pair cannot be an ontological principle, since left and right (width) are not really opposites, as there are also top and bottom (height), back and front (depth), and center to form a set of seven elements in four categories. In the field of visual design, this set of seven elements is known as "form" (as opposed to "function"), as addressed in *FPHARCH*. (WK 2011nn)

Although these six examples (criteria) are not exhaustive, they illustrative the selectiveness of any ontological principle to be existentially dialectic.

Further Clarifications on Classification

Lest any misunderstanding carelessly occurs, some further clarifications on classification should be made here (and summarized in *Table 4.8*).

(a) Firstly, the total number of ontological principles is unknown, to be discovered later, as our knowledge of the world becomes more advanced.

Consequently, the principles as introduced in my books are not exhaustive, with new ones being added, whenever more of them are discovered in later research.

At least, future generations can pick up where I leave off and continue the discovery.

In this sense, my work should be treated as a pioneering effort for the development of a systematic, comprehensive analysis of a new general ontology for the future of knowledge.

(b) Secondly, the principles are not rigidly classified, as they can be reclassified in a different way.

For instance, the preciseness-vagueness principle is classified under the category of "structure" but can be reclassified under the category of "method," although in so doing, it has a different meaning in the context of method.

The same logic applies to the same-difference principle under the category of "outcome," which can be reclassified under the category of "structure," although in so doing, once more, it has a different meaning in the context of structure.

(c) Thirdly, the principles are to be understood together, not that each prinicple is to be treated independently of others. After all, the principles are mutually constraining, in that they work together as a whole.

(d) Fourthly, the principles are relevant to all subject matters, but some principles are, relatively speaking of course, more useful to some subject matters than others under certain conditions—as implied in the symmetry-asymmetry principle.

(e) And fifthly, the principles constitute only two levels of analysis, this time, at the ontological and methodological levels—while other levels of analysis (from the perspectives of the mind, nature, society, and culture) are also needed, in order to understand reality in its totality (as explained in the section on "sophisticated methodological holism," so as to avoid the dual dangers of reductionism and reverse-reductionism).

Ontomethodology

With these selection criteria and clarifications in mind—the syntax of existential dialectics can be divided into two main smaller sections for analysis, since they are related to each other in the context of some fundamental principles, that is, (a) ontomethodology and (b) ontologic, to be addressed hereafter, respectively.

In the context of ontomethodology, some ontological principles are proposed here in relation to the nature of methodology, which however is tied up with the general issue of ontology in the context of knowledge.

In this sense, ontomethodology (in the context of these ontological principles in the category of method) are related with ontologic in general (in the context of ontological principles in general).

Some good instances of ontological principles involving ontomethodology include the formalness-informalness principle, the absoluteness-relativeness principle, the partiality-totality principle, the predictability-unpredictability principle, the explicability-inexplicability principle, and the fiction-reality principle, which will be introduced in conjunction with ontological principles in the ontologic of existential dialectics in general hereafter.

Ontologic

With this clarification of the relationship between ontomethodology and ontologic in mind (which is comparable to the relationship between methodology and ontology as described in *Sec. 1.3* of *FPHK*)—the first three principles in existential dialectics were introduced in *BCPC*. And more principles were later added to the logic in my subsequent books (like *FC*, *FAE*, and others), as explained below.

More specifically, in *BCPC*, I proposed three principles (based on previous works of mine, not just something out of the blue) for the ontological logic, namely, (a) the regression-progression principle on the direction of history, (b) the symmetry-asymmetry principle on the relationships among existents, and (c) the change-constancy principle on the alteration of things—or in a more elegant term, the dynamics of space-time (in *FC*).

In *The Future of Complexity* (*FC*), I added three more principles, on top of the three aforementioned, namely, (d) the partiality-totality principle on the relationships between parts and whole, (e) the order-chaos principle on the pattern of things, and (f) the predictability-unpredictability principle on the occurrence of things—as part of the ontological logic of existential dialectics.

In *The Future of Aesthetic Experience* (or simply *FAE*), another principle is added, namely, (g) the evolution-transformation principle (on the multiple kinds of agency).

Also in *FAE*, I then reclassified the 7 principles into four categories, namely, (i) in relation to method, for the partiality-totality principle and the predictability-unpredictability principle, (ii) in relation to process, for the change-constancy principle and the order-chaos principle, (iii) in relation to agency, for the symmetry-asymmetry principle and the evolution-transformation principle, and (iv) in relation to outcome, for the regression-progression principle.

In *The Rise of Authoritarian Liberal Democracy* (*ALD*), I added two more principles, namely, (h) the softness-hardness principle on the force of agency (which is to be classified under the category about agency) and (i) the same-difference principle on the metamorphosis of change (which is to be classified under the category about outcome).

In *The Future of Information Architecture* (*FIA*), I introduce three additional principles, that is, (j) the simpleness-complicadness principle on the interconnection among things, (k) the preciseness-vagueness principle on the refinement of things, and (l) the slowness-quickness principle on the speed of change—with the first two in relation to structure and the third in relation to process.

In *The Future of Post-Human Unconsciousness* (*FPHU*), I added three more principles, that is, (m) the openness-hidden principle on the detection of things, in relation to structure, (n) the explicability-inexplicability principle on the underlying mechanisms of things, in relation to method, and (o) the expansion-contraction principle on the growth of things, in relation to process.

In *The Future of Post-Human Knowledge* (*FPHK*), I added one more principle, that is, (p) the absoluteness-relativeness principle on the multiplicity of entities, in relation to method.

In *The Future of Post-Human Mathematical Logic* (*FPHML*), I added one more principle, that is, (q) the formalness-informalness principle on the formal requirements of systems, in relation to method.

In *The Future of Post-Human Engineering* (*FPHE*), I added one more principle, that is, (r) the theory-praxis principle on the duality of knowledge, in relation to agency.

In *The Future of Post-Human Creative Thinking* (*FPHCT*), I added one more principle, that is, (s) the convention-novelty principle on the nature of creative thinking, in relation to agency.

In *The Future of Post-Human Geometry* (*FPHG*) here, I added one more principle, that is, (t) the finiteness-transfiniteness principle on the nature of numbers, in relation to structure.

In *The Future of Post-Human Urban Planning* (*FPHUP*), I added one more principle, that is, (u) the denseness-emptiness principle on the distribution of entities in space, in relation to structure.

In *The Future of Post-Human Humor* (*FPHH*), I added one more principle, that is, (v) the seriousness-playfulness principle on the extent of seriousness, in relation to agency.

In *The Future of Post-Human Literature* (*FPHLIT*), I added one more principle, that is, (w) the fiction-reality principle on the different faces of reality, in relation to method.

In *The Future of Post-Human Chemistry* (*FPHCHEM*), I added one more principle, that is, (x) the stability-reaction principle on the feedback mechanism for further changes, in relation to outcome.

In *The Future of Post-Human Transportation* (*FPHTRAN*), I added one more principle, that is, (y) the activeness-inactiveness principle on the extent of activeness, in relation to agency.

In *The Future of Post-Human Culiinary Art* (*FPHCA*), I added two more principles, that is, (z) the selfness-otherness principle on the magnitude of self-centeredness in relation to agency and (aa) the functionality-nonfunctionality principle on the presence of function in relation to outcome.

And in *The Future of Post-Human Waste* (*FPHW*), I add one more principle, that is, (bb) the intentionality-nonintentionality principle on the planning for outcomes in relation to outcome.

With this update, there are twenty-eight principles so far in existential dialectics, that is:

- six principles in the category of "method"—viz., the formalness-informalness principle, the absoluteness-relativeness principle, the partiality-totality principle, the predictability-unpredictability principle, the explicability-inexplicability principle, and the fiction-reality principle
- five principles in the category of "structure"—viz., the finiteness-transfiniteness principle, the simpleness-complicatedness principle, the preciseness-vagueness principle, the openness-hiddenness principle, and the denseness-emptiness principle
- four principles in the category of "process"—viz., the change-constancy principle, the slowness-quickness principle, the order-chaos principle, and the expansion-contraction principle
- eight principles in the category of "agency"—viz., the theory-praxis principle, the convention-novelty principle, the symmetry-asymmetry principle, the evolution-transformation principle, the softness-hardness principle, the seriousness-playfulness principle, the activeness-inactiveness principle, and the selfness-otherness principle
- five principles in the category of "outcome"—viz., the regression-progression principle, the same-difference principle, the stability-reaction principle, the functionality-nonfunctionality principle, and the intentionality-nonintentionality principle

There can be more than twenty-eight principles in existential dialectics, but they are what I have discovered so far (as summarized in *Table 4.4*).

The Principles as Short Cuts

Yet, these principles should be treated with caution, lest misunderstanding occurs, since they do not constitute rigid dualities (or dichotomies).

The reason is that each pair in an ontological principle consists of two opposites, which are, however, merely short cuts both for multiple variations and degrees, as well as for different interactions with multiple other entities unlike them.

In this light, each pair can end up having hundred (or even thousand, if not more) different versions, which interact with hundred (or even thousand, if not more) other entities—as summarized in *Table 4.5*.

There are two clarifications here.

Firstly, this conception of shortcuts is not mutually exclusive nor absolute, in that the opposites can come in all shapes and sizes, with different degrees. For instance, by analogy, just as there are different degrees of the two colors "white" and "black"—there are likewise different degrees of the opposites in each ontological pair, to the extent that there can be multiple entities (not only two) interacting with each other in each pair.

And secondly, this conception of shortcuts do not ignore other possible entities in interacting with the two opposites (with their different versions) in each pair. For instance, by analogy, there are not only the two colors "white" and "black" as opposites, since there are other colors too besides them like "yellow," "green," "purple," or else. The same logic can be applied to each ontological pair, in that they also interact with other entities, not with only two of them (with their different versions).

Therefore, with these two clarifications in mind—each pair in an ontological principle serves only as an abbreviation for something more complicated and, therefore, although it contains two opposites, it should not be confused as a duality (dualism).

Existential dialectics rejects any dualism (or dichotomy) as too rigid and instead allows the multiplicity of entities, to the extent that between the two opposites in each pair exist many other alternatives to choose from. They are named in that short form for aesthetic elegance only, instead of listing all possible entities between the two opposites in the title (which would be too cumbersome and sound extremely awkward).

To be dialectic is to go beyond any rigid dichotomy and transcend into something different altogether in the long haul. One may be tempted to call the dialectic logic here with a different name like existential "multilectics" (instead of "dialectics"), but this naming is incorrect (or even misleading), for the two reasons aforestated.

After all, all the possible entities between any two opposites in an ontological pair are still about the two opposites, although each ontological pair come in all shapes and sizes (with different degrees) and do not exist by themselves but also interact with other entities unlike them. The virtue of revealing an ontological pair is to show how they relate within themselves (in multiple versions) and also interact with others unlike them (also in multiple versions).

This will be clear in a later sub-section titled "the dialectic constraints imposed by the principles."

The Principles as Family Resemblances

In addition, each ontological principle is generic, with some other comparable ontological pairs to be put in the same family (like a *family resemblance*)—as summarized in *Table 4.6.*

Ludwig Wittgenstein (1953) once suggested the idea of "family resemblance" in explaining different games classified under the same family called *games.*

Why should, for instance, playing football and chess as playing "games," when it is well understood that football is not the same as chess? The answer is that, although each game is different and has different rules, many of them (though not all) share, more or less, some commonalites (e.g., scoring as necessary for winning).

And this is so, even though some games share more than some others in any given selection of critieria, and no two games are exactly identical. So, Wittgenstein's point here is that there is no essential core which is common to all games, and the best that one can look for is some characteristics which are common to many (but not all) games. (A. Biletzki 2006)

(a) Appropriate Family Resemblances

In the context of existential dialectics, I want to distinguish two types of family resemblances, and, in the absence of better words, let me call them (a) *appropriate* family resemblances and (b) *inappropriate* family resemblances.

In regard to appropriate family resemblances—good examples include some of the pairs as already explained in an earlier section on selection criteria, like the flexibility-inflexibility pair, which can be put in the family resemblance of the simpleness-complicatedness principle, although the two pairs are not exactly identical. Likewise, the directness-indirectness pair can be put in the family resemblance of the preciseness-vagueness principle, although, again, the two pairs are not exactly identical.

In this sense, which specific pair in a family should be used to designate the name of the family can be at times a bit arbitrary, but with good reason.

For illustration, in the context of method, the partiality-totality principle can take the different form like individualisticness-holisticness, just as the explicability-inexplicability principle can take the different form like underlyingness-regularness—although each two pairs are not exactly identical and have slightly different meanings and usages.

A more comprehensive listing of this family resemblance for all other principles is shown in *Table 4.6* on the syntax of existential dialectics.

In fact, this section on family resemblances overlaps with the semantics of existential dialectics, which will be introduced later in the next section (and summarized in *Table 4.11*).

(b) Inappropriate Family Resemblances

However, there are some family resemblances which are not appropriate, and let me distinugish them by calling them, in the absence of better words, (b1) *inadequate* family resemblances, (b2) *compound* family resemblances, (b3) *one-sided* family resemblances, and (b4) *vague* family resemblances (as summarized in *Table 4.10*).

(b1) In regard to "inadequate" family resemblances—there are some pairs which are not adequate, because they fail the test of selection criteria (as already pointed out in an earlier section), especially in relation to the criterion of not being *trivial* (or *parochial*)—as if it is only one component competing with other ones for inclusion in a set.

For instance, the *consistency-inconsistency* pair is inadequate, because it is only one component competing with other ones for inclusion in the larger set of the formalness-informalenss principle (like the *completeness-incompleteness* pair and the *soundness-unsoundness* pair, as discussed in my book titled *The Future of Post-Human Mathematical Logic*, or *FPHML* in short).

The same is true for the *cognition-noncognition* pair, which is related with both the completeness-incompleteness pair and the soundness-unsoundness pair in the larger set of the formalness-informalenss principle.

(b2) In regard to "compound" (or non-atomic) family resemblances—there are some pairs which are not appropriate for a different reason, because they fail the test of selection criteria (as already pointed out in an earlier section), especially in relation to the criterion of not *overlapping* with other principles, in that it is too closely related with other ones.

For instance, the *normality-abnormality* pair is compound (and thus inappropriate), because it overlaps with several other principles, like the symmetry-asymmetry principle, the density-emptiness principle, the convention-novelty principle, and the same-difference principle, for instance (as already discussed in *FPHP*).

The normality-abnormality pair overlaps with the symmetry-asymmetry principle, because being "abnormal" (say, for those with mental illnesses) is often treated not equally by others who are "normal." So, the relationship reveals the symmetry-asymmetry principle.

The normality-abnormality pair also overlaps with the density-emptiness principle, because being "abnormal" (say, for those with mental illnesses) is often not as dense in the overall (not local) distribution of a population as those who are "normal." So, the relationship reveals the density-emptiness principle.

The normality-abnormality pair also overlaps with the convention-novelty principle, because being "abnormal" (say, for those with mental illnesses) often exhibits non-conventional (non-conformist) ideas and behaviors when contrasted with those who are "normal." So, the relationship reveals the convention-novelty principle.

The normality-abnormality pair also overlaps with the same-difference principle, because being "abnormal" (say, for those with mental illnesses) shares some similarities in outcome when compared with those who are "normal" (in that both contribute to the control and regulation of human thoughts and behaviors in society, albeit in different ways). Yet, being "abnormal" also has something different from being "normal," in that abnormal thoughts and behaviors follow a different pattern when contrasted with those which are normal. So, the relationship reveals the same-difference principle.

(b2.1) There is a special case of compound family resemblances (as summarized in *Table 4.10*), which can be called, in the absence of better words, *one-sided and compound* family resemblances.

For instance, the extroversion-introversion pair is a good example of *one-sided and compound* family resemblances, because, firstly, it is "one-sided," in being part of the side (category) of "normality" in the normality-abnormality pair—and, secondly, the normality-abnormality pair is in turn "compounded," because it is related to the symmetry-asymmetry principle, the density-emptiness principle, the convention-novelty principle, and the same-difference principle, as already explained above.

(b3) In regard to "one-sided" family resemblances—there are some pairs which are not appropriate for another different reason, because they fail the test of selection criteria (as already pointed out in an earlier section), especially in relation to the criterion of not being *one-sided*, that is, part of one side (category) of a dialectic pair.

For instance, the extroversion-introversion pair cannot be an ontological principle, since it is part of the side (category) of "normality" in the normality-abnormaltiy pair (because both extroversion and introversion are two different "normal" personality traits in psychology).

By the same logic, the vision-blindness pair and the sound-silence pair cannot be ontological principles, because the pairs are part of the side (category) of "asymmetry" in the symmetry-asymmetry principle (since being able to see and hear for some individuals and being unable to see or hear for others are two contrastive abilities in the world of animals and humans, for instance).

These illustrations of indequate, compound, and one-sided family resemblances show how difficult it is to come up with an ontological principle in existential dialectics out of the blue, because many of them violate the selection criteria in one way or the other. It is all too easy to join two opposing words and simply make them into a pair, but it is all too difficult to pass the test of the selection criteria to be qualified as an ontological principle in existential dialectics.

(b4) And in regard to "vague" family resemblances—there are some pairs which are not appropriate for still another different reason, because the terms in the pairs are vague in meaning, so that they are inappropriate family resemblances if defined in some particular ways, although they can be appropriate family resemblances if defined in other particular ways.

For instance, such pairs like shortness-longness, narrowness-broadness, smallness-largeness, smallness-greatness, microness-macroness, and littleness-muchness can be understood in at least three different meanings and thus have three different possibilities. The first possibility is that they are one-sided as part of finiteness (in the finiteness-transfiniteness principle), if both opposites are defined on a finite scale. The second possibility is that they are one-sided as part of transfiniteness (in the finiteness-transfiniteness principle) if both opposites are defined on a transfinite scale. And the third possibility is that they are appropriate family resemblances in relation to the finiteness-transfiniteness principle if one opposite is defined on a finite scale and the other opposite is defined on a transfinite scale.

Other examples of inappropriate family resemblances are shown in *Table 4.10* on "The Syntax of Existential Dialectics VII: Types of Inappropriate Family Resemblances."

The Dialectic Constraints Imposed by the Principles

The principles, as they constitute the syntax of existential dialectics, are dialectic in character, such that, when they are applied, they impose dialectic constraints on how reality is to be understood.

Consider, say, the symmetry-asymmetry principle as an illustration here, in order to summarize two main characters of the dialectic constraints in question (as summarized in *Table 4.7*).

Firstly, to be dialectic here is to go beyond the narrow dichotomies (and, for that matter, any rigid multi-dimensional classificatory scheme), be they about "self" vs. "world," "freedom" vs. "unfreedom," "barbarity" vs. "civilization," "individuality" vs. "communality," and so on.

One way to do so (to go beyond) is to consider them all in terms of co-existence (without heavily favoring one over the rest). For instance, my theory of "post-civilization" (to be summarized later in the section on the pragmatics of existential dialectics) is to go beyond barbarity and civilization in terms of understanding barbarity and civilization as being co-existent.

And the same logic can be said in relation to my theories of "post-democracy," "post-capitalsm," and others (also to be introduced later in the section on the pragmatics of existential dialectics), in regard to freedom vs. unfreedom, equality vs. inequality, communality vs. individuality, spirituality vs. materiality, and so on.

But to consider them all (in the dichotomies—and, for that matter, in any rigid multi-dimensional classifactory scheme) as co-existent is not the same as to imply that the opposing categories in any classificatory scheme are all equal, since, in accordance to the symmetry-asymmetry principle (as an illustration here), if they are equal in terms of being considered as co-existent, they are asymmetric in terms of being unequal in dialectic interaction (e.g., X can be more dominant than Y in case A, or Y is more dominant than X in case B), although in some special cases, they can be relatively equal (e.g., X and Y are relatively equal in case C).

For this reason, there are different versions of "post-democracy" and "post-capitalism" in my theories. As an illustration, in version I of the theory of post-democracy, freedom is more dominant than equality, whereas in version II of the theory of post-democracy, equality is more so than freedom.

But this "X more than Y" has to be understood in the context of dialectic logic (not in symbolic logic, as conventionally understood), in that both "X" and "Y" are important in post-democracy (in the context of dialectic logic), but in an asymmetry way.

By contrast, in symbolic logic, it often favors one over the other—be it in regard to privileging freedom over equality (in Fascism), favoring freedom relatively more than equality (in Liberal Democracy), or favoring equality relatively more than freedom (in Socialist Democracy). In regard to the latter two cases (about Liberal Democracy and Socialist Democracy), the difference between dialectic logic and symbolic logic can be one in degree, not in kind—in this sense, albeit not in other senses.

The same reasoning can be said about the relatonships between individuality and communality, between spirituality and materiality, and between formal legalism and informal legalism in the different versions of my theory of "post-capitalism."

Secondly, to be dialectic is to go beyond the narrow dichotomies (and, for that matter, any rigid multi-dimensional classificatory scheme) in another way, this time, in a transcendent way, that is, in exploring other possibilities or even other issues not considered within the narrow confines of narrow dichotomies (and, for that matter, any rigid multi-dimensional classificatory scheme).

As an analogy, in this second meaning, to go beyond the narrow dichotomy of "black" vs. "white" is not just to choose both "black" and "white" (as in the first meaning) but also to explore other options (e.g., "green," "purple," "blue," etc.—and, alternatively, "shade," "line," "curve," etc.).

By the same reasoning, to go beyond "democracy" is to transcend democracy (as in version III of my theory of "post-democracy") and to explore other possibilties of lifeforms (e.g., "floating consciousness," "hyper-spatial consciousness," etc., to live beyond the narrow obsession with freedom and equality).

This dialectic character of the principles in existential dialectics has important implications for the pragmatics of existential dialectics (as will be clear shortly, in the section on the pragmatics of existential dialectics).

The Semantics of Existential Dialectics (or Its Ontosemantics)

The syntax of existential dialectics so understood in terms of ontological principles only gives us the structure of ontology in the world, in an abstract (general) sense. These principles by themselves do not tell us the specific meanings in a given context.

In order to grasp the specific meanings of the principles in a given context, it is necessary to study the semantics of existential dialectics (as summarized in *Table 4.11*).

And the analysis of the ontological principles as family resemblances in the previous section on syntax is only a starting point (and thus overlaps a bit with the semantics of existential dialectics here).

For this reason, I have often gone to great lengths in my previous books on different subjects to explain the specific meanings of the principles when applied in different contexts.

Consider two examples below for illustration, which are not exhaustive.

In *FPHST*, for instance, I used the first three principles (i.e., the change-constancy principle, the regression-progression principle, and the symmetry-asymmetry principle) to propose "the perspectival theory of space-time," for a better way to understand space and time—especially, though not exclusively, in relation to future post-human history (as summarized in *Table 1.3*, *Table 1.4*, *Table 1.5*, and *Table 1.6*).

In so doing, I had to introduce concepts and theories specific to the field of physics and other related fields (e.g., "absolute space" and "absolute time" in "classical mechanics" and "relative space-time" in "the theory of relativity").

And in *BNN*, I also exploited the three principles to propose the "transcendent" approach to the study of genes and memes as a new way to understand the interaction between nature and nurture.

In so doing, I had to explore concepts and theories in the world of evolutionary biology (e.g., "mutation," "variation," "adaptation," "selection," and "inheritance" in classical evolutionary theory) and neural biology (e.g., "chromosome," "gene," "DNA," "RNA," "protein," "neuron," "neural network," and "behavior").

Other examples, besides these two here, are given in other books of mine on different topics.

Therefore, this semantic dimension of existential dialectics is important, since, in each of my books when the ontological principles are used, each principle becomes relevant in a different meaning when contextualized within a given case study in question.

In this book, for instance, the principles are relevant in relation to a different subject matter (as summarized in the concluding chapter), and they can be shown in all shapes and sizes (since they are also to be treated as "short cuts," as indicated before), without, however, being reduced to "specific ontology" (as opposed to "general ontology"), as explained in *Table 4.9*.

In the technical jargons of semantics, these exercises perform two essential functions of translation (or interpretation), namely, (a) "intensive" interpretation (about the properties or relations of the principles in existential dialectics), and (b) "extensive" interpretation (about the classes or cases to which the properties and relations of the principles in existential dialectics are applied). (WK 2011oo)

But there are limits of translation (or interpretation), and an excellent example is that "there is a problem with the 'objectivity' of translation, as it is relative to different cultural (or sub-cultural) contexts. For instance, one of the still unresolved problems in linguistics is: 'Is there an objective gauge for the quality of translation?'" (*FPHCOM* & WK 2011pp) And the problems of "radical translation" by W. V. Quine and "radical interpretation" by Donald Davidson are another well-known examples, as already discussed in *FPHSEM*.

This is one serious problem about the logic of translation and interpretation that I already addressed in *Sec. 2.5.2* of *FPHSEM*, *Sec. 3.5.1* of *FPHCOM* (and also, for that matter, in *FHC*, in the context of hermeneutics).

The Pragmatics of Existential Dialectics (or Its Ontopragmatics)

Even the semantics of existential dialectics is not enough to understand reality, since semantics by itself does not tell us the nature of intentionality and interests in relation to the matrix of power formations and conflicting interests when existential dialectics as an ontology is applied.

Surely, one cannot completely separate the pragmatics of existential dialectics from its semantics (and for that matter, its syntax), but the point here is to give each a more elaborated analysis.

For this reason, the pragmatics of existential dialectics needs to be studied too, in special relation to the inclusion of power formation and conflicting interests in the application of the principles of existential dialectics (as summarized in *Table 4.12*).

It is thus no wonder that all my previous books have shown in different ways how and why these ontological principles and their theoretical applications can reveal the future world to come in a way that humans have never known, in a totally different battleground for competing human (and later post-human) interests by myriad groups fighting for their own versions of hegemony.

Perhaps a summary of my previous works in relation to existential dialectics and their contributions to my new theories as proposed over the years is deemed revealing here, for the further understanding of the pragmatics of existential dialectics (in close relation to the semantics of existential dialectics too).

That stressed—the following summary in four sub-sections is something that I often do in my previous books too, as a way to introduce the intellectual background of my ideas.

So, for those readers who had read my previous works, the following sub-sections serve as a reminder, as they are repeated hereafter *often verbatim* (with only some updated changes, with the rest kept intact).

These sub-sections are, namely, (a) the two-way street connecting theory with meta-theory, (b) the theoretical application of existential dialectics, (c) direct and indirect applications of existential dialectics, and (d) multiple applications of existential dialectics—to be summaried herefafter, respectively.

The Two-Way Street Connecting Theory and Meta-Theory

However, lest any misunderstanding occurs, it is important to remember that the pragmatics of existential dialectics is not a one-way street (that is, using the ontological principles for theoretical insights in praxis) but a two-way one, namely, (a) from meta-theory to theory, and (b) from theory to meta-theory.

(a) On one side of the street, the ontological principles can inspire some theoretical insights in praxis, that is, in relation to some specific fields.

(b) On the other (opposing) side of the street, however, the study of a subject matter in the specific fields in turn reveals some more hitherto unknown ontological principles to be discovered and identified.

For this reason, three new principles were added in *FC*, one in *FAE*, two in *ALD*, and three in *FIA*, on top of the original three in *BCPC*, for example (as there are more)—after some research on the specific subject matters.

Both theory and meta-theory enrich each other in all my works over the years.

The Theoretical Application of Existential Dialectics

This two-way street connecting theory and meta-theory has contributed to my original construction of many theories in numerous books of mine on numerous fields in all domains of knowledge, which can be classified in terms of specific "meta-theories" (about epistemology, methodology, logic, ontology, etc.) and specific "theories" (about the mind, nature, culture, society, and the rest—as will be further explained later in the sub-section on "the holistic organization of an inquiry" in the section about "sophisticated methodological holism").

The final aim of my numerous new specific theories (on both theory and meta-theory) is to converge all of them (as proposed in my numerous books) into a grand project about the future of intelligent life, both here on earth and in deep space.

In this sense, all these theories of mine serve as a foundation to unify all domains of knowledge for an overarching understanding of all forms of reality (by way of some ontological principles and the comprehensive perspectives of the mind, nature, society, and culture), while allowing for heterogeneity, conflict, subjectivity, and complexity, for example.

With this grand project in mind—the summary of my specific theories on meta-theory are given in this section (on "existential dialectics") and the next (on "sophisticated methodological holism"), whereas the specific theories on theory will be summarized in the following categories of (a) society, (b) culture, (c) the mind, (d) nature, and (e) the rest, as shown below, in that order.

(a) Society

The category of "society" here can be divided into four sub-categories, i.e., (a1) social organizations, (a2) social institutions, (a3) social structure, and (a4) social systems (or trends).

(a1) Social Organizations

In relation to social organizations, I proposed a new thesis on the ambivalent nature of virtual organizations in the future, known as "the theory of e-civic alienation" in *FCD* (and summarized in *Table 4.19*).

And in *FPHO*, another theory called "the combinational theory of organization" was worked out, although it was first proposed in Ch.6 of *ALD*.

(a2) Social Institutions

There are different social institutions, be they about (a2.1) politics, (a2.2) economics, (a2.3) military affairs, (a2.4) mass media, (a2.5) law, (a2.6) education, and (a.7) gambling, for instance.

(a2.1) In relation to politics, I offered different visions on (a2.1.1) political theory, (a2.1.2) comparative politics, and (a2.1.3) international relations, to cite three main ones.

(a2.1.1) In the context of political theory, my vision of a path-breaking political system in future times concerns the different forms of "post-democracy" to supersede democracy unto the post-human age.

My vision of post-democracy was called "the theory of post-democracy" (as summarized in *Table 4.22*, *Table 4.23*, and *Table 4.24*), whereas *Table 4.25* explains the distinctions among democracy, non-democracy, and post-democracy.

And *Table 4.26* stresses the multiple causes of the emergence of post-democracy, while *Table 4.27* clarifies some possible misunderstandings in regard to post-capitalism and post-democracy.

(a2.1.2) In the context of comparative politics, I also showed, in *ALD*, how and why authoritarianism can be an advanced form of liberal democracy (just as it has its own counterpart in non-liberal democracy), or what I originally called "authoritarian liberal democracy" (as summarized in *Table 4.21*), before it is to be superceded by "post-democracy."

(a2.1.3) In the context of international relations (both at the institutional and structural levels), I provided, in *BWT*, "the theory of cyclical progression of empire-building," for a better way to understand the logic of empire-building on earth and beyond—with the rise of what I originally envisioned as "hyper-empires" (like "the Chinese Union" and "the Indian Union"), "meso-empires" (like "the European Union" and "the North American Union") and "micro-empires" (like "the Middle Eastern Union" and "the Latin American Union") in the coming "post-post-Cold War era" as a visionary illustration, as summarized in *Table 4.20*.

This is also related to the issues of "social structure" in (a3), "migration" in (a4,6), "identity" in (b11), and "geography" in (d4), for instance.

In addition, in *FPHWP*, I proposed "the multifaceted theory of war and peace" to understand war and peace, in a small sense—or aggression and pacificity, in a broad sense.

(a2.2) In relation to economics, I also offered another vision, this time to understand the future forms of economic systems, that is, the different forms of "post-capitalism" to supersede capitalism unto the post-human age.

My vision of post-capitalism was referred to (in *FCD*, *BDPD* and *BCPC*) as "the theory of post-capitalism" (as summarized in *Table 4.28*, *Table 4.29*, *Table 4.30*, *Table 4.31*, and *Table 4.32*). And *Table 4.33* clarifies the differences among capitalism, non-capitalism, and post-capitalism, whereas *Table 4.34* shows multiple causes of the emergence of post-capitalism.

This vision on "post-capitalism" constitutes the third theoretical application of existential dialectics, this time, in that there is no wealth without poverty (or the wealth/poverty dialectics).

(a2.3) In relation to military affairs, I also proposed "the multifaceted theory of war and peace" (in *FCD*, although it was first so labeled in *FPHK*) for a new way to understand the causes of war—and, for that matter, the rise and fall of different forms of warfare over time.

This theory was further expanded in *FPHWP*, where I suggested a new way to understand war and peace, in a small sense—or aggression and pacificity, in a broad sense.

(a2.4) In relation to mass media, I also proposed "the totalistic theory of communication" in *The Future of Post-Human Mass Media* (or in short, *FPHMM*).

(a2.5) In relation to law, I also proposed "the reconstruction theory of law" in *The Future of Post-Human Law* (or in short, *FPHLAW*). And in relation to criminality, I proposed "the reflective theory of criminality" in *FPHCRIM* for a different way to understand heroes and villains.

(a2.6) In relation to education, I proposed "the heterodox theory of education" in *The Future of Post-Human Education* (or in short, *FPHEDU*).

(a2.7) And in relation to gambling, I proposed "the detached theory of gambling" in *The Future of Post-Human Gambling* (or in short, *FPHGAM*).

(a3) Social Structure

In relation to social structure, I first proposed two additional theories, that is, "the theory of the cyclical progression of hegemony" in *FCD* (and also in *FPHC* and *BDPD*), and "the theory of the cyclical progression of empire-building" (in *BWT*, as already indicated above)—with important implications for the symmetry-asymmetry princple in existential dialectics, just to cite one example.

In addition, this is also related to the issues of "migration" in (a4,6), "identity" in (b11), and "geography" in (d4), for instance.

Then, in *BNN*, I proposed another theory, namely, "the theory of contrastive advantages" to understand the persistence of social structure by gender, race, ethnicity, or any other groupings.

(a4) Social Systems (or Trends)

In relation to social systems, there are several sub-categories for illustration, which can be about (a4.1) urbanization, (a4.2) technology, (a4.3) demography, (a4.4) the environment, (a4.5) social trends as a whole (which also include the current phenomenon of globalization), and (a4.6) migration.

(a4.1) In relation to urbanization, I proposed "the contingent theory of urban planning" in *FPHUP* for a new way to understand density and void—which is also related to the issues of "social trends" (especially in relation to globalization) in (a4.5), "migration" in (a4.6), and "identity" in (b11).

(a4.2) In relation to technology, I worked out "the ambivalent theory of technology" in *FPHE* for a new way to understand the nature of technology.

Some other books (like *FPHMM*, *FAE*, *FCD*, and *FHC*) also deal with technology, and a good example concerns "the synthetic theory of information architecture" in *FIA* on taxonomy, network, and intelligence, in conjunction with the rise of technology.

(a4.3) In relation to demography, I proposed, first in *FHC* (and later in other books like *FCD*, *FPHC*, *FPHU*, etc.), "the theory of post-humanity," in relation to the "post-human" age at some distant point of "after-postmodernity," long after human extinction, to be eventually superseded by "post-humans" of various forms.

Some good candidates of the post-human lifeforms include, for instance, "thinking robots," "thinking machines," "cyborgs," "genetically altered superior beings," "floating consciousness," "hyper-spatial consciousness," "unfolding unconsciousness," "the hyper-sexual body," and "the hyper-martial body." This post-human vision of mine was first originally worked out in *FHC* and further elaborated in both *FCD*, *FPHC*, *FPHU*, *FPHS*, and *FPHMA*, for instance.

Lest any confusion occurs, I need to emphasize two clarifications here about the term "post-human" as a neologism in my works.

Firstly, the word "post-human" here should *not* be confused with another term which looks similar but has a totally different meaning in the literature of postmodernism, namely, "post-humanism"—which constitutes a critique of "humanism" as tradionally understood (especially, though not exclusively, in relation to the idea of progress in science and reason in the Enlightenment project). (WK 2008)

And secondly, the word "post-human" here should also *not* be confused with a similar term which is used to champion the ideology of technology for the future co-existence between humans and cyborgs in "trans-humanism." (WK 2008a)

Instead, my term "post-human" in relation to "posthuman-ism" also rejects "transhumanism" (especially, though not exclusively, in relation to the promises of technology) and refers to something else altogether, that is, the future extinction of humans and its post-human successors in deep space and beyond unto multiverses.

Later, in *BEPE*, I went further and provided a more comprehensive critique of "transhumanism," together with other issues.

With these two clarifications in mind (as summarized in *Table 1.31*)—in the end, a most fundamental question about intelligent life now has an answer, in that, if asked, "What is the future of human civilization?"—my answer in *FCD* (89) is thus: "As addressed in Ch.7 of *FHC*, a later epoch of the age of after-postmodernity (that is, at some point further away from after-postmodernity) will begin, as what I called the 'post-human' history (with the term 'post-human' originally used in my doctoral dissertation at M.I.T., which was finished in November 1995, under the title *After Postmodernity*, still available at M.I.T. library, and was later revised and published as *FHC*). The post-human history will be such that humans are nothing in the end, other than what culture, society, and nature (with some luck) have shaped them into, to be eventually superseded by post-humans (e.g., cyborgs, thinking machines, genetically altered superior beings, and others), if humans are not destroyed long before then."

(a4.4) In relation to the environment, I also addressed the issue of sustainability, especially in *FHC*, *FCD*, and *FPHUP*. For instance, in *FPHUP*, I specifically proposed a new theory, "the contingent theory of urban planning," on the need to go beyond the contemporoary obsession with sustainability, in the context of not only the environment but also other issues.

I also proposed "the interventive-reshaping theory of geography" in *FPHGEOG*—which is related to the issues of "migration" in (a4.6), "geography" in (d4), and "geology" in (d5).

In addition, I propose "the transfigurative theory of waste" in *FPHW*.

(a4.5) In relation to social trends as a whole, I proposed in Ch.9 of *FCD* (367-8), that "civilizational history will continue into the following cyclical progression of expansion, before it is to be superseded (solely as a high probability, since humans might be destroyed sooner either by themselves or in a gigantic natural calamity) by posthumans at some distant point in after-postmodernity (which I already discussed in *FHC*)" unto multiverses (different constellations of universes):

Local → Regional → Global →Solar →
Galactic → Clustery…→ Multiversal

The current phenomeon of "globalization" lies in the third category on the line above and was extensively analyzed in *FCD*, in close connection with "globalism" at the cultural level—in the context of "postmodernity" (which will be introduced later in the section on history).

In *BDPD*, this thesis of mine was referred to as "the theory of the cyclical progression of system integration and fragmentation," at the systemic level—in close relation to another thesis analyzed in Ch.9 of *FCD*, which was called, in the absence of better words, "the multifaceted theory of war and peace" in *FPHK* (as already indicated above).

In *FC*, I also proposed "the dialectic theory of complexity" to account for the factor of uncertainty, which can have a major role to play in the world.

All this is also related to the issues of "urbanization" in (a4.1), "the environment" in (a4.4), "migration" in (a4,6), "identity" in (b11), "geography" in (d4), and "geology" in (d5).

(a4.6) And in relation to migration, I proposed "the resettlement theory of geology" in *FPHGEOL* and "the theory of the cyclical progression of migration" in *FPHMIG*—which are also related to the issues of "social structure" in (a3), "urbanization" in (a4.1), "the environment" in (a4.4), "social trends" in (a4.5), "identity" in (b11), "geography" in (d4), and "geology" in (d5).

(b) Culture

The category of "culture" here can also be divided into different sub-categories like (b1) aesthetics, (b2) religion, (b3) popular (mass) culture, (b4) logic, (b5) civilization as a whole, (b6) ethics, (b7) epistemology (methodology), (b8) metaphysics (ontology), (b9) ideology, (b10) language, and (b11) identity.

(b1) Aesthetics

In relation to aesthetics, I proposed "the transformative theory of aesthetic experience" in *FAE* about some great transformations of aesthetic experience in future history.

In *FPHLIT*, I further proposed "the comparative-impartial theory of literature" for a new way to understand fiction and non-fiction.

In *FPHPA*, I further proposed "the transdisciplinary theory of performing arts" for a new way to understand the body and its presence—which is also related to the issues of "popular (mass) culture" in (b3) and of "the human body" in (c1.2). And the aesthetic issue also overlaps with (d8) on architecture.

And in *FPHCA*, I further proposed "the inquisitive theory of culinary art" for a new way to understand ingredients and techniques.

(b2) Religion

In relation to religion, I proposed "the comparative theory of religion" in *FHC* (and also in *FCD*, although it was first so labeled in *FPHK*).

And in *FPHR*, I provided a solid foundation of this theory with a better name, that is, "the comparative-substitutive theory of religion" for a new way to understand spirituality.

(b3) Popular (Mass) Culture

In relation to popular (mass) culture, there are some sub-categories for illustration, which can be, for example, (b3.1) mass culture in general and (b3.2) sports (like chess)—both of which are also related to the issues of "aesthetics" in (b1) and of "the human body" in (c1.2).

(b3.1) In relation to mass culture in general, I proposed, in *FHC*, a new way to understand mass culture in terms of some fundamental dilemmas, which is called, in the absence of better words, "the dualistic theory of mass culture" in *FPHO*.

(b3.2) In relation to sports (like chess), I proposed, in *FPHCHESS*, "the mediative-variative theory of chess" for a new way to understand tactics and strategy (in different traditions).

(b4) Logic

In relation to logic, I also worked out, in *FPHML*, "the contrastive theory of rationality" for a new way to understand rationality and knowledge (in the context of mathematical logic).

(b5) Civilization as a Whole

In relation to civilization as a whole, I also argued, in *FHC*, no freedom without unfreedom (as shown in *Table 4.13*) and no equality without inequality (as shown in *Table 4.14*), especially in relation to the seven dimensions of life existence first worked out in *FHC* (i.e., the technological, the everyday, the true, the holy, the sublime/beautiful, the good, and the just).

Even more interestingly, in *BDPD* and later in *BCIV*, I further revealed a theoretical refinement of existential dialectics, this time, in arguing that there is no civilization without barbarity, with human civilization to be eventually superseded by what I originally analyzed as "post-human post-civilization" (which should not be confused with "post-human civilization"), in the context of the freedom/unfreedom and equality/inequality dialectics.

In the final analysis, civilization cannot live without barbarity and has to learn to co-exist with it in ever new ways. Preposteorus as this may seem to many contemporaries—it is no more imperative to preserve civilization than necessary to destroy barbarity, and the ideal of civilization is essentially bankrupt, to be eventually superseded by "post-civilization" (as summarized in *Tables 4.35-4.43*).

This freedom/unfreedom and equality/inequality dialectics holds in pre-moderny, modernity, postmodernity, and what I orignally called "after-postmodernity" in *FHC* and *FCD*—especially, though not exclusively, in the context of the emergence of "post-humans."

There are two specific theories about "after-postmodernity" here. The first theory about "after-postmodernity" is about the trinity of pre-modernity, modernity, postmodernity, and after-postmodernity at the cultural level—and was called, in *BDPD*, "the theory of the trinity of modernity to its after-postmodern counterpart," although it was first analyzed in *FHC* and later in *FCD* and also *BCIV* (as summarized in *Table 1.34*, *Table 1.35*, *Table 1.36*, and *Table 1.37* and will be elaborated further in the later sub-section on "the rest" about future history).

The second theory about "after-postmodernity" is about pre-modernity, modernity, postmodernity, and after-postmodernity at the historical level. For instance, a conclusion in *FHC*, which is shocking indeed for many contemporaries, is none other than that "[t]he post-human history will therefore mark the end of human history as we know it and, for that matter, the end of human dominance and, practically speaking, the end of humans as well. The entire history of human civilization, from its beginning to the end, can be summarized by four words, linked by three arrows (as already discussed in *FHC*)":

Pre-Modernity → Modernity →
Post-Modernity → After-Postmodernity

In *BDPD*, this thesis of mine was specifically called "the theory of the evolution from pre-modernity to after-postmodernity," at the historical level—as shown in *Table 1.34*, *Table 1.35*, *Table 1.36*, and *Table 1.37* (and will be elaborated further in the later sub-section on "the rest" about future history).

Therefore, "[t]he end of humanity in the coming human extinction is the beginning of post-humanity. To say an untimely farewell to humanity is to foretell the future welcome of post-humanity." (P. Baofu 2002: 89) This thesis of mine was known in *BDPD* as "the theory of post-humanity," at the systemic level—as already indicated in (a4.3) above.

(b6) Ethics

In relation to ethics, I proposed, in *BEPE*, "the theory of post-ethics" for a new way to understand morality and immorality.

(b7) Epistemology (Methodology)

In relation to epistemology (methodology), I proposed "the complex theory of data analysis" in *FPHDA* for a different way to understand quantitative and qualitative research.

In addition, I already proposed "the critical-dialectic theory of formal science" in *FPHFS* and "the theory of methodological holism" in several books (like Ch.1 *FCD*, Ch.1of *FPHC*, *Sec.2.1* and *Sec.2.5* of *BCPC*, *FC*, *FPHK*, etc.) for a different way to do methodology.

(b8) Metaphysics (Ontology)

I also proposed "the theory of existential dialectics" in several books (like *FHC*, *FCD*, *FPHC*, *BDPD*, *FC*, *FAE*, *ALD*, *FIA*, *FPHK*, etc,) and "the contrastive theory of rationality" in *FPHML* for a different way to understand ontology.

(b9) Ideology

The issue about ideology was already discussed under the categories of "social institutions" and of "civilization as a whole," especially in relation to my theories of "post-democracy," "post-capitalism," "post-civilization," and so on.

(b10) Language

I already proposed "the multilogical theory of learning" in *FPHL* for a new way to understand the nature of language structure and context .

And I also worked out, in *FPHSEM*, "the interactive theory of semantics" for a new way to understand internality and externality (in the context of semantics).

(b11) Identity

I propose here "the theory of the cyclical progression of migration" in *FPHMIG* for a new way to understand the nature of sameness, otherness, and identity—which is also related to the issues of "social structure" in (a3), "urbanization" in (a4.1), "social trends" (especially in relation to globalization) in (a4.5), "migration" in (a4.6), religion in (b2), and ideology in (b9).

(c) The Mind

The important vision about "post-humans" (as summarized earlier) brings us to the category of "the mind" here, which, as is often understood in everyday's usage, refers to the "brain" and the "body" (together with their interactions) and has different levels of consciousness which can be both "spiritual" and non-spiritual.

Therefore, the "mind" already presupposes the "brain," "body," and "spirit" (or "soul"), so that one does not fall into the intellectual trap of the "mind-body dualism" as debunked in modern (and especially, postmodern) philosophy (which I already went to great lengths to explain in *The Future of Human Civilization*).

That said—the category of "the mind" can be divided into two sub-categories, namely, in relation to (c1) biology and (c2) psychology

(c1) Biology

In relation to biology, there are two concerns here, namely, (c1.1) genes and intelligence and (c1.2) the human body.

(c1.1) In relation to genes and intelligence, I worked out "the theory of contrastive advantages" (which was originally proposed in *FCD* and further worked out in *BNN*, as summarized in *Table 1.24*), to show the nature-nurture interactions of multiple levels in action (e.g., the biological, the psychological, the structural, the systemic, the cultural, and others) for humans and post-humans on earth and beyond, to the extent that different groups, be they on the basis of race, gender, ethnicity, class, age, or else, are not equal, on average, and have not been, nor will they be, even if there can be exceptions.

(c1.2) In relation to the human body, for instance, in *FPHMA*, I suggested "the expansive-contractive theory of martial arts," in which two great future transformations of the martial body were envisioned, that is, both "with the hyper-martial body" and "without the body" (as summarized in *Table 1.32*)—although this issue can also be classified under the category of "culture."

In *FPHS*, I proposed "the theory of virtual sexuality," in which three great future transformations of the sexual body were envisioned, that is, "virtual pleasure," "the hyper-sexual body," and "the post-sexual floating mind" (as summarized in *Table 1.33*).

And in *FPHPA*, I further proposed "the transdisciplinary theory of performing arts" for a new way to understand the body and its presence, with seven great future transformations of the body in performing arts—which is also related to the issues of "aesthetics" in (b1) and "popular (mass) culture" in (b3).

(c2) Psychology

In relation to psypchology (which also overlaps with biology too, however), I also proposed some greater transformations of the mind to come in the future, especially though not exclusively in the post-human age.

These greater transformations of the mind in psychology can be further sub-divided into different categories, namely, (c2.1) consciousness, (c2.2) unconsciousness (together with subconsciousness), (c2.3) creativity, (c2.4) learning, (c2.5) aggression and pacificity, (c2.6) personality, (c2.7) humor, and (c2.8) artificial intelligence.

(c2.1) In relation to consciousness, I analyzed, in *FPHC*, the nature of consciousness in relation to the conceptual dimensions of consciousness (as shown in *Table 1.12*), the theoretical levels of consciousness (as shown in *Table 1.13*), the thematic issues of consciousness (as shown in *Table 1.14*), and the different dimensions of human existence (viz., having-ness, belonging-ness, and being-ness) in relation to consciousness (as shown in *Table 1.15*, *Table 1.16*, *Table 1.17*, and *Table 1.18*).

In *BCPC*, the nature of consciousness was further explored in relation to cognition, emotion, and behavior, which can be in consciousness and other mental states (e.g., subconsciousness and unconsciousness), in the context of existential dialectics, as shown in *Table 1.19*, *Table 1.20*, *Table 1.21*, *Table 1.22,* and *Table 1.23*.

In fact, in *FCD*, I already proposed three different theories in relation to the limits of cognition, emotion, and behavior, namely, "the theory of cognitive partiality," "the theory of emotional non-neutrality," and "the theory of behavioral alteration."

Then, I argued that the existence of human consciousness will be superseded one day with "floating consciousness" (as first proposed in *FCD*) and "hyper-spatial consciousness" (as first proposed in *FPHC*) as a climax of evolution in consciousness, after the future extinction of human consciousness (as shown below):

Primordial consciousness → Human consciousness →
Post-human consciousness (with floating consciousness and hyper-spatial consciousness as a climax in the evolution of consciousness)

In *BDPD*, these latest theses of mine were called "the theory of floating consciousness" and "the theory of hyper-spatial consciousness," both at the cosmological and psychological levels (as summarized in *Table 1.26* and *Table 1.25*).

(c2.2) In relation to unconsciousness, for instance, in *FPHU*, the unconsious mind will undergo tremendous change, to the rise of what I originally called "unfolding unconsciousess" in relation to anomalous experience, as a climax of evolution in unconsciousness, in conjunction with the conscious mind (both human now and post-human later), especially at the psychological level, as shown below:

Primordial unconsciousness → Human unconsciousness →
Post-human unconsciousness (with unfolding unconsciousness as a climax in the evolution of unconsciousness)

In *FPHU*, this thesis of mine is called "the theory of unfolding unconsciousness" in a small sense, or "the unfolding theory of anomalous experience," in a larger one, as shown in *Table 1.27* and *Table 1.28*.

(c2.3) In relation to creativity, for instance, in *FPHCT*, I also proposed "the comprehensive theory of creative thinking" to explore the role of creative thinking, in relation to technology and other factors, together with my original proposal of important creative techniques and traits (as shown in *Table 1.29*) and its double nature of desirability and undesirability (as shown in *Table 1.30*).

(c2.4) In relation to learning, for instance, in *FPHL*, I worked out "the multilogical theory of learning" for a new way to understand the nature of structure and context in language, in relation to learning and its future development in the context of the mind.

(c2.5) In relation to aggression (and pacificity), I proposed "the multifaceted theory of war and peace" in *FPHWP* (as mentioned above).

(c2.6) In relation to personality, I worked out “the contrarian theory of personality” in *FPHP* for a new way to understand normality and abnormality.

(c2.7) In relation to humor, I worked out “the metamorphic theory of humor” in *FPHH* for a new way to understand joking and laughing.

(c2.8) And in relation to artificial intelligence, I proposed “the supersession theory of computing” in *FPHCOM* for a new way to understand hardware, software, and the mind.

In addition, I also proposed “the complex theory of data analysis” in *FPHDA* for a different way to understand quantitative and qualitative research in the context of the evolution of new minds.

(d) Nature

The category of “nature” here refers to the state of nature, which can be divided into some sub-categories like (d1) physics, (d2) cosmology, (d3) biology, (d4) geography, (d5) geology, (d6) acoustics, (d7) chemistry, (d8) architecture, and (d9) transportation.

But since biology in (d3) overlaps with biology in (c1)—only the rest of the categories are addressed hereafter.

(d1) Physics

In relation to physics, I also proposed “the perspectival theory of space-time” in *FPHST* (based on *FPHC*) for a new way to understand space and time, as shown in *Table 1.3*, *Table 1.4*, *Table 1.5*, and *Table 1.7*.

In *FPHG* (based on *FPHC* and *FPHST*), I proposed “the selective theory of geometry” for a new way to understand geometry, with implications for the understanding of space-time in the context of infinity, symmetry, and dimensionality for future lifeforms that our world has never known (as shown in *Table 1.6*, *Table 1.7*, *Table 1.8; Table 1.9*, *Table 1.10,* and *Table 1.11*).

And in *FC*, I proposed “the dialectic theory of complexity” for a new way to understand order and chaos in the state of nature and beyond.

(d2) Cosmology

In relation to cosmology, my theories on physics (as indicated above) already have important implications for understanding the nature and the future of the universe (which was also addressed in *FHC*, among other works aforcited).

I also proposed “the theory of hyper-spatial consciousness” in *FPHC* (and also in *FPHG*) for a new way to understand the future evolution of the mind to enter into different dimensions of space-time that our current world has never known.

In addition, in *BCOS*, I proposed "the theory of post-cosmology" for a new way to understand the beginnings and ends of the cosmos and beyond.

(d4) Geography

In relation to geography, I proposed "the theory of the geopower of nature" in *ALD* for a different way to understand the relationships between geography and other entities (like the mind, culture, and society).

I further refined this issue of geography by proposing "the interventive-reshaping theory of geography" in *FPHGEOG*—which is related to the issues of "the environment" in (a4.4), "migration" in (a4.6), and "geology" in (d5).

(d5) Geology

In relation to geology, I proposed "the resettlement theory of geology" in *FPHGEOL* for a different way to understand the relationships between statics and dynamics—which is related to the issues of "the environment" in (a4.4), "migration" in (a4.6), and "geography" in (d4).

(d6) Acoustics

In relation to acoustics, I proposed "the multilateral theory of acoustics" in *FPHA* for a different way to understand the relationships between sound and silence.

(d7) Chemistry

In relation to chemistry, I proposed "the creational theory of chemistry" in *FPHCHEM* for a different way to understand substances and their changes.

(d8) Architecture

In relation to architecture, I proposed "the transcendent theory of architecture" in *FPHARCH* for a different way to understand form and function.

(d9) Transportation

And in relation to transportation, I proposed "the panoramic theory of transportation" in *FPHTRAN* for a different way to understand networks and operations.

(e) The Rest

The category of "the rest" here refers to what cannot be strictly classified in any of the previous categories (i.e., society, culture, the mind, and nature). A good illustration of "the rest" refers to (e1) history.

(e1) History

The category of history can be divided into some sub-categories like (e1.1) empirical history and (e.1.2) meta-history.

(e1.1) In relation to empirical history, I proposed "the theory of the evolution from pre-modernity to after-postmodernity" (as first worked out in *FHC* and later in *FCD* and *FPHC*), which overlaps—as a reminder from (b5)—with the issue concerning "civilization as a whole" in (b5).

For instance, in both *FCD* and *FPHC*, I worked out the structure of "post-human civilization" in terms of the trinity of after-postmodernity (i.e., "free-spirited after-postmodernity," "post-capitalist after-postmodernity," and "hegemonic after-postmodernity").

Both conceptually and theoretically, the trinity of after-postmodernity is a sequential extension of the trinity of modernity (i.e., "free-spirited modernity," "capitalist modernity,"and "hegemonic modernity") and the trinity of postmodernity (i.e., "free-spirited postmodernity," "capitalist postmodernity," and "hegemonic postmodernity") as first proposed in *FHC*.

And the trinity of pre-modernity (i.e., "pre-free-spirited pre-modernity," "pre-capitalist pre-modernity" and "hegemonic pre-modernity") was later conceived in *BCIV* to complete the historical set from pre-modernity to after-postmodernity.

In *BDPD*, this thesis about the trinity of pre-modernity, modernity, postmodernity, and after-postmodernity was collectively known as "the theory of the trinity of modernity to its after-postmodern counterpart," at the cultural level—as already so mentioned in (b5) earlier, and summarized in *Table 1.34*, *Table 1.35*, *Table 1.36*, and *Table 1.37*.

At the structural level, all these trinities are subject to the existential constraints (e.g., the freedom/unfreedom and equality/inequality dialectics in the context of "the cyclical progression of hegemony"), be the historical epoch in pre-modernity, modernity, postmodernity, or after-postmodernity in future times. Each of the historical epochs has its ever new ways of coming to terms with the ever new (different) mixtures of freedom/unfreedom and equality/inequality.

This is importantly so, not because, as is falsely assumed in conventional wisdom, one certain way is superior (or better) than another in terms of achieving more freedom and less unfreedom, or more equality with less inequality.

On the contrary, indeed, in each of the historical epochs, each increase of unfreedom greets each freedom achieved, just as each increase of inequality welcomes each equality achieved, albeit in ever new (different) ways. In *BDPD*, this thesis of mine was labeled as "the theory of the cyclical progression of hegemony," at the structural level, though it was first analyzed in *FCD*.

In *BDPD*, more theoretical applications of existential dialectics were further examined, in relation to five main features, in the context of the duality of oppression, namely, (a) that each freedom/equality achieved is also each unfreedom/inequality created, (b) that the subsequent oppressiveness is dualistic, both by the Same against the Others and itself and by the Others against the Same and themselves, (c) that both oppression and self-oppression can be achieved by way of downgrading differences (between the Same and the Others) and of accentuating them, (d) that the relationships are relatively asymmetric among them but relatively symmetric within them, even when the Same can be relatively asymmetric towards itself in self-oppression, and the Others can be likewise towards themselves, and (e) that symmetry and asymmetry change over time, with ever new players, new causes, and new forms, be the locality here on Earth or in deep space unto multiverses—as summarized in *Table.4.13*, *Table 4.14*, *Table 4.15*, and *Table 4.16*.

The same logic also holds both in relation to wealth and poverty (as addressed in *BCPC* and summarized in *Table 4.17* on the wealth/poverty dialectics) and in relation to civilization and barbarity (as addressed in *BCIV* and summarized in *Table 4.18* on the civilization/barbarity dialectics).

In *BDPD*, this thesis on existential dialectics was labeled as "the theory of existential dialectics."

(e.1.2) And in relation to meta-history, I proposed "the multifold theory of history" in *FPHHIST* for a different way to understand universality and relativity (together with other debates).

Direct and Indirect Applications of Existential Dialectics

A different way to appreciate the usefulness of existential dialectics is by way of the analysis of its (a) direct and (b) indirect applications.

(a) In direct applications, on the one hand, the logic of existential dialectics can shed some theoretical insights on diverse phenomena in the world, and good instances are the usage of the principles of existential dialectics for the theoretical insights on the freedom/unfreedom dialectics, the equality/inequality dialectics, and the wealth/poverty dialectics—as introduced earlier in (b5) and (e) in the previous sub-section.

For instance, my books like *FPHST*, *BNN*, *FAE*, *FC*, *FIA*, *FPHU*, *FPHK*, *FPHML*, *FPHE*, *FPHMM*, *FPHCT*, *FPHG*, *FPHUP*, *FPHL*, *FPHO*, *FPHMA*, *FPHS*, *FPHLAW*, *FPHFS*, *FPHWP*, *FPHP*, *BEPE*, *BCOS*, *FPHGEOL*, *FPHCHESS*, *FPHR*, *FPHEDU*, *FPHH*, *FPHA*, *FPHGAM*, *FPHCOM*, *FPHLIT*, *FPHCHEM*, *FPHDA*, *FPHGEOG*, *FPHPA*, *FPHSEM*, *FPHARCH*, *FPHCRIM*, *FPHHIST*, *FPHMIG*, *FPHTRAN*, *FPHCA*, and *FPHW* use the principles to reveal some theoretical insights on the perspectives of space and time (as in *FPHST*), of nature and nurture (as in *BNN*), of beauty and ugliness (as in *FAE*), of order and chaos (as in *FC*), of taxonomy and network (as in *FIA*), of the unconscious mind (as in *FPHU*), of the nature of knowledge (as in *FPHK*), of mathematical logic (as in *FPHML*), of engineering (as in *FPHE*), of mass media (as in *FPHMM*), of creative thinking (as in *FPHCT*), of geometry (as in *FPHG*), of urban planning (as in *FPHUP*), of language (as in *FPHL*), of organization (as in *FPHO*), of martial arts (as in *FPHMA*), of sexuality (as in *FPHS*), of law (as in *FPHLAW*), of formal science (as in *FPHFS*), of aggression and pacificity (as in *FPHWP*), of normality and abnormality (as in *FPHP*). of morality and immorality (as in *BEPE*), of the beginnings and ends of the universe (as in *BOCS*), of statics and dynamics (as in *FPHGEOL*), of tactics and strategy (as in *FPHCHESS*), of secularness and sacredness (as in *FPHR*), of teaching and learning (as in *FPHEDU*), of joking and laughing (as in *FPHH*), of sound and silence (as in *FPHA*), of risk and caution (as in *FPHGAM*), of hardware and software (as in *FPHCOM*), of fiction and non-fiction (as in *FPHLIT*), of substances and their changes (as in *FPHCHEM*), of quantitative and qualitative research (as in *FPHDA*), of environments and their interactions (as in *FPHGEOG*), of the body and its presence (as in *FPHPA*), of internality and externality (as in *FPHSEM*). of form and function (as in *FPHARCH*), of heroes and villains (as in *FPHCRIM*), of universality and relativity (as in *FPHHIST*), of Sameness and Otherness (as in *FPHMIG*), of networks and operations (as in *FPHTRAN*), of ingredients and techniques (as in *FPHCA*), and of uselessness and usefulness (as in *FPHW*).

(b) In indirect applications, on the other hand, the theoretical insights can further be used to reveal other phenomena directly from them (viz., the theoretical insights) and therefore indirectly from the principles themselves.

A good illustration is of course the use of the theoretical insights on the freedom/unfreedom and equality/inequality dialectics for the understanding of the civilization/barbarity dialectics.

This distinction between direct and indirect applications may sound a bit academic, since even in indirect applications, the phenomena under study can still be directly related back to the principles themselves.

In the previous example, as an illustration, the civilization/barbarity dialectics can be directly related to the principles of existential dialectics without the intermediate role of the freedom/unfreedom and equality/inequality dialectics.

Multiple Levels of Application

There is another issue to be clarified, however. In other words, the theoretical insights can be applied to multiple levels of analysis—even though, in a given example, it may refer to one level or a few only.

For instance, in the example concerning the freedom/unfreedom dialectics, it can be used at the structural level (e.g., in relation to the theory of the cyclical progression of hegemony), but it can also be exploited for other levels (e.g., the theory of post-capitalism at the institutional level).

All these levels of application should not be misleadingly construed, as I stress this before, as a one-way street (that is, to use the ontological principles for theoretical application at multiple levels of specific analysis) but a two-way one, in which theoretical insights in praxis, when studied in more specific contexts, can in turn refine the nature of existential dialectics (for example, with the addition of new principles).

Sophisticated Methodological Holism

The summary of my original meta-theory on methodology is provided below, again, to be repeated *often verbatim* from my previous works—as this is something I regularly do in each new book to introduce my past works, which are in conversation with this current one.

This means that, for those readers who have not read any of my previous books, the summary below is helpful.

That clarified—this original contribution of mine to the study of methodology takes the form of an approach known as "the theory of methodological holism" or "methodological holism" in short, as already worked out in *FPHC* (2004).

I have made good use of this methodological approach of mine for all of my previous works.

The Confusion of Holism with Systems Approach

Yet, it is imperative to stress at the outset, as I often emphasized this in all my previous books, that my approach of "methodological holism" does not oppose or exclude "methodological individualism" (as some readers may be tempted to assume, as is conventionally understood) but actually includes it.

The confusion here has to do with equating holism with systems approach in general (be it about systems theory, chaos theory, complexity theory, or else), which I debunked, while learning something from it, as already elaborated in *The Future of Complexity* (2007)—and *The Future of Post-Human Formal Science* (2010) is another addition to this critique of the intellectual fad of systems approach.

In other words, my meta-theory of holism, or sophisticated methodological holism in short, makes a distinction between doing holistic work and doing systems work, for two reasons.

The first reason is that holism, in my theory, does not imply systems approach, since it rejects systems approach because of the myriad problems that the literature on systems approach has been unable to resolve, while learning something from it.

And the second reason is that my theory is related to the other theory of mine, that is, existential dialectics, which does not exclude methodological individualism (but includes it, because it simply goes beyond both reductionism and reverse-reductionism in a dialectic way).

For these reasons (and others too, as summarized in *Table 4.1*), my version of methodological holism is *sophisticated*—not *vulgar* as sometimes used by inapt scholars employing the same term.

The Ontological Constraints

Sophisticated methodological holism is subject to some ontological constraints, and some good examples include "the formalness-informalness principle," "the absoluteness-relativeness principle," "the partiality-totality principle," "the predictability-unpredictability principle," "the explicability-inexplicability principle," and "the fiction-reallity principle" in existential dialectics.

Of course, there can be more than six ontological principles in relation to method, but the examples here are sufficient to illustrate the point in question.

With this caveat in mind—the first two were already analyzed in previous books of mine like *FC* and *FAE*, whereas the third one was addressed in *FPHU*, the fourth and fifth ones were accounted for in *FPHK* and *FPHML*, respectively, and the last one was addressed in *FPHLIT*.

For now, it suffices to show that sophisticted methodological holism, because of these ontological constraints on methodology, targets two major sins of methodology, namely, what I call, in the absence of better words, (a) *reductionism* and (b) *reverse-reductionism*—both of which come in all shapes and sizes, to be summarized below (and also shown in *Table 4.2*).

Against the Varieties of Reductionism

There are four versions of reductionism to be summarized here, which sophisticated methodological holism rejects, namely, (a1) conceptual, (a2) theoretical, (a3) methodological, and (a4) ontological forms of reductionism.

(a1) In conceptual reductionism, a good case in point concerns myriad dualities like mind vs. body, self vs. world, democracy vs non-democracy, and the like (as already addressed in *FHC*, *FPHC*, and *BDPD*, for instance).

(a2) In theoretical reductionism, an illuminating case study is best exemplified by what I originally called "the foundation fallacy" in *FPHST*, in any attempt to naively understand space-time from the physical perspective as the foundation and, consequently, to dangerously dismiss (or belittle) other perspectives.

In *FAE*, I elaborated further these versions of reductionism in the literature on aesthetics (e.g., form vs. content, representation vs. expression, critics vs. artists, and externalism vs. internalism).

In *FIA*, I revealed other forms of reductionism in the literature on information architecture (e.g., the constructivist argument and the representational argument).

In *FPHU*, I showed the persistent legacy of reductionism, this time, in the literature on anomalous experience (e.g., the obsession with physics, chemistry, and biology for explaining anomalous experience).

In *FPHE*, I examined another case of reductionism in action, in the context of engineering (e.g., technical constraints vs. normative constraints).

In *FPHMM*, I elaborated one more version of reductionism, in relation to the three domains of communication (e.g., the competing views on sending, connecting, and receiving).

In *FPHCT*, I explored another version of reductionism, in relation to invention and innovation (e.g., the bio-psychological argument vs. the socio-cultural arguments).

In *FPHG*, I identified another version of reductionism, in relation to infinity, symmetry, and dimensionality (e.g., the Euclidean argument vs. the non-Euclidean arguments).

In *FPHUP*, I analyzed another version of reductionism, in relation to density and void (e.g., the engineering argument and the ecology argument).

In *PFHL*, I examined another version of reductionism, in relation to structure and context (e.g., the structuralist argument and the contextualist argument).

In *PFHO*, I explored another version of reductionism, in relation to communcation, decision-making, and leadership (e.g., the rational-system argument vs. the natural-system argument vs. the open-system argument).

In *PFHMA*, I revealed another version of reductionism, in relation to the martial body and spirit (e.g., the spiritual argument vs. the materialist argument vs. the defensive argument).

In *PFHS*, I analyzed another version of reductionism, in relation to the sexual body and spirit (e.g., the naturalist argument vs. the constructivist argument).

In *PFHLAW*, I examined another version of reductionism, in relation to law (e.g., the necessity argument vs. the contengency argument).

In *FPHWP*, I show another version of reductionism, in relation to war and peace (e.g., the aggressivist argument vs. the pacifist argument).

In *BEPE*, I analyzed another version of reductionism, in relation to morality and immorality (e.g., the objectivist argument vs. the non-objectivist argument vs. the skeptical argument).

In *BCOS*, I examined another version of reductionism, in relation to the contested beginnings and speculative ends of the universe (e.g., the scientific argument vs. the religious argument vs. the esoteric argument vs. the metaphysical argument).

In *FPHP*, I showed another version of reductionism, in relation to normality and abnormality (e.g., the natural argument vs. the social argument vs. the cultural argument vs. the mental argument).

In *FPHGEOL*, I scrutinized another version of reductionism, in relation to statics and dynamics (e.g., the catastrophe argument vs. the uniformity argument vs. the revision argument).

In *FPHCHESS*, I showed another version of reductionism, in relation to tactics and strategy (e.g., the natural argument vs. the social argument vs. the cultural argument vs. the mental argument).

In *FPHR*, I revealed another version of reductionism, in relation to secularness and sacredness (e.g., the critical argument vs. the skeptical argument vs. the theist argument).

In *FPHEDU*, I examined another version of reductionism, in relation to teaching and learning (e.g., the teacher-centered argument vs. the student-centered argument vs. the balanced argument).

In *FPHH*, I explored another version of reductionism, in relation to joking and laughing (e.g., the natural argument vs. the social argument vs. the cultural argument vs. the mental argument).

In *FPHA*, I revealed another version of reductionism, in relation to sound and silence (e.g., the natural argument vs. the social argument vs. the cultural argument vs. the mental argument).

In *FPHGAM*, I revealed another version of reductionism, in relation to risk and caution (e.g., the natural argument vs. the social argument vs. the cultural argument vs. the mental argument).

In *FPHCOM*, I showed another version of reductionism, in relation to hardware and software (e.g., the natural argument vs. the social argument vs. the cultural argument vs. the mental argument).

In *FPHLIT*, I proposed another version of reductionism, in relation to fiction and non-fiction (e.g., the natural argument vs. the social argument vs. the cultural argument vs. the mental argument).

In *FPHCHEM*, I proposed another version of reductionism, in relation to substances and their changes (e.g., the natural argument vs. the social argument vs. the cultural argument vs. the mental argument).

In *FPHDA*, I proposed another version of reductionism, in relation to quantitative and qualitative research (e.g., the natural argument vs. the social argument vs. the cultural argument vs. the mental argument).

In *FPHGEOG*, I proposed another version of reductionism, in relation to environments and their interactions (e.g., the natural argument vs. the social argument vs. the cultural argument vs. the mental argument).

In *FPHPA*, I proposed another version of reductionism, in relation to the body and its presence (e.g., the natural argument vs. the social argument vs. the cultural argument vs. the mental argument).

In *FPHSEM*, I proposed another version of reductionism, in relation to internality and externality (e.g., the natural argument vs. the social argument vs. the cultural argument vs. the mental argument).

In *FPHARCH*, I proposed another version of reductionism, in relation to form and function (e.g., the natural argument vs. the social argument vs. the cultural argument vs. the mental argument).

In *FPHCRIM*, I proposed another version of reductionism, in relation to heroes and villains (e.g., the natural argument vs. the social argument vs. the cultural argument vs. the mental argument).

In *FPHHIST*, I proposed another version of reductionism, in relation to universality and relativity (e.g., the natural argument vs. the social argument vs. the cultural argument vs. the mental argument).

In *FPHMIG*, I proposed another version of reductionism, in relation to Sameness and Otherness (e.g., the natural argument vs. the social argument vs. the cultural argument vs. the mental argument).

In *FPHTRAN*, I proposed another version of reductionism, in relation to networks and operations (e.g., the natural argument vs. the social argument vs. the cultural argument vs. the mental argument).

In *FPHCA*, I proposed another version of reductionism, in relation to ingredients and techniques (e.g., the natural argument vs. the social argument vs. the cultural argument vs. the mental argument).

And in *FPHW*, I propose another version of reductionism, in relation to uselessness and usefulness (e.g., the natural argument vs. the social argument vs. the cultural argument vs. the mental argument).

(a3) In methodological reductionism, a good illustration can be the debate between different versions of qualitative and quantitative methods (as already analyzed in *FC* and also *FHC*).

In *FPHML*, I examined similar reductionism, this time, in the literature on mathematical logic (e.g., the obsession with consistency, soundness, and completeness).

And in *FPHFS*, I explored the problems of reductionism in the context of formal science (e.g., the analytical argument, as opposed to the synthetic argument).

(a4) In ontological reductionism, an excellent instance involves another debate, this time between emergentism and reductionism in complexity theory (as addressed in *FC*) and also in psychology (as elaborated in *FPHC*, in the context of Being and Belonging).

But to target against the varieties of reductionism constitutes only a side of the same coin, and the other side concerns the varieties of reverse-reductionism.

Against the Varieties of Reverse-Reductionism

The other side of the same coin is the reverse version of reductionism, which is what I want to call, in the absence of better words, *reverse-reductionism*.

My sophisticated version of methodological holism targets against the varieties of reverse-reductionism (just as it also rejects the ones of reductionism).

Perhaps nothing expresses better the popularity of reverse-reductionim than the "anything-goes" mentality in postmodernism of our time, as shown in the following four versions.

(b1) In conceptual reverse-reductionism, any concept of "art" (e.g., fine art, applied art, outsider art, junk art) is welcome in postmodernism (as already analyzed in Ch.4 of *FHC*—and also in *FAE*).

(b2) In theoretical reverse-reductionism, a variety of art and literary theories co-exist. Take the case of literary studies, as there are now Literary Structuralism, Marxist Literary Criticism, New Criticism, Phenomenology, Hermeneutics, Language-Game Literary Criticism, Feminist Literary Criticism, Reception Theory, Reader Response

Criticism, Poststructuralism, Semiotics, Pyschoanalytic Literary Criticism, just to cite some well-known ones, with no one being said to be better than any others (as detailedly analyzed in Ch.4 of *FHC*). (S. Raman 1997)

In *BNN*, I even introduced "the compromise fallacy" as another good illustration of theoretical reverse-reductionism, in misleadingly treating both genetic and environmental approaches as equally valid.

(b3) In methodological reverse-reductionism, multiple methodologies are deemed as acceptable in postmodernism (e.g., doing art without praxis, doing art with praxis, and doing art by sublation), as analyzed in Ch.4 of *FHC*.

And in *FPHFS*, I exposed the problems of reverse-reductionism in the context of formal science in relation to systems theory (e.g., the synthetic argument).

(b4) In ontological reverse-reductionism, no privileged ontology is allowed, and the door is open practically for anything in postmodernism (e.g., the equal status of the ontology of Being vs. that of Becoming, as already addressed in Ch.4 of *FHC*—and also in *FPHC*).

In *FAE*, I also introduced another version of reverse-reductionism, that is, "the pluralist fallacy," in the context of understanding aesthetic experience, for instance—although this fallacy has been committed not exclusively in relation to the ontological level (but also at the conceptual, theoretical, and methodological ones).

These dual dangers against reverse-reductionism (in this sub-section) and reductionism (in the previous sub-section) are something that sophisticated methodological holism rejects. And in *FPHFS*, I addressed these two problems further.

This point about the dual dangers is important enough, since many scholars often favor one against the other, but it is vital to target both.

The Holistic Organization of an Inquiry

With these dual dangers against reductionism and reverse-reductionism in mind—sophisticated methodological holism suggests that an inquiry of any given phenomenon is more complete, if treated in the context of a comprehensive analysis at all relevant levels, which challengingly encompass all the fields of human knowledge, ranging from the natural sciences and formal sciences through the social sciences to the humanities.

This section is something that I had already stressed in all my previous books and repeat (*often verbatim*) hereafter. But, for those readers who had read my previous books, this serves as a reminder.

With this reminder in mind—there are multiple ways to engage in a holistic inquiry with all relevent levels of analysis. Over the years, I have proposed different ways to fulfill this holistic methodological requirement, as repeated in my previous books.

Hereafter is a summary of four major ways, namely, (a) by discipline, (b) by domain, (c) by subject, and (d) by meta-analysis—all of which fulfill the holistic requirement of sophisticated methodological holism on a given issue.

(a) In a holistic organization by discipline, a good illustration is the multiple levels of analysis, namely, (a1) the micro-physical, (a2) the chemical, (a3) the biological, (a4) the psychological, (a5) the organizational, (a6) the institutional, (a7) the structural, (a8) the systemic, (a9) the cultural, (a10) the macro-physical (cosmological), and (a11) other relevant levels, which are either a combination of all these levels or the practical applications with a combination of them.

This holistic organization by discipline is used in *FCD* and *BCPC*, for instance.

(b) In a holistic organization by domain, some good candidates include the classification of different perspectives of analysis in relation to (b1) nature, (b2) the mind, (b3) society, and (b4) culture, as already worked out in *BCIV*, *FPHST*, *FC*, *FAE*, *FPHU*, *FPHK*, *FPHE*, and *FPHFS*, just to cite some examples.

Culture in (b4), in this re-classification, is the same as culture in (a9) and can be further divided into different sub-categories, with good examples like (a9i) epistemology (methodology), (a9ii) aesthetics, (a9iii) ethics, (a9iv) metaphysics (ontology), and (a9v) religion. For clarification, epistemology (methodology) in (a9i) and metaphysics (ontology) in (a9iv) overlap with the category of "meta-theory."

Society in (b3) includes the organizational in (a5), the institutional in (a6), the structural in (a7), and the systemic in (a8).

The mind (b2) has more to do with the chemical in (a2), the biological in (a3), and the psychological in (a4), although it can overlap with (a1), (b3) and (b4), for instance.

It should be stressed that the "mind" here also includes the categories like the "brain," "body," and the "spirit" (or the "soul")—whenever needed in a given context, so that one does not fall into the intellectual trap of the "mind-body dualism" in the history of modern (and especially, postmodern) philosophy, which I already went to great lengths to explain in *The Future of Human Civilization.* After all, in everyday's usage, the "mind" already implies the "brain" and the "body" (together with their interactions) and has different levels of consciousness which can be both "spiritual" and non-spiritual.

And nature in (b1) refers to the micro-physical in (a1) and the macro-physical (cosmological) in (a10), although it can also overlap with (a2), (a3) and (a4).

(c) In a holistic organization by subject, some excellent examples concern the analysis of consciousness in *FPHC* (i.e., on Having, Belonging, and Being); of civilization in *FHC* (i.e., the True, the Holy, the Everyday, the Technological, the Beautiful/Sublime, the Good, and the Just); of international politics in *BWT* (i.e., Hyper-Empires, Meso-Empires, and Micro-Empires); of nature and nurture in *BNN* (i.e., genes and memes); of information architecture in *FIA* (i.e., taxonomy and network); of mathematical logic in *FPHML* (i.e., consistency, soundness, and completeness); of communication in *FPHMM* (i.e., sending, connecting, and receiving); of creative thinking in *FPHCT* (i.e., invention and innovation); of geometry in *FPHG* (i.e., infinity, symmetry, and dimensionality); of urban planning in *FPHUP* (i.e., density and void), of language in *FPHL* (i.e., structure and context), of organization in *FPHO* (i.e., communication, decision-making, and leadership), of martial arts in *FPHMA* (i.e., the martial body and the martial spirit), of sexuality in *FPHS* (i.e., the sexual body and the sexual spirit), of law in *FPHLAW* (i.e., necessity and contingency), of war and peace in *FPHWP* (i.e., aggression and pacificity), of personality in *FPHP* (i.e., normality and abnormality), of ethics in *BEPE* (i.e., morality and immorality), of cosmology in *BCOS* (i.e., the beginnings and ends of the universe), of geology in *FPHGEOL* (i.e., statics and dynamics), of chess in *FPHCHESS* (i.e., tactics and strategy), of religion in *FPHR* (i.e., secularness and sacredness), of education in *FPHEDU* (i.e., teaching and learning), of humor in *FPHH* (i.e., laughing and joking), of acoustics in *FPHA* (i.e., sound and silence), of gambling in *FPHGAM* (i.e., risk and caution), of computing in *FPHCOM* (i.e., hardware and software), of literature in *FPHLIT* (i.e., fiction and non-fiction), of chemistry in *FPHCHEM* (i.e., substances and their changes), of data analysis in *FPHDA* (i.e., quantitative and qualitative research), of geography in *FPHGEOG* (i.e., environments and their interactions), of performing arts in *FPHPA* (i.e., the body and its presence), of semantics in *FPHSEM* (i.e., internality and externality), of architecture in *FPHARCH* (i.e., form and function), of criminality in *FPHCRIM* (i.e., heroes and villains), of history in *FPHHIST* (i.e., universality and relativity), of migration in *FPHMIG* (i.e., Sameness and Otherness), of transportation in *FPHTRAN* (i.e., networks and operations), of culinary art in *FPHCA* (i.e., ingredients and techniques), and of waste in *FPHW* (i.e., uselessness and usefulness)—by incorporating all the relevent levels of analysis as cited above in each.

(d) And in a holistic organization by meta-analysis, two good cases in point involve the works on post-capitalism in *BCPC* and on authoritarian liberal democracy in *ALD*, with the classification of analysis in theory and meta-theory.

The essential point to remember here, as I thus stress in all my previous books, is that the multiple levels of analysis can be reorganized in many different ways, insofar as none of the levels (if relevant to an inquiry in question) is ignored or dismissed, to avoid the danger of reductionism (and for that matter, the one of reverse-reductionism, in the opposite direction).

Three Clarifications

Three clarifications are needed here, however, to avoid any hasty misunderstanding.

Firstly, the entities in each classification are not mutually exclusive. For instance, in the holisitc organization by domain, nature in (b1) can also be linked to the chemical in (a2), the biological in (a3), and the systemic in (a8). And the mind (b3) can alternatively be related to the micro-physical in (a1) and the chemical in (a2), for example.

Secondly, there is also the important factor of luck (or uncertainty in conventional usage), but it is already implied or allowed in each of the organizations of an inquiry—especially in relation to the predictability-unpredictability principle and the order-chaos principle in existential dialectics. This topic about uncertainty was already addressed in *FHC*—and more extensively, in *FC*, in the context of order and chaos (and later, to an extent, in *FPHK*, *FPHFS*, and *FPHGAM*, for example).

And lastly, the comparison in a classification is not absolute, but relative, as there are often some shades of gray, not exactly black or white (figuratively speaking).

Some further clarifications and qualifications of sophisticated methodological holism are shown in *Table 4.2*.

Some Distinctive Usefulness

In the end, sophisticated methodological holism—when applied, especially though not exclusively, as illustrated in my numerous works—can enrich the understanding of reality in some distinctive ways. Here are four examples (as revealed in each of my books).

Firstly, it provides a comprehensive analysis of a subject matter, from which much can be learned about reality, both in relation to the perspectives of the mind, nature, society, and culture.

Secondly, it offers a new classification of the subject matter.

Thirdly, it suggests some visions of the future in relation to the subject matter in question.

And fourthly, it proposes some insights on meta-theory (e.g., methodology and ontology) in general—with the clear understanding, however, of the dilemma of specific vs. general ontology (as shown in *Table 4.9* on the syntax of existential dialectics in the context of the dilemma of ontology).

For this very reason, all of these ways are important, without reducing one into the analysis of another. (*FPHCT*)

Chapter Outline

With this summary of my two original meta-theories (on methodology and ontology, *almost verbatim* from my previous works) in mind—this book is organized in four main parts, namely, (a) Part I on the introduction to this book, (b) Part II on uselessmess, (c) Part III on usefulness, and (d) Part IV on the conclusion of the book.

The book is thus organized in four chapters.

The introductory chapter, that is, Chapter One titled "Introduction—The Existence of Waste," begins with a story of two practices, provides a definition and history of waste, and then introduces uselessness and usefulness in the study of waste, as well as the diverse aspects of waste—together with the theoretical debate on waste, my transfigurative theory of waste, the relationship between theory and meta-theory, existential dialectics, sophisticated methodological holism, and some further clarifications.

Chapter Two titled "Uselessness and its Dual Poles" examines the nature of uselessness in waste from the four perspectives of the mind, nature, society, and culture, with a deconstructive critique of each.

Chapter Three titled "Usefulness and its Double Sides" explores the nature of usefulness in waste from the four perspectives of the mind, nature, society, and culture, with a constructive assessment of each.

The last chapter titled "Conclusion—The Future of Waste" summarizes the analysis in the book in the dialectic context of uselessness and usefulness in waste with twenty-nine major theses, namely, (a) the first thesis: the formalness-informalness principle, (b) the second thesis: the absoluteness-relativeness principle, (c) the third thesis: the partiality-totality principle, (d) the fourth thesis: the predictability-unpredictability principle, (e) the fifth thesis: the explicability-inexplicability principle, (f) the sixth thesis: the fiction-reality principle, (g) the seventh thesis: the finiteness-transfiniteness principle, (h) the eighth thesis: the preciseness-vagueness principle, (i) the ninth thesis: the simpleness-complicatedness principle, (j) the tenth thesis: the openness-hiddenness principle, (k) the eleventh thesis: the denseness-emptiness principle, (l) the twelfth thesis: the change-constancy principle, (m) the thirteenth thesis: the order-chaos principle, (n) the fourteenth thesis: the slowness-quickness principle, (o) the fifteenth thesis: the expansion-contraction principle, (p) the sixteenth thesis: the theory-praxis principle, (q) the seventeenth thesis: the convention-novelty principle, (r) the eighteenth thesis: the evolution-transformation principle, (s) the nineteenth thesis: the symmetry-asymmetry principle, (t) the twentieth thesis: the softness-hardness principle, (u) the twenty-first thesis: the seriousness-playfulness principle, (v) the twenty-second thesis: the activeness-inactiveness principle, (w) the twenty-third thesis: the selfness-otherness principle, (x) the twenty-fourth thesis: the regression-progression principle, (y) the twenty-fifth thesis: the sameness-difference principle, (z) the twenty-sixth thesis: the stability-reaction principle, (aa) the twenty-seventh thesis: the functionality-nonfunctionality principle, (bb) the twenty-eighth thesis: the intentionality-nonintentionality principle, and (cc) the twenty-ninth thesis: the post-human transfiguration—to be elaborated in the rest of the book, with a summary in the concluding chapter.

This seminal project here, if successful, will fundamentally change the way that we think about uselessness and usefulness in the study of waste (together with other debates as will be discussed in the rest of the book), from the combined perspectives of the mind, nature, society, and culture, with enormous implications for the human future and what I originally called its "post-human" fate.

Some Clarifications

However, some additional (last) clarifications are needed.

These clarifications, lest any potential misunderstanding by the readers occurs, refer to (a) the conversation with my previous books, (b) the illustrative selection of case studies and examples, (c) detailed analysis versus overall analysis, (d) three distinctive features of using quotations, (e) two difficulties with understanding, (f) repetition, and (g) the use of neologisms.

These clarifications are something that I also regularly repeat in my previous books too, so these points are repeated hereafter, *often verbatim*—albeit in a different context this time.

In Conversation with My Previous Books

Firstly, my *inquisitive theory of culinary art* is constructed on the theoretical foundation of my previous books (i.e., *FHC*, *FCD*, *FPHC*, *BDPD*, *BCPC*, *BCIV*, *FPHST*, *BNN*, *BWT*, *FC*, *FAE*, *ALD*, *FIA*, *FPHU*, *FPHK*, *FPHML*, *FPHE*, *FPHMM*, *FPHCT*, *FPHG*, *FPHUP*, *FPHL*, *FPHO*, *FPHMA*, *FPHS*, *FPHLAW*, *FPHFS*, *FPHWP*, *FPHP*, *BCOS*, *BEPE*, *FPHGEOL*, *FPHCHESS*, *FPHR*, *FPHEDU*, *FPHH*, *FPHA*, *FPHGAM*, *FPHCOM*, *FPHLIT*, *FPHCHEM*, *FPHDA*, *FPHGEOG*, *FPHPA*, *FPHSEM*, *FPHARCH*, *FPHCRIM*, *FPHHIST*, *FPHMIG*, *FPHTRAN*, and *FPHCA*—as explicated in "The List of Abbreviations").

For this reason, many of my books have some familiar words in their titles like "The Future of Post-Human ____" or "Beyond ____to ____," as well as some tables for background information—as they constitute a series of books with the final aim to converge all of my original theories on numerous fields into a unified theory of everything (as already summarized in the section on theory and meta-theory in Chapter One).

Just consider the following two illustrations, namely, (a) in relation to other related theoretical debates and (b) in relation to different visions as worked out in my previous books.

(a) The first illustration is that the theoretical debate here also involves a few other related theoretical issues.

These other related theoretical issues have to do with some deeply contested theoretical debates, which, however, I had already gone to great lengths to analyze in my numerous (previous) books, as summarized below. It should also be stressed, lest misunderstanding occurs, that each of the theoretical debates (as listed below) does not exhaust the issues covered in the books, as each book covers more issues than what is listed below (for illustration only).

• On the Environment
  –Ex: *The Future of Post-Human Geography*
  –Ex: *The Future of Post-Human Geology*
  –Ex: *The Future of Post-Human Urban Planning*
  –Ex: *The Future of Post-Human Migration*
  –Ex: *The Future of Capitalism and Democracy*
  –Ex: 2 volumes, *The Future of Human Civilization*

- On Food
  - Ex: *The Future of Post-Human Culinary Art*
- On Construction
  - Ex: *The Future of Post-Human Architecture*
- On Chemistry
  - Ex: *The Future of Post-Human Chemistry*
- On Technology
  - Ex: *The Future of Post-Human Engineering*
  - Ex: 2 volumes, *The Future of Human Civilization*
  - Ex: *The Future of Post-Human Computing*
- On Engineering
  - Ex: *The Future of Post-Human Engineering*
- On Space Colonization
  - Ex: *The Future of Post-Human Geography*
  - Ex: *The Future of Post-Human Migration*
  - Ex: *The Future of Capitalism and Democracy*
  - Ex: 2 volumes, *The Future of Human Civilization*
- On Economics
  - Ex: *The Future of Capitalism and Democracy*
  - Ex: 2 volumes, *The Future of Human Civilization*
  - Ex: *Beyond Capitalism to Post-Capitalism*
- On Ethics
  - Ex: *Beyond Ethics to Post-Ethics*
  - Ex: 2 volumes, *The Future of Human Civilization*
- On Spirituality
  - Ex: 2 volumes, *The Future of Human Civilization*
  - Ex: *The Future of Capitalism and Democracy*
  - Ex: *The Future of Post-Human Religion*
  - Ex: *Beyond Civilization to Post-Civilization*
  - Ex: *The Future of Post-Human Sexuality*
  - Ex: *The Future of Post-Human Martial Arts*
  - Ex: *The Future of Post-Human Unconsciousness*
  - Ex: *The Future of Post-Human Consciousness*
  - Ex: *Beyond Ethics to Post-Ethics*
- On Aesthetics
  - Ex: *The Future of Aesthetic Experience*
  - Ex: *The Future of Post-Human Culinary Art*
  - Ex: *The Future of Post-Human Architecture*
  - Ex: *The Future of Post-Human Martial Arts*
  - Ex: *The Future of Post-Human Performing Art*
  - Ex: *The Future of Post-Human Literature*

- On Globalization
  - –Ex: *The Future of Capitalism and Democracy*
  - –Ex: 2 volumes, *The Future of Human Civilization*
- On Civilization
  - –Ex: *Beyond Civilization to Post-Civilization*
  - –Ex: 2 volumes, *The Future of Human Civilization*
- On Epistemology and Ontology
  - –Ex: *The Future of Post-Human Knowledge*
  - –Ex: *The Future of Post-Human Consciousness*
  - –Ex: 2 volumes, *The Future of Human Civilization*
- On Rationality
  - –Ex: *The Future of Post-Human Mathematical Logic*
  - –Ex: *The Future of Post-Human Humor*
- On Freedom
  - –Ex: 2 volumes, *The Future of Human Civilization*
  - –Ex: *The Future of Capitalism and Democracy*
  - –Ex: *Beyond Capitalism to Post-Capitalism*
  - –Ex: 2 volumes, *Beyond Democracy to Post-Democracy*
  - –Ex: *Beyond Civilization to Post-Civilization*
- On Communication
  - –Ex: *The Future of Post-Human Mass Media*
  - –Ex: *The Future of Post-Human Organization*
- On Learning
  - –Ex: *The Future of Post-Human Language*
  - –Ex: *The Future of Post-Human Education*
- On Psychology
  - –Ex: *The Future of Post-Human Unconsciousness*
  - –Ex: *The Future of Post-Human Consciousness*
  - –Ex: *The Future of Capitalism and Democracy*
  - –Ex: *The Future of Post-Human Creative Thinking*
  - –Ex: *The Future of Post-Human Personality*
- On Cognition
  - –Ex: *The Future of Post-Human Unconsciousness*
  - –Ex: *The Future of Post-Human Consciousness*
  - –Ex: *The Future of Capitalism and Democracy*
  - –Ex: *The Future of Post-Human Creative Thinking*
  - –Ex: *The Future of Post-Human Personality*
- On Intuition
  - –Ex: *The Future of Post-Human Unconsciousness*
  - –Ex: *The Future of Post-Human Consciousness*
  - –Ex: *Beyond Capitalism to Post-Capitalism*

–Ex: *The Future of Capitalism and Democracy*
- On Emotion
  –Ex: *The Future of Post-Human Unconsciousness*
  –Ex: *The Future of Post-Human Consciousness*
  –Ex: *The Future of Aesthetic Experience*
  –Ex: *Beyond Capitalism to Post-Capitalism*
  –Ex: *The Future of Capitalism and Democracy*
  –Ex: *The Future of Post-Human Personality*
- On Behavior
  –Ex: *The Future of Post-Human Language*
  –Ex: *The Future of Post-Human Unconsciousness*
  –Ex: *The Future of Post-Human Consciousness*
  –Ex: *Beyond Capitalism to Post-Capitalism*
  –Ex: *The Future of Capitalism and Democracy*
  –Ex: *The Future of Post-Human Sexuality*
  –Ex: *The Future of Post-Human Personality*
  –Ex: *The Future of Post-Human Martial Arts*
- On Being and Belonging
  –Ex: *The Future of Post-Human Consciousness*
  –Ex: 2 volumes, *The Future of Human Civilization*
- On Different Dimensions of Human Existence
  –Ex: 2 volumes, *The Future of Human Civilization*
- On Modernity, Post-Modernity, and After-Postmodernity
  –Ex: 2 volumes, *The Future of Human Civilization*
  –Ex: *Beyond Civilization to Post-Civilization*
- On Pre-Modernity
  –Ex: *Beyond Civilization to Post-Civilization*

Therefore, those who are interested in these other theoretical issues (and debates) which are related to the current context can consult these other books of mine for more details.

(b) And the second illustration is that one of the major theses of this book concerns the future, be it human and later post-human, which is closely linked with the different visions that I had worked out in my previous books, in relation to the mind, nature, society, and culture.

Of course, for the convenience of the readers, a summary of some of these visions will be provided whenever necessary, as shown in the two sections on meta-theory above (*often verbatim*) and in different tables at the end of this chapter and the concluding chapter, for instance.

Yet, I still expect the readers to read my previous books directly for more details—as this book is written in conversation with my previous ones.

Case Studies and Examples

Secondly, the case studies and examples in the chapters are not exhaustive but solely illustrative. Nor are they necessarily mutually exclusive, as they can be reclassified in a different way. And exceptions are allowed.

One is tempted to assume, however, that, since the case studies and their examples are not exhaustive, more case studies and examples are needed. But the problem here is that just adding more of them does not necessarily change the conclusions to be drawn and may even lead to redundancy.

After all, the case studies and examples used in this book are chosen with care and deemed sufficient for the purpose at hand, even though they are solely illustrative.

Detailed Analysis versus Overall Synthesis

Thirdly, there is a distinction between detailed analysis and overall synthesis, in that the former is, especially though not exclusively, more for the specialized viewpoint of a specific discipline, whereas the latter is, especially though not exclusively, more for the broad horizon of an interdisciplinary approach—relatively speaking, of course.

Or, by analogy, there is a distinction between depicting a "tree" (with tiny details) in a "forest" and viewing the entire "forest" itself (with the broader horizon).

This book, like all previous ones of mine, tries to strike a balance between the two forms of understanding, although this does not imply that the two forms should be exactly equally offered, in an exact 50-50 balance.

Instead, the book, more often than not, does not allow the tiny details to obscure the larger picture in a given case study, because the entire book is inter-disciplinary in nature and tries to look at the entire "forest," not being bogged down with the detailed analysis of an individual "tree" in the forest, by analogy.

In this sense, although detailed analysis can be provided whenever needed for a given case study, the important point to remember is the overall horizon with its broadness of scope.

Three Distinctive Features of Using Quotations

Fourthly, the way that quotes are used in this book (and others of mine) has three distinctive features, as explained below.

The first distinctive feature is that, at the end of each quotation, sometimes there are more than one reference—in which case the first reference is for the original citation, and the rest of the references are for the convenience of the reader for more information about the issue(s) under quotation.

The second distinctive feature is that quotations are used as often as possible, so as to let others speak for themselves, without the nuisance of putting my own words into their mouths. Even more importantly, this allows me to use the original text as an evidence for critical analysis at the end of each section.

And the third distinctive feature is that all these quotations are used for my own purposes, in that I often creatively (or originally) re-organize them for a new way of analyzing the issue(s) in question.

Two Difficulties with Understanding

Fifthly, to understand this book (or any other book of mine) is difficult for two main reasons.

In regard to the first difficulty, this book, just like many other books of mine, is in conceptual and theoretical conversation with other books of mine (54 books, as of July 2012), so this requires a vast background information on numerous subject matters.

And in regard to the second difficulty, this book also analyzes the subject matter in a way very different from the expectation of a conventional specialist reader, since it examines reality outside the specialist's interest (or outside his or her "comfort zone"), from the four broad perspectives of the mind, nature, society, and culture, such that it covers all domains of knowledge (in the social sciences, the natural sciences, the formal sciences, and the humanities). In other words, those who are truly interested in my project for a unified theory of everything would find many books of mine highly stimulating, but those who are specialists in a subject matter (like physics, geography, chess, and so on, in a strictly specialist sense) will not find them suitable to their specialist taste (or expectation).

Not surprisingly, a conventional specialist reader would find these two difficulties (with understanding) almost practically impossible to overcome, as they almost require the mind of a polymath. Even then, the best advice is that one should still try one's best to understand it.

Repetition

Sixthly, many of my books have something in common, and the reason is to show "the unified theory of everything" in all domains of knowledge.

For this reason, there is certain repetition in all books (especially in regard to the background information, together with the tables, in the introductory chapter and the concluding chapter). Therefore, these books are all titled "The Future of Post-Human ____" (or at times "Beyond ____ to ____") and constitute the "post-human" series.

The Use of Neologisms

And finally, I use different neologisms in my books, mostly to introduce my original concepts and theories, and good instances include those here or elsewhere in my previous books (e.g., "the transfigurative theory of waste," "post-democracy," "hyper-spatial consciousness," "post-capitalism," and whatnot). Needless to say, they are used here solely for our current intellectual convenience, as they will be renamed differently in different ways in future history.

As I thus wrote in *FCD* (508-9), "all these terms 'post-capitalism,' 'post-democracy'...and other ones as introduced in...[the] project (e.g.,...'posthuman elitists,' and 'posthuman counter-elitists,' just to cite a few of them) are more for our current intellectual convenience than to the liking of future humans and post-humans, who will surely invent more tasteful neologisms to call their own eras, entities, and everything else, for that matter. But the didactic point here is to use the terms to foretell what the future might be like, not that its eras and entities must be called so exactly and permanently. After all, William Shakespeare (1995: Act II, Scene II, Line 47) well said long ago: 'What is in a name? That which we call a rose by any other name would smell as sweet.'"

As I had also stressed time and again before, each of the neologisms can be re-written as a different "X," only to be re-named differently by the powers that be in different eras of future history.

With these clarifications for thought in mind—I now proceed to Chapter Two on uselessness and its dual poles.

Table 1.1. Diverse Aspects of Waste

- **Uselessness and Usefulness**
  - —Uselessness
    - Ex: waste prevention
    - Ex: waste disposal
  - —Usefulness
    - Ex: waste reuse
    - Ex: waste recycle
    - Ex: waste-to-energy

- **Waste Types**
  - —Ex: "industrial waste"
  - —Ex: "electronic waste (e-waste)"
  - —Ex: "construction and demolition waste"
  - —Ex: "commercial waste"
  - —Ex: "solid wastes"
  - —Ex: "animal by-products"
  - —Ex: "biomedical waste"
  - —Ex: "chemical waste"

- **Waste Management Topics**
  - —Ex: "biodrying"
  - —Ex: "carbon management"
  - —Ex: "industrial wastewater treatment"
  - —Ex: "packaging and labelling"
  - —Ex: "recycling"
  - —Ex: "sewage collection and disposal"
  - —Ex: "windrow composting"

Notes: The categories are solely illustrative (not exhaustive), and the comparison is relative (not absolute), nor are they necessarily mutually exclusive. And some can be easily re-classified elsewhere. As generalities, they allow exceptions.

Source: From *Sec. 1.3* & *Sec. 1.4* of *FPHW*. See book for citations.

Table 1.2. Nine Great Future Transformations of Waste (Part I)

- **In Relation to the Mind**
  —*New Goals*
  - This search search for "new goals" will give future advanced humans and later their successors, namely, the post-humans, different motivations to treat waste in a way that current humans cannot—in relation to my previous visions on the future "post-human" supersession of the human mind like "the supersession computing," "metamorphic humor," "the contrarian personality," "the hyper-martial body," "the hyper-sexual body," "floating consciousness," "hyper-spatial consciousness," "thinking machines," "unfolding unconsciousness," "thinking robots," "genetically altered superior beings," "cyborgs," etc. For instance, most advanced forms of future intelligent life in deep space until multiverses (e.g., "floating consciousness") do not need to have the physical bodies that current humans have and therefore do not necessarily produce the kinds of waste that current humans do.

  —*New Means*
  - This quest for "new means" will give future advanced humans and later their successors, namely, the post-humans, the capacity to find novel options to achieve the goals of waste treatment in a way that current humans cannot.

- **In Relation to Nature**
  —*New Comparisons*
  - This search for "new comparisons" will give future advanced humans and later their successors, namely, the post-humans, the ability to learn from animals, plants, and the universe, through comparison and contrast, about the future of waste in a way that current humans cannot.

(continued on next page)

Table 1.2. Nine Great Future Transformations of Waste (Part II)

- **In Relation to Nature (*cont'd*)**
  —*New Integrations*
  - This search for "new integrations" will give future advanced humans and later their successors, namely, the post-humans, the ability to live and interact with nature for the future of waste in a way that current humans cannot. This is especially so, in the context of the great future transformations of geology (in *FPHGEOL*), of geography (in *FPHGEOG*), of migration (in *FPHMIG*), and of transportation (in *FPHTRAN*)—in relation to my previous visions on the future "post-human" adventures into different dimensions of space-time unto multiverses like "the alteration of space-time" (*Table 1.3* & *Table 1.4*), "multiverses" (*Table 1.7* & *Table 1.8*), "hyperspace" (*Table 1.9* & *Table 1.25*), "time travel" (*Table 1.10* & *Table 1.11*), and others.

    This extension is consistent with my other theory, namely, "the theory of the cyclical progression of system integration and fragmentation" in *FCD* (2002)—at the social-systemic level (as already clarified in *Sec. 1.8* and shown below):

    Local → Regional → Global →Solar →
    Galactic → Clustery…→ Multiversal

(continued on next page)

Table 1.2. Nine Great Future Transformations of Waste (Part III)

- **In Relation to Society**
  —*New Praxes*
  - This search for "new praxes" will give future advanced humans and later their successors, namely, the post-humans, the ability to engage in practice (like activism, politics, personal experience, etc.) for the future of waste in a way that current humans cannot.

  —*New Technologies*
  - This search for "new technologies" will give future advanced humans and later their successors, namely, the post-humans, the ability to come up with different techniques, materials, and experiments for the future of waste in a way that current humans cannot.

  —*New Stratifications*
  - This emergence of "new stratifications" will lead future advanced humans and later their successors, namely, the post-humans, to face new inequalities in the future of waste in a way that current humans cannot.

- **In Relation to Culture**
  —*New Sub-Fields*
  - This search for "new sub-fields" will give future advanced humans and later their successors, namely, the post-humans, the ability to explore new specialties for the future of waste in a way that current humans cannot.

  —*New Interpretations*
  - This search for "new interpretations" will give future advanced humans and later their successors, namely, the post-humans, the ability to come up with different evaluations and views about waste in a way that current humans cannot.

Notes: The categories and examples are solely illustrative (not exhaustive), and the comparison is relative (not absolute), nor are they necessarily mutually exclusive. And some can be easily re-classified elsewhere. As generalities, they allow exceptions.
Source: From *Ch.4* of *FPHW*. See text for citations.

Table 1.3. The Theoretical Debate on Space-Time (Part I)

- **Isaac Newton's Absolutist (Substantivist) Theory of Space-Time**
  —space and time are independent from each other. The structure of space-time is *E3×E1* (with the structure of space, *P*, as the set of spatial locations in a three-dimensional Euclidean space, *E3*, and the structure of time as the set of temporal moments, *T*, in the one-dimensional real time, *E1*).
  —space and time are also independent from the effects of matter and events. The existence of space and time is possible even in a world absent of matter (and, for that matter, even in a world absent of events), as if they were material objects but with their total unchangingness thorough time.

- **Albert Einstein's Relativist Theory of Space-Time**
  —space and time are interchangeable (not absolute), just as matter and energy are equivalent (not independent) with the famous equation, $E = mc^2$ (as in the special theory of relativity in 1905).
  —space-time and matter-energy are also relative in a grand union (as in the general theory of relativity in 1915). Thus, each pair affects the other pair, as "matter 'tells' spacetime how to curve [in a non-Euclidean geometry] and curved spacetime 'tells' matter how to behave….Space contracts near mass and dilates away from it. Time dilates near mass and contracts away from it. Clocks positioned farther away from the mass of the earth run faster than clocks closer to the earth." (L. Shlain 1991: 328-330)

(continued on next page)

Table 1.3. The Theoretical Debate on Space-Time (Part II)

- **Peter Baofu's Perspectival Theory of Space-Time**

—space and time can be understood from multiple perspectives, be they in relation to culture, society, nature, and the mind, with each perspective revealing something about the nature of space-time and simultaneously delimiting its view. This is subject to "the regression-progression principle" in existential dialectics.

—each perspective of space and time exists in society and culture with good reasons, with some being more successful and hegemonic (dominant) than others. This is subject to "the symmetry-asymmetry principle" in existential dialectics.

—space and time will not last, to be eventually superseded (altered) by post-humans in different forms (e.g.,stretching/shrinking space-time, engineering more dimensions of space-time, and manipulating multiverses), be they here in this universe or in multiverses. Thus, even the physical existence of space-time cannot last forever, with ever more transformations in the process. This is subject to "the change-constancy principle" in existential dialectics.

—the conventional wisdom (especially by physicists) of treating the physical perspective of space and time as the foundation of all other perspectives (of space and time) and of regarding them as much less important is a form of reductionism,committing what I call *the foundation fallacy*, in misleadingly dismissing the multiple perspectives of space and time in relation to culture, society, nature, and the mind.

Notes: The examples in each category are solely illustrative (not exhaustive), and the comparison is relative (not absolute), nor are they necessarily mutually exclusive. Some can be easily re-classified elsewhere. As generalities, they allow exceptions.

Source: A summary of *Sec.1.1*, *Sec.1.2*, and *Sec.1.3* of *FPHST*—and, for that matter, the rest of *FPHST*

Table 1.4. Main Reasons for Altering Space-Time

- **The Need to Make New Energy-Matter**
  —Ex: manipulating molecular bonds for new materials
  —Ex: creating nanotechnologies on the atomic scale
  —Ex: engineering the atomic nucleus
  —Ex: restructuring most elementary particles
  —Ex: inventing new forms of matter and energy

- **The Need to Create New Space-Time**
  —Ex: creating "warp drive" (as in science fiction) for space travel
  —Ex: creating "pocket universes"

- **The Need to Conquer the Cosmos unto Multiverses**
  —Ex: spreading floating consciousness and hyper-spatial consciousness, besides other forms that humans have never known, in the cosmos and beyond unto multiverses for ultimate conquest

Notes: The examples in each category are solely illustrative (not exhaustive) nor necessarily mutually exclusive, and the comparison is relative (not absolute). As generalities, they allow exceptions. Also, it should be stressed that the three reasons are all related, in that they all contribute to the evolution of intelligent life in the cosmos unto multiverses in the most distant future beyond our current knowledge.
Sources: A summary of *Sec.6.2* of *FPHST*. See also *FHC*, *FCD*, *FPHC*, *FC*, *FPHU*, and *FPHG*, for example.

Table 1.5. The Technological Frontiers of the Micro-World

- **Type I-Minus**
  —Ex: building structures and mining

- **Type II-Minus**
  —Ex: playing with the genetic makeups of living things

- **Type III-Minus**
  —Ex: manipulating molecular bonds for new materials

- **Type IV-Minus**
  —Ex: creating nanotechnologies on the atomic scale

- **Type V-Minus**
  —Ex: engineering the atomic nucleus

- **Type VI-Minus**
  —Ex: restructuring most elementary particles

- **Type Ω-Minus**
  —Ex: altering the structure of space-time

Notes: As already indicated in *Sec.4.4.2.2* of *FPHC*, the problem with this micro-classification (from Barrow's work) is that the civilization types (with the exception of Type Ω-Minus, for example) are not quite distinct, since many of them can be achieved more or less in a civilization, to the extent that Type II-minus and Type III-minus, just to cite two plausible types, can be historically contemporaneous, relatively speaking, unlike the vast historical distance between, say, Type O and Type I (or Type I and Type II) civilizations. In other words, the micro-classification here is not very useful to understand civilization types but is revealing to see the technological frontiers of the micro-world.
Sources: A reconstruction from J.Barrow (1998:133), as originally shown in *Table 4.7* of *FPHC*. See *FPHC* for more info.

Table 1.6. Finity, Transfinity, and Infinity (Part I)

• **Infinity**

—The term "infinity," which is "symbolically represented with ∞," derives from the Latin *infinitas* (or "unboundedness"), to refer to "several distinct concepts—usually linked to the idea of 'without end'—which arise in philosophy, mathematics, and theology." (WK 2008c)

—In the context of mathematics, "infinity" is often "treated as if it were a number (i.e., it counts or measures things: 'an infinite number of terms') but it is a different type of 'number' from the real numbers. Infinity is related to limits,…large cardinals,…projective geometry, extended real numbers and the absolute Infinite," for instance. (WK 2008c)

• **Transfinity**

—To avoid confusion, my usage of the word "transfinite" (as used here in the ontological principle) is not the same as the one which was originally coined by Georg Cantor to refer to "cardinal numbers or ordinal numbers that are larger than all finite numbers, yet not necessarily absolutely infinite" (in the mind of God), as opposed to "relative" infinity (in the mind of Man). (WK 2008d) In other words, for Cantor, there are two kinds of infinity, "absolute" and "relative"—with "relative" infinity or "transfinity" (in the mind of Man) being dependent on "absolute" infinity (in the mind of God) for its existence.

—By contrast, my usage of the word "transfinite" here differs radically from the one used by Cantor (and other mathematicians) and is more limited, in light of the problems confronting any attempt to understand the idea of infinity, be it by intuition, imagination, and conception (as already shown in the prevous three sub-sections).

—With this clarification in mind, I allow numbers which can be many times larger—or smaller, for that matter—than the finite things that we encounter in daily life, but they do not have to be related to the idea of infinity at all (which may not exist).

(continued on next page)

Table 1.6. Finity, Transfinity, and Infinity (Part II)

• **Transfinity** ***(cont'd)***

—Of course, there may be some *borderline* cases, in which it is not clear whether the number in question is transfinite (in my usage) or simply a mathematical convenience. A good example of a borderline case is the Planck unit of length for "the smallest space possibly measured in nature," which is "less than billionths of trillionths of trillionths of an inch" (or something like 1.6×10^{-35} meters). (P. Baofu 2006a; N. McAleer 1987: 219; WKV 2008; D. Corbett 2008)

—The ideas concerning symmetry and dimensions in the examples above were addressed in Chapter Three and Chapter Four of *FPHG*. For now, it suffices to further add that it is mind-boggling to even conceive or imagine extreme numbers like the Planck unit of length (or others). Does it really exist at all? Or is it merely a mathematical convenience to describe a physical situation that the human mind does not understand properly?

—In any event, "transfinity" (in my usage) can be used as a further extension (of number) to the finite numbers in daily life—albeit with some borderline cases in mind.

• **Finity**

—It refers to numbers which are bounded (that is, with an end), especially (though not exclusively) in relation to things in everyday life.

Source: From *Sec. 1.4* and *Sec. 2.2.3* of *FPHG*

Table 1.7. Theoretical Speculations of Multiverses

- **"Baby Universes" (Ex: Andre Linde and others)**
  —Ex: In a flat universe theory, "even if our part of it eventually collapses,…some spots in the cosmos would suddenly start inflating on their own, creating brand-new 'baby universes.'" (P. Baofu 2000: 623)

- **"Parallel Universes" (Ex: Stephen Hawking and others)**
  —Ex: In quantum cosmology, there allows the existence of infinite numbers of parallel universes, with tunneling among them. (M. Kaku 1994: 256) Hawking later revised his views on this.

- **"Pocket Universes" (Ex: Alan Guth)**
  —Ex: "As the pocket universes live out their lives and recollapse or dwindle away, new universes are generated to take their place….While life in our pocket universe will presumably die out, life in the universe as a whole will thrive for eternity." (A. Guth 1997: 248; P. Baofu 2002: 482)

- **"Brane Worlds" (Ex: Warren Siegel, Lisa Randall, and others)**
  —Ex: Our universe is stuck on a membrane of space-time embedded in a larger cosmos, with different brane worlds connecting and/or colliding with each other.

Notes: These examples are solely illustrative (not exhaustive), and some of the items can be reclassified somewhere else. Nor are they always mutually exclusive. Since they are generalities, exceptions are expected.
Source: From *Table 4.8* of *FPHC*

Table 1.8. The Confusion between "Many Worlds" and "Multiverse" (Part I)

• **First Problem**

—The first problem in this second confusion concerns a debate between the "many-worlds" interpretation of quantum mechanics by Hugh Everett (1957) and others, and the Copenhagen interpretation by Niels Bohr and Werner Heisenberg on the effects of quantum states after interacting with an external environment (e.g., by a measuring observer).

—On the one hand, the Copenhagan interpretation of quantum mechanics is more indeterministic in its concern with the probabilities of the outcomes of a wave function after interacting with an external environment (e.g., by a measuring observer using light to detect, for example, "an electron in a particular region around the nucleus at a particular time." (WK 2008e) Each wave function for the state of a particle refers to "a mathematical representation used to calculate the probability for it to be found in a location, or state of motion." (WK 2008f) And the "wavefunction collapse" means that "the act of measurement causes the calculated set of probabilities to 'collapse' to the value defined by the measurement." (WK 2008f)

—On the other hand, the Everett's interrpetation is more deterministic, in rejecting "the objective reality of wavefunction collapse" and "instead explaining the subjective appearance of wavefunction collapse with the mechanism of quantum decoherence." (WK 2008g) The words "quantum decoherence" here refers to "the mechanism by which quantum systems interact with their environments to exhibit probabilistically additive behavior," in such a way that "the quantum nature of the system is simply 'leaked' into the environment." (WK 2008g) For example, "[a]s a result of an interaction, the wave functions of the system and the measuring device become entangled with each other. Decoherence happens when different portions of the system's wavefunction become entangled *in different ways* [italic added] with the measuring device...." (WK 2008g) Thus, for Everett, the appearance of "wavefunction collapse" can be explained by "quantum decoherence" for quantum states to branch out in different independent worlds after interacting with an external environment.

(continued on next page)

Table 1.8. The Confusion between "Many Worlds" and "Multiverse" (Part II)

- **First Problem *(cont'd)***
  —Yet, a major criticism against Everett's interpretation is that these different independent worlds in Everett's sense "will never be accessible to us," because of the interference by the measuring observer in an external environment. (WK 2008g)
  —For illustration, "once a measurement is done, the measured system becomes entangled with both the physicist who measured it and a huge number of other particles, some of which are photons flying away towards the other end of the universe; in order to prove that the wave function did not collapse one would have to bring all these particles back and measure them again, together with the system that was measured originally. This is completely impractical, but even if one can theoretically do this, it would destroy any evidence that the original measurement took place (including the physicist's memory)." (WK 2008g)
  —But then, the critics ask, If these independent worlds in Everett's sense are inaccessible to us, how do we know much of anything really about them?

- **Second Problem**
  —What is relevant at the quantum mechanical level may not necessarily apply to the world at the physicsl level.
  —In fact, the persistent inability among contemporoary physicists to unifiy the theory of relativity for large objects at the physical level and quantum mechanics for tiny objects at the quantum mechanical level should warn us about the danger of reductionism, by reducing the understanding of different universes ("parallel universes") at the physical level from a generalization of "many worlds" at the quantum mechanical level.

- **Third Problem**
  —There is no empirical evidence to suggest that there are infinite (or extremely many) possible worlds in the cosmos as the many-worlds model by Everett would have us to believe.

Source: From *Sec.4.4.1* of *FPHG*

Table 1.9. Hyperspace and Its Challenge (Part I)

• **First Problem**

—Hyperspace for intra- and inter-universal travel is assumed to be possible, on the basis of four main assumptions; although there may be others, these are the often cited ones.

—The first assumption concerns the existence of "entry" and "exit" points in hyperspace, which is easier said than done to enter, go through, and then exit hyperspace.

—For instance, using hyperspace "requires complex calculations…of single or multiple hyperspace jumps and the control of the jump….Peturbations such as those experienced by ship in space from the gravitional field around an object such as a planet or even a star are exacerbated in hyperspatial travel, since mass in real space distorts hyperspace in an equal measure. 'Jumping' near to a gravitational mass is likely to make resulting exit from hyperspace to be highly uncertain, with the level of improbability *i* increasing with the square of the distance to the nearest gravitional 'well.'" (WK 2008h)

—Besides, "[o]nly spaceships equipped with a special force field can enter hyperspace, because exposure to hyperspace even for short period of time is hazardous to unprotected humans." (WK 2008h)

• **Second Problem**

—The second assumption concerns the faster-than-light speed. It is well known that Einstein argued for the speed of light as the ultimate limit, but nowadays phyiscists do allow the violation of this limit under certain conditions (like the hypothetical existence of tachyons, as will be explained further in a later sub-section on time travel).

—Even then, there is a challenge to explain in science fiction "why ships can travel faster than light in hyperspace," and two good arguments are that "hyperspace may be smaller than real space and therefore a star ship's propulsion seems to be greatly multiplied, or else the speed of light in hyperspace is not a barrier as it is in real space. Whatever the reasoning, the general effect is that ships traveling in hyperspace seem to have broken the speed of light, appearing at their destinations much more quickly and without the shift in time that the Theory of Relativity would suggest." (WK 2008h)

(continued on next page)

Table 1.9. Hyperspace and Its Challenge (Part II)

• **Third Problem**
—The third assumption concerns the speed of time (to be measured, say, by a clock), but this is elaborated in a different section on the related issue of "time travel" (as shown in a different table on time travel).

• **Fourth Problem**
—The fourth assumption concerns spatial shortcut, as it is possible, because of the curvature of space, in that "the manifold of ordinary three-dimensional space is curved in four or more 'higher' spacial dimensions (a 'hyperspace' in the geometric sense…). This curvature causes certain widely separated points in three-dimensional space to nonetheless be 'adjacent' to each other four-dimensionally. Creating an aperture in 4D space (a wormhole) between these locations can allow instantaneous transit between the two locations; a common comparison is that of a folded piece of paper, where a hole punched through two folded sections is more direct than a line drawn between them on the sheet. This idea probably arose out of certain popular descriptions of General Relativity and/or Riemannian manifolds, and may be the original form from which later concepts of hyperspace arose." (WK 2008h)
—But this does not mean that hyperpsace is always safe. In fact, "in some science fiction, the danger of hyperspace travel is due to the chance that the route through hyperspace may take a ship too close to a celestial body with a large gravitational field, such as a star. In such scenarios, if a starship passes too close to a large gravitational field while in hyperspace, the ship is forcibly pulled out of hyperspace and reverts to normal space. Therefore, certain hyperspace 'routes' may be mapped out that are safe, not passing too close to stars or other dangers." (WJ 2008h)

Source: From *Sec.4.4.1* of *FPHG*

Table 1.10. The Problems of Time Travel into the Future (Part I)

- **First Problem**

—The first problem concerns "time dilation under the Theory of Special Relativity" (which separates space-time from mass-energy as special cases). Suppose you are on a moving train and an observer outside the train is watching you along the tracks. As your train moves along at the speed of light (or something close to it), "time, as measured by your watch, ticked along at a slower pace than time measured by the observer. Not only that, distance changed, too. For the observer, a one-foot ruler whizzing by on the train would have measured less than a foot." (NO 2000)

—In fact, you may not even notice that this slower clicking of the clock; in fact, everything looks normal to you inside the train as it was before: "The weird thing is that, for you on the train, time wouldn't seem to be moving slower and your ruler wouldn't be shorter—all would appear normal." (NO 2008) By contrast, you instead think that "time on the rest of the Earth would appear to be ticking along slower and its rulers would be shorter." (NO 2000) So, both of you insist that the other clock (not his) is clicking more slowly—assuming, of course, an inertial frame of reference.

—Now, change the story a bit, in that, this time, instead of an "inertial reference frame," acceleration is allowed, and you are on your way to board a spaceship. As an illustration, suppose "[y]ou board a spaceship and take off for deep space. The ship approaches the speed of light. Time for you seems to pass as it always has. It takes you about five seconds to tie your shoe. But to an observer on Earth (assuming he or she could watch you), you are moving at a snail's pace. It takes hours to tie your shoe." (NO 2000) In any event, "you continue on your journey. You slow down, stop, and accelerate back to Earth. You arrive home. You have aged two years during your flight. Two hundred years have passed on Earth. You have successfully travelled forward through time." (NO 2000) In this case, both the observer (if he is still alive) and you agree that you have aged less.

(continued on next page)

Table 1.10. The Problems of Time Travel into the Future (Part II)

• **First Problem** ***(cont'd)***

—In other words, when you are back to Earth, a lot of things have changed since, and you may find rather strange to be in a new environment which you hardly recognize, when compared with what you used to think as your city, or the Earth (two years ago, well, in accordance to your time of measurement). This means that you are in the future (two hundred years later), even though you are still relatively young (only two years older). But, can you go back? Well, no. The reason is that "[a]ccording to relativity, you can only move through time in one direction." (NO 2000)

• **Second Problem**

—The second problem concerns "time dilation under the Theory of General Relativity" (which unifies space-time with mass-energy in a general theory), the same result of time dilation occurs.

—For instance, when "one clock is deeper in a gravity well than the other," a result is that "the clock deeper in the well" will "tick…more slowly; this effect must be taken into account when calibrating the clocks on the satellites of the Global Positioning System, and it could lead to significant differences in rates of aging for observers at different distances from a black hole." (WK 2008i)

—More precisely, by way of another example, "it has been calculated that, under general relativity, a person could travel forward in time at a rate four times that of distant observers by residing inside a spherical shell with a diameter of 5 meters and the mass of Jupiter. For such a person, every one second of their 'personal' time would correspond to four seconds for distant observers. Of course, squeezing the mass of a large planet into such a structure is not expected to be within our technological capabilities in the near future." (WK 2008i; R. Gott 2002)

—The point here, however, is that time travel into the future, in this sense, is allowed, theoretically speaking, but in a very limited way, from the theory of relativity. But one can also ask, What if the theory of relativity were wrong? Well, in that case, there would be a need to find a better theory to explain time travel into the future.

Source: From *Sec.4.4.2* of *FPHG*

Table 1.11. The Problems of Time Travel into the Past (Part I)

- **First Problem**

—The first problem concerns the faster-than-light-speed, since, in accordance to the theory of special relativity, when an object goes faster than the speed of light, something weird can happen, in that, instead of having the clock clicking more slowly, it is moving backward, in relation to some inertial frame of reference—which then raisies the issue of whether time travel into the past is possible at all.

—But the problem here is about the violation of the law of casuality. In everyday language, an excellent illustration of causality violation is that, suppose you travel back into the past and then kill your parents, this then means that you could not have been born. But this is not true, since you are alive.

—Yet, "in the case of a hypothetical signal moving faster than light, there would always be some frames in which the signal was received before it was sent, so that the signal could be said to have moved backwards in time." The violation of casuality exists in this case.

—The first part of the problem is that, in accordance to the theory of special relativity, "it would take an infinite amount of energy to accelerate a slower-than-light object to the speed of light"—let alone the energy needed to propel an object to go faster than the speed of light. (WK 2008i) Although some suggest the possibility of "negative energy," it remains to be seen to what extent this is true.

—And the second part of the problem is that, "although relativity does not forbid the theoretical possibility of tachyons which move faster than light at all times, when analyzed using quantum field theory it seems that it would not actually be possible to use them to transmit information faster than light and there is no evidence for their existence." (WK 2008i; S. Chase 1993)

(continued on next page)

Table 1.11. The Problems of Time Travel into the Past (Part II)

• **Second Problem**

—The second problem concerns cosmic string and black holes, since it is the theory of general relativity which "extends the special theory to cover gravity, illustrating it in terms of curvature in spacetime caused by mass-energy and the flow of momentum." (WK 2008i)

—On the other hand, "[g]eneral relativity describes the universe under a system of field equations…and there exist solutions to these equations that permit what are called 'closed time-like curves,' and hence time travel into the past….The first of these was proposed by Kurt Gödel, a solution known as the Gödel metric, but his (and many others') example requires the universe to have physical characteristics that it does not appear to have." (WK 2008i; K. Thorne 1994)

—But then, the problem now becomes: "Whether general relativity forbids closed time-like curves for all realistic conditions is unknown." (WK 2008i)

• **Third Problem**

—The third problem concerns wormholes and related ones, and "[w]ormholes are a hypothetical warped spacetime which are also permitted by the Einstein field equations of general relativity, although it would be impossible to travel through a wormhole unless it was what is known as a traversable wormhole." (WK 2008i; M. Visser 1995) In other words, "a wormhole is a hypothetical topological feature of spacetime that is fundamentally a 'shortcut' through space and time. Spacetime can be viewed as a 2D surface, and when 'folded' over, a wormhole bridge can be formed. A wormhole has at least two mouths which are connected to a single throat or tube. If the wormhole is traversable, matter can 'travel' from one mouth to the other by passing through the throat." (WK 2008j)

—But the problem here is three-fold, as shown in what follows.

(continued on next page)

Table 1.11. The Problems of Time Travel into the Past (Part III)

• **Third Problem** ***(cont'd)***
—The first part of the problem is that "there is no observational evidence for wormholes." (WK 2008j)
—The second part of the problem is that it is not clear how exactly wormholes (and black holes, for that matter) could allow time travel into the past at all.
—And the third part of the problem is that if using wormholes (and black holes, for that matter) to travel into the past requires something to go faster than light too, then this raises another issue concerning the two problems as previously indicated in the sub-section on the faster-than-light-speed.

• **Fourth Problem**
—The fourth problem is that some physicists have argued that, even should time travel into the past be possible, by way of these three methods, the person still would not return to the same past history that he exactly experienced before, but in a different one. (WK 2008i)
—In other words, "strictly speaking," time travel into the past is not "really" possible, in light of these criticisms, so there is no need to worry that one could travel into the past and kill one's parent, as a way to violate the law of casuality.

Source: From *Sec.4.4.2* of *FPHG*

Table 1.12. The Conceptual Dimensions of Consciousness (and Other Mental States)

- **On Heredity and Time**
  —Heredity and the Environment
  —The Past and the Present

- **On Layers of Mental States and Abnormality**
  —Consciousness, Unconsciousness, and Preconsciousness
  —Normality and Paranormality

- **On Organicity and Motivation**
  —Mechanicity and Organicity
  —Primary Motivations and Multiple Motivations

- **On Other (Mostly Epistemic) Considerations**
  —Synthesis and Analysis
  —Situation and the Subject
  —Process and Outcome
  —Reasoning and Other Modes of Thinking
  —Meta-Conceptual Nominalism and Realism

Notes:: These examples are solely illustrative (not exhaustive), and some of the items can be reclassified somewhere else. Nor are they always mutually exclusive. Since they are generalities, exceptions are expected.
Sources: From *FPHC*. A re-construction, but with my own contribution, originally from G.Lindzey & C.Hall, *Introduction to Theories of Personality* (NY: John Wiley & Sons, 1985).

Table 1.13. The Theoretical Levels of Consciousness (and Other Mental States) (Part I)

- **At the Micro-Physical Theoretical Level**
  —Quantum-Mechanics
  —Electromagnetism

- **At the Chemical Theoretical Level**
  —Biochemistry

- **At the Biological Theoretical Level**
  —Evolutionary Biology
  —Neuroscience
  —Artificial Intelligence

- **At the Psychological Theoretical Level**
  —Psychodynamic Psychology
    - Psychoanalytical Psychology
    - Analytical Psychology
    - Socially Oriented Psychology

  —Experimental (Behavioral) Psychology
    - Operant Reinforcement Theory
    - Stimulus-Response Theory
    - Social Learning Theory

  —Cognitive (Gestalt) Psychology
    - Humanist Psychology
    - Existential Psychology
    - Field Theory

  —Psychometric Psychology
    - Idiographics
    - Constitutional Psychology

  —Social Psychology
    - Symbolic Interactive Theory
    - Social Exchange Theory

(continued on next page)

Table 1.13. The Theoretical Levels of Consciousness (and Other Mental States) (Part II)

- **At the Organizational Theoretical Level**
  —Managerial-Bureaucratic Theory
  —Oligarchic Theory
  —Network Theory

- **At the Institutional Theoretical Level**
  —Functionalist Theory
  —Anomic Theory

- **At the Structural Theoretical Level**
  —Conflict Theory
    - Marxian Theory
    - Critical Theory
    - Weberian Theory

  —Games Theory (in Formal Theory)
  —Feminist Theory
    - Feminist Concerns
    - Feminist Hopes

- **At the Systemic Theoretical Level**
  —Equilibrium Theory
  —System Theory
  —Chaos Theory

- **At the Cosmological Theoretical Level**
  —Superluminal Model
  —The Theory of Floating Consciousness

(continued on next page)

Table 1.13. The Theoretical Levels of Consciousness (and Other Mental States) (Part III)

- **At the Cultural Theoretical Level**
  - —Substantive Theories
    - Structuralist Theory
    - Post-Structuralist Theory (in Postmodernism)
  - —Meta-Theories
    - Epistemic Objectivism vs. Epistemic Historicism
    - Epistemic Subjectivism vs. Epistemic Non-Subjectivism
      - –Phenomenology
      - –Ethnomethodology
      - –Hermeneutics
    - Epistemic Relativism vs. Epistemic Absolutism
    - Epistemic Reductionism vs. Epistemic Emergencism

- **At Other Levels**
  - —Structuration Theory
  - —Reflexive Socioanalysis

Notes: These examples are solely illustrative (not exhaustive), and some of the items can be reclassified somewhere else. Nor are they always mutually exclusive. Since they are generalities, exceptions are expected.
Source: From many different sources as indicated in *FPHC*

Table 1.14. The Thematic Issues of Consciousness (and Other Mental States)

- **The Factor of History**
  —Pre-Modernity
  —Modernity
  —Postmodernity
  —After-Postmodernity

- **The Importance of Needs**
  —Having (e.g., the everyday, the technological)
  —Belonging (e.g., the just and the good)
  —Being (e.g., the true, the holy, the sublime/beautiful, and the good)

Notes: These examples are solely illustrative (not exhaustive). Nor are they always mutually exclusive. Since they are generalities, exceptions are expected. Also, some categories overlap with others; for instance, the good is also related with being-ness, just as the everyday can be connected with belong-ness.
Sources: From *FPHC*, based on *FHC* and *FCD*

Table 1.15. Having, Belonging, and Being in Consciousness

- **Having (e.g., About the Everyday and Technological)**
  —Ex: Physiological needs (e.g., thirst, hunger, sex)
  —Ex: Safety (e.g., stability, freedom from fear and anxiety)

- **Belonging (e.g., About the Good and Just)**
  —Ex: Social interaction
  —Ex: Friendship, acquaintance
  —Ex: Love, family
  —Ex: Self-respect, respect from others

- **Being (e.g., About the True, Holy, Beautiful/Sublime, and Good)**
  —Ex: Understanding of reality about self and world
  —Ex: Spiritual quest for holiness, beauty, sublimity, and goodness

Notes: These examples are solely illustrative (not exhaustive), and some of the items can be reclassified somewhere else. Nor are they always mutually exclusive. Since they are generalities, exceptions are expected.
Sources: From *FPHC*, based on *FHC* and *FCD*

Table 1.16. The Having-Ness of Consciousness (and Other Mental States)

- **At the Micro-Physical Level**
  —Ex: The finer physics of nutrition

- **At the Chemical Level**
  —Ex: Food guide pyramid

- **At the Bio-Psychological Level**
  —Ex: The hierarchy of needs

- **At the Institutional Level**
  —Ex: The production of commodity fetishism

- **At the Organizational Level**
  —Ex: Mass standardization (and customization); bureaucratic rationality

- **At the Structural Level**
  —Ex: The divide between the haves and the have-nots

- **At the Cultural Level**
  —Ex: The bondage of greater expectations

- **At the Systemic Level**
  —Ex: The technological transformation of poverty with transvaluations

- **At the Cosmological Level**
  —Ex: Floating consciousness in deep space without the human physiological needs

Notes: These examples are solely illustrative (not exhaustive), and some of the items can be reclassified somewhere else. Nor are they always mutually exclusive. Since they are generalities, exceptions are expected.
Sources: From *FPHC*. See also *FCD* and *FHC*.

Table 1.17. The Belonging-Ness of Consciousness (and Other Mental States)

- **At the Micro-Physical Level**
  —Ex: Statistical physics and social networks
- **At the Chemical Level**
  —Ex: Olfaction and social relations
- **At the Bio-Psychological Level**
  —Ex: Dual human nature on relationships
- **At the Institutional Level**
  —Ex: Private property and social cleavage
- **At the Organizational Level**
  —Ex: Legal formalism and iron cage
- **At the Structural Level**
  —Ex: Class conflicts and social alienation
- **At the Cultural Level**
  —Ex: Ruthless society and the culture of law
- **At the Systemic Level**
  —Ex: Human extinction and post-human relations
- **At the Cosmological Level**
  —Ex: The transcendence of equality in the cosmos and beyond

Notes: These examples are solely illustrative (not exhaustive), and some of the items can be reclassified somewhere else. Nor are they always mutually exclusive. Since they are generalities, exceptions are expected.
Sources: From *FPHC*. See also *FCD* and *FHC*.

Table 1.18. The Being-Ness of Consciousness (and Other Mental States) (Part I)

- **At the Micro-Physical Level**
  —Ex: Space-time and matter-energy in classical mechanics, quantum mechanics, and the theory of relativity

- **At the Chemical Level**
  —Ex: Drugs and well being

- **At the Biological Level**
  —Ex: Neuroscience and mystical experiences

- **At the Psychological Level**
  —Ex: Reductionism and emergencism

- **At the Institutional Level**
  —Ex: The proliferation of cults and sects, and the corporate-art industry

- **At the Organizational Level**
  —Ex: The organization of ideas in the scheme of things

- **At the Structural Level**
  —Ex: Disciplinary power and control in the order of things

- **At the Systemic Level**
  —Ex: Mutualism and ecological psychology

- **At the Cultural Level**
  —Ex: The recurrence of competing spirits, and floating consciousness as a climax of evolution

(continued on next page)

Table 1.18. The Being-Ness of Consciousness (and Other Mental States) (Part II)

- **At the Cosmological Level**
  —Ex: The search for elsewherewhen, and the evolution of other consciousnesses (e.g., the hyper-spatial consciousness)

Notes: These examples are solely illustrative (not exhaustive), and some of the items can be reclassified somewhere else. Nor are they always mutually exclusive. Since they are generalities, exceptions are expected.
Sources: From *FPHC*. See also *FCD* and *FHC*.

Table 1.19. Cognitive Partiality in Different Mental States

- **The Conscious**
  - —*Biased*
    - Ex: Self-overratingness
    - Ex: Self-modesty
    - Ex: Self-fulfilling prophesy
  - —*Shallow*
    - Ex: Attribution error
    - Ex: Heuristics
    - Ex: Illusory thinking
  - —*Short-term*
    - Ex: Deindividuation
  - —*Materialistic*
    - Ex: Hierarchy of needs

- **The Subconscious**
  - Ex: mental reconstruction
  - Ex: stereotyping

- **The Unconscious**
  - Ex: automatic processing
  - Ex: cognitive intuition

Notes: These examples are solely illustrative (not exhaustive), and some of the items can be reclassified somewhere else. Nor are they always mutually exclusive. Since they are generalities, exceptions are expected. And the comparison is relative, not absolute.
Source: From *BCPC*

Table 1.20. Emotional Non-Neutrality and Behavioral Alteration in Different Mental States

- **The Emotional**
  - —*Conscious*
    - Ex: Feelings
  - —*Subconscious*
    - Ex: Moods
  - —*Unconscious*
    - Ex: Empathy
    - Ex: Instincts

- **The Behavioral**
  - —*Conscious*
    - Ex: Foot-in-the-door
    - Ex: Role playing
    - Ex: Saying-is-believing
  - —*Subconscious*
    - Ex: Discrimination
  - —*Unconscious*
    - Ex: Non-Verbal Communication

Notes: These examples are solely illustrative (not exhaustive), and some of the items can be reclassified somewhere else. Nor are they always mutually exclusive. Since they are generalities, exceptions are expected. And the comparison is relative, not absolute.
Source: From *BCPC*

Table 1.21. The Limits of Intuition in Unconsciousness

- **Cognitive Intuition (or Intuitive Thinking)**
  —*Features*
    Ex: mental schemas
    Ex: expertise
    Ex: blindsight
  —*Problems*
    Ex: inaccurate
    Ex: non-explanatory

- **Emotional Intuition**
  —*Features*
    Ex: empathy
    Ex: instincts
  —*Problems*
    Ex: over-generalized
    Ex: non-explanatory

- **Behavioral Intuition**
  —*Features*
    Ex: prosemics
    Ex: kinesics
  —*Problems*
    Ex: unreliable
    Ex: non-explanatory

Notes: These examples are solely illustrative (not exhaustive), and some of the items can be reclassified somewhere else. Nor are they always mutually exclusive. Since they are generalities, exceptions are expected. And the comparison is relative, not absolute.
Source: From *BCPC*

Table 1.22. The Wealth/Poverty Dialectics in Different Mental States: The Case of Cognition

- **The Conscious**
  - —*Biased*
    - Ex: If richer in self-serving bias, then poorer in understanding others
  - —*Shallow*
    - Ex: If richer in heuristic thinking, then poorer in comprehending reality
  - —*Short-term*
    - Ex: If richer in short-term tendency, then poorer in long-term planning
  - —*Materialistic*
    - Ex: If richer in physiological obsession, then poorer in spiritual enlightenment

- **The Subconscious**
  - Ex: If richer in mental reconstruction, then poorer in accuracy of recall
  - Ex: If richer in stereotyping, then poorer in understanding groups

- **The Unconscious**
  - Ex: If richer in automatic processing, then poorer in depth of analysis
  - Ex: If richer in cognitive intuition, then poorer in reliability of judgments

Notes: These examples are solely illustrative (not exhaustive), and some of the items can be reclassified somewhere else. Nor are they always mutually exclusive. Since they are generalities, exceptions are expected. And the comparison is relative, not absolute.
Source: From *BCPC*

Table 1.23. The Wealth/Poverty Dialectics in Different Mental States: The Case of Emotion and Behavior

- **Emotion**
  - —*Conscious*
    - Ex: If richer in relying on familiarity and similarity for social attraction, then poorer in having diverse social relations
  - —*Subconscious*
    - Ex: If richer in moodiness, then poorer in sober thinking
  - —*Unconscious*
    - Ex: If richer in empathy, then poorer in task-oriented efficiency
    - Ex: If richer in instincts, then poorer in critical thinking

- **Behavior**
  - —*Conscious*
    - Ex: If richer in foot-in-the-door technique, then poorer in instant Gratification
    - Ex: If richer in role playing, then poorer in identity stability
  - —*Subconscious*
    - Ex: If richer in discriminating, then poorer in social harmony
  - —*Unconscious*
    - Ex: If richer in kinesics and proxemics, then poorer in verbal skills, relatively speaking.

Notes: These examples are solely illustrative (not exhaustive), and some of the items can be reclassified somewhere else. Nor are they always mutually exclusive. Since they are generalities, exceptions are expected. And the comparison is relative, not absolute.
Source: From *BCPC*

Table 1.24. The Theoretical Debate on Nature vs. Nurture (Part I)

- **The Environmental Approach**
  —*Thesis*: It focuses, relatively speaking, more on the environment (culture and society) in explaining the achievement gap among individuals and for that matter, countries or regions, when contrasted with the natural factors.
  —*Discourse*: Examples include Jose Ortega y Gasset ("Man has no nature; what he has is history"), Ashley Montagu ("Man is man because he has no instincts, because everything he is and has become he has learned from his culture, from the man-made part of the environment, from other human beings"), Stephen Jay Gould ("[The] brain [is] capable of a full range of behaviors and predisposed to none"), and Jesse Jackson (who blames white racism for the failure of blacks to close the achievement gap between whites and blacks in America). The works on dependency theory in international political economy (with a Marxian influence) and on the Protestant work ethic (by Max Weber) also point to this environmental direction.

- **The Genetic Approach**
  —*Thesis*: It focuses instead, relatively speaking again, on hereditory factors (e.g., genes and evolution) in explaining the achievement gap among individuals and for that matter, countries or regions, when contrasted with the envrionmental factors.
  —*Discourse*: Examples are Hans Eysenck and William Sheldon (in constitutional psychology), Konrad Lorenz (in his work on innate aggressive human nature), Gary Marcus (on the complexities of human thought by a tiny number of genes) and Robert Plomin, together with Michael Owen and Peter McGuffin (on the genetic basis of complex human behaviors).

(continued on next page)

Table 1.24. The Theoretical Debate on Nature vs. Nurture (Part II)

- **The Compromise Approach**
  —*Thesis*: It seeks the middle-of-the-road argument in regard to nature and nurture and regards all differences among individuals and groups as the result of the mixture of both nature and nurture, more or less equally.
  —*Discourse*: Examples include C. Murray and R. Herrnstein ("It seems highly likely to us that both genes and environment have something to do with this issue") and Dan Dennett ("Surely 'everyone knows' that the nature-nurture debate was resolved long ago, and neither side wins since everything – is – a – mixture – of – both – and - it's – all – very -complicated…."

- **The Transcendent Approach**
  —*Thesis*: It goes beyond both nature and nurture (without, however, committing the compromise fallacy) in showing their closely intertwined interactions in producing the behavioral differences as often seen in individual human endeavors on the micro scale, and for that matter, in country (or regional) endeavors on the macro one—in the context of my five theses, namely, (a) the compromise fallacy, (b) no oppression without self-oppression, (c) no success without failure, (d) the factor of randomness, and (e) the post-human vision, to be elaborated in Chapter Six.
  —*Discourse*: Peter Baofu proposed this approach on the basis of his "theory of contrastive advantages" (as an original theoretical contribution to the debate, which was first proposed and analyzed in *The Future of Capitalism and Democracy*). In the end, the human genes will not last, to be eventually superseded by post-human life forms, so the debate between genes and memes has obscured something profoundly important about the future that the world has never known. And the debate is also misleading and faulty in its dichotomy.

Source: From *BNN*. See the book for citations and details.

Table 1.25. Physical Challenges to Hyper-Spatial Consciousness

- **The Understanding of a Higher-Dimensional World of Space-Time**
  —Ex: 4 for traditional aspects of space-time (e.g., length, width, breadth and time) plus 6 more new dimensions in theory of hyper-space, with profound implications for practical applications to new forms of consciousness.

- **The Mastering of Dark Matter and Dark Energy**
  —Ex: "ordinary matter" (e.g., atoms, molecules) as a mere 4.4% of the universe, with 23% made of "cold dark matter" and the rest (about 73%) of mysterious "dark energy," with fundamental significance to questions about the limit of the speed of energy (or info), the availability of energy for use, and the nature of space-time, just to cite some examples.

- **The Exploration of Multiverses**
  —Ex: theoretical speculation of other universes (e.g., "baby universes," "gateways" in black holes, "wave function of the universe," "many worlds," "brane worlds"), with potentially seminary discoveries of different physical laws in relation to matter-energy and space-time, and vital differences to the future of post-human conquest of other universes (for the emergence of new forms of consciousness).

Notes: These examples are solely illustrative (not exhaustive), and some of the items can be reclassified somewhere else. Nor are they always mutually exclusive. Since they are generalities, exceptions are expected. The point here is to give a rough picture of the evolution of consciousness to the hyper-spatial consciousness and others totally unknown to current earthlings. As a note of clarification, it makes no difference to my argument as to whether or not the hyper-spatial consciousness may emerge before, during, and after floating consciousness.
Source: From *Table 4.5* of *FPHC*

Table 1.26. The Theory of Floating Consciousness (Part I)

• **At the Micro-Physical Level**
—Ex: intelligent life without the human physical-chemical system

• **At the Chemical Level**
—Ex: space radiation and toxins

• **At the Bio-Psychological Level**
—Ex: exo-biological evolution in deep space
—Ex: genetic engineering of new beings

• **At the Institutional Level**
—Ex: post-capitalism
—Ex: post-democracy

• **At the Organizational Level**
—Ex: less legal-formalistic routines

• **At the Structural Level**
—Ex: alien forms of violence

• **At the Cultural Level**
—Ex: transcending freedom
—Ex: transcending equality

(continued on next page)

Table 1.26. The Theory of Floating Consciousness (Part II)

• **At the Cosmological Level**
—Ex: parallel universes
—Ex: pocket universes

• **At the Systemic Level**
—Ex: space habitats (in zero-gravity)

Notes: Each example draws from the works of different scholars in the field. For instance, at the cosmological level, the idea of parallel universes is from the theoretical speculation in quantum cosmology by Stephen Hawking and others, while the one of pocket universes comes from the theoretical work of Allan Guth at MIT. And at the institutional level, I proposed post-capitalism and post-democracy in *FCD* (and later, from *BDPD* and *BCPC*). In addition, the examples are solely illustrative (not exhaustive), and some of the items can be reclassified somewhere else. Nor are they always mutually exclusive. Since they are generalities, exceptions are expected.
Sources: From *FPHC*—and, originally, from *FCD*

Table 1.27. The Potential of Unfolding Unconsciousness

- **Superior Senses**
  - —*Through space*
    - Ex: clair-sentience (in feeling)
    - Ex: clair-voyance (in seeing)
    - Ex: clair-audience (in hearing)
    - Ex: clair-austance (in tasting)
    - Ex: clair-alience (in smelling)
    - Ex: clair-cognizance (in knowing)
  - —*Through time*
    - Ex: pre-cognition (in knowing), pre-sentiment (in feeling)
    - Ex: retro-cognition (in knowing), retro-sentiment (in feeling)

- **Superior Contacts**
  - —*With living minds*
    - Ex: telepathy
  - —*With the dead*
    - Ex: mediumship
  - —*With other-living entities*
    - Ex: telekinesis

Notes: The categories and examples are solely illustrative (not exhaustive), and the comparison is relative (not absolute), nor are they necessarily mutually exclusive. And some can be easily re-classified elsewhere. As generalities, they allow exceptions.
Source: A summary of Ch.5 of *FPHU*

Table 1.28. The Future Exploration of Unfolding Unconsciousness

- **In the Context of Nature**
  - —Ex: the higher-dimensional space-time in hyperspace theory
  - —Ex: the particle-wave interactions in quantum mechanics
  - —Ex: the study of brain waves in electromagnetic theory

- **In the Context of the Mind**
  - —Ex: the unconscious fantasies in psychoanalysis
  - —Ex: the mind-altering drugs in biochemistry
  - —Ex: the manipulation of neural activity in neurobiology

- **In the Context of Culture**
  - —Ex: the correlation between anomalous experience and moral/religious/aesthetic interests in culture studies
  - —Ex: the role of myths and language in epistemology

- **In the Context of Society**
  - —Ex: different organizational agendas in anomalous research
  - —Ex: the scholarly divide on anomalous study
  - —Ex: interests and propaganda in institutioanl research
  - —Ex: the transformative power of technology in research on social systems
  - —Ex: the natural and non-natural selection of other worlds in exo-biology
  - —Ex: the role of the post-human conscious mind in qualitative demography

Notes: The examples in each category are solely illustrative (not exhaustive), and the comparison is relative (not absolute), nor are they necessarily mutually exclusive. And some can be easily re-classified elsewhere. As generalities, they allow exceptions.
Source: A summary of Chs.2-5 of *FPHU*, in conjunction with the rest of the book

Table 1.29. Creative Techniques and Traits

- **Creative Techniques**
  - —*Comprehensiveness*
    - Ex: Broadness of Scope
    - Ex: In-Depth Analysis
    - Ex: Multifaceted Taxonomy
    - Ex: Numerous Theories
  - —*Arrangement*
    - Ex: Separation (or Subtraction)
    - Ex: Combination (or Addition)
    - Ex: Replacement
    - Ex: Accommodation
    - Ex: Reversal
    - Ex: Stretching
  - —*Serendipity*
    - Ex: Play
    - Ex: Chance

- **Creative Traits**
  - —*Openness / Risk*
    - Ex: Ability to welcome new ideas or to do new things, with risk
  - —*Discipline*
    - Ex: Ability to sacrifice whatever it takes to succeed
  - —*Resilience / Confidence*
    - Ex: Ability to take failures to mind, not to heart
  - —*Trust*
    - Ex: Ability to cooperate with others for team work, if necessary

Notes: The examples in the categories are solely illustrative (not exhaustive), and the comparison is relative (not absolute), nor are they necessarily mutually exclusive. And some can be easily re-classified elsewhere. As generalities, they allow exceptions.
Source: From *Sec.4.2* of *FPHCT*. For more details, also read the rest of the book.

Table 1.30. The Desirability of Creativity, and Its Dark Sides (Part I)

- **In Relation to Instrumental Rationality**
  —*Desirability*
  - Ex: For the sake of efficiency and effectiveness in this capitalist era of our time.

  —*Undesirability*
  - Ex: What Max Weber (1930) called the "iron cage" of capitalism, where live the "sensualists without heart, specialists without spirit"—or something which the Frankfurt School has forcefully asked us to be freed from.

- **In Relation to Substantive Rationality**
  —*Desirability*
  - Ex: For the sake of God, the King, Motherland, or other comparable ideals.

  —*Undesirability*
  - Ex: Domination and oppression of various forms—something which the moderns had struggled to free themselves from in the first place, since the dawn of modernity.

- **In Relation to Autonomous Rationality**
  —*Desirability*
  - Ex: For the sake of itself, the autonomy of creative endeavor.

  —*Undesirability*
  - Ex: The decadence of its own degeneration (e.g., the donwsides of postmodernism in our postmodern times, or what Nietzsche once called the unbearable "unreality" and "falseness" of the autonomous artist's "innermost existence"—and, in other cases, the painful suffering from different mental illnesses for those highly creative individuals).

(continued on next page)

Table 1.30. The Desirability of Creativity, and Its Dark Sides (Part II)

Notes: The examples in the categories are solely illustrative (not exhaustive), and the comparison is relative (not absolute), nor are they necessarily mutually exclusive. And some can be easily re-classified elsewhere. As generalities, they allow exceptions.
Also, I already went to great lenghts to explain the desirability and its dark sides of all these three rationalities (although there may be others too) in the two volumes of *FHC*—and, for that matter, *FCD*.
Source: From *Sec.4.9* of *FPHCT*. For more details, also read the rest of the book.

Table 1.31. Posthuman-Ism, Post-Humanism, and Trans-Humanism

- **Post-Humanism**
  —The neologism "post-human" used in my books should *not* be confused with another term which looks similar but has a totally different meaning in the literature of postmodernism, namely, "post-humanism"—which constitutes a critique of "humanism" as tradionally understood (especially, though not exclusively, in relation to the idea of progress in science and reason in the Enlightenment project). (WK 2008)
  —My works reject the project of "postmodernism" and propose the future world of what I orignally called "after-postmodernity" in *FHC* and *FCD*, for instance.

- **Trans-Humanism**
  —Also, the neologism "post-human" used in my books should *not* be confused with a similar term which is used to champion the ideology of technology for the fturue co-existence between humans and cyborgs in "trans-humanism." (WK 2008a)
  —Instead, my term "post-human" in relation to "posthuman-ism" also rejects "trans-humanism" (especially, though not exclusively, in relation to the promises of technology) and refers to something else altogether, that is, the future extinction of humans and its post-human successors in deep space and beyond unto multiverses.
  —Chapter Two of my book on ethics (*BEPE*) provides a more comprehensive critique of trans-humanism.

- **Posthuman-Ism**
  —With these two clarifications in mind, the "post-humans" (as envisioned in my books) can take different forms, and I proposed, in my different books, some of them, such as "unfolding unconsciousness," "floating consciousness," "hyper-spatial consciousness," "thinking machines," "thinking robots," "genetically altered superior beings," "cyborgs," and others (as already analyzed in my previous works).
  —They will evolve in conjunction with other visions of mine in relation to nature, society, and culture.

Sources: From *Sec.1.6* of *FPHE* (and also *BEPE*). For more details, also read the rest of the books (and other books of mine).

Table 1.32. Three Great Future Transformations of the Martial Body (Part I)

• **Virtual Battle**
  —Ex: virtual fight
  —Ex: virtual memory

• **The Hyper-Martial Body**
  —*In relation to the physical dimension*
  • Ex: striking (e.g., punching, kicking, trapping, acupressure-striking, and so on in an amazing way)
  • Ex: grapping (e.g., throwing, pinning, joint-locking, and so on in an amazing way)
  • Ex: running and walking (e.g., on different surfaces)
  • Ex: jumping (e.g., through walls, trees, etc.)
  • Ex: flying (e.g., through space)
  • Ex: enduring pain (e.g., absorbing punches, kicks, and so on in an amazing way)
  • Ex: using extremely advanced hi-tech weapons
  • Ex: healing (e.g., with special energy, etc.)
  • Ex: entering (e.g., into different dimensions of space)
  • Ex: sensing (e.g., in extraordinary ways through space and time, etc.)
  • Ex: wearing hi-tech suits for defensive and offensive purposes (e.g., like a hi-tech armor)
  —*In relation to the mental dimension*
  • Ex: the ability to relax the martial body in a way that current humans cannot.
  • Ex: the ability to concentrate the martial mind in a way that current humans are not accustomed to.
  • Ex: the ability to breathe and be timely in preparing the martial body for fighting in a way that current humans are not good at.

• **The Post-Martial Floating Mind**
  —Ex: the mind without the human body
  —Ex: the quest for spirituality beyond martial arts

(continued on next page)

Table 1.32. Three Great Future Transformations of the Martial Body (Part II)

Notes: The examples in the categories are solely illustrative (not exhaustive), and the comparison is relative (not absolute), nor are they necessarily mutually exclusive. And some can be easily re-classified elsewhere. As generalities, they allow exceptions.
Source: From *Sec. 2.5.1* of *FPHMA*. See text for citations.

Table 1.33. Three Great Future Transformations of the Sexual Body

- **Virtual Pleasure**
  —Ex: virtual Sex
  —Ex: virtual Recall

- **The Hyper-Sexual Body**
  —*In relation to the physical dimension*
  - Ex: the ability to perform differenet sexual positions without the various physical limitations that humans currently have.
  - Ex: the ability to produce unusual amount of sexual energy to endure in sexual acts in a way that humans currently cannot.
  - Ex: the ability to engage in multiple sexual experiences comparable to (but different from) non-human sexuality (e.g., hermaphrodite, monoecious, sex-switching, and many other types) that current humans cannot.

  —*In relation to the mental dimension*
  - Ex: the ability to relax the sexual body in a way that current humans are not accustomed to.
  - Ex: the ability to concentrate the sexual mind in a way that current humans are not good at.
  - Ex: the ability to breathe and be patient in preparing the sexual body for sexuality in a way that current humans are not able to.

- **The Post-Sexual Floating Mind**
  —Ex: the mind without the human body
  —Ex: the quest for spirituality beyond sexuality

Notes: The examples in the categories are solely illustrative (not exhaustive), and the comparison is relative (not absolute), nor are they necessarily mutually exclusive. And some can be easily re-classified elsewhere. As generalities, they allow exceptions.
Source: From *Sec. 3.4.2* of *FPHS*. See text for more info.

Table 1.34. The Trinity of Pre-Modernity

• **Pre-Free-Spirited Pre-Modernity (Pre-Modernism) and Its Internal Split**
—Competing worldviews and values both within and between linear centric (e.g., Islamic, Christian, Judaic, Imperial Roman) and cyclical-centric (e.g., Confucian, Taoist, Hindu, and Buddhist) orientations
—Compare modernism with pre-modernism here in relation to the seven dimensions of human existence like the true and the holy (e.g., different versions of epistemic dogmas and religious superstitions), the everyday and the technological (e.g.,different versions of non-technophilism and non-consumerism), the beautiful/sublime (e.g., different versions of aesthetic non-autonomy), and the good and the just (e.g., different versions of moral particularism).

• **Pre-Capitalist Pre-Modernity (Pre-Modernization) and Its Own Discontents**
—Competing versions of societal arrangements (e.g., feudalism, monarchism, and the holy order)

• **Hegemonic Pre-Modernity and Its Countervailing Forces**
—Different power centers and their enemies (e.g., the Roman Empire and the "barbarian hordes," the "Holy Crusades" and the Muslims, the Middle Kingdom and the invading tribes, different social castes in India, and warring Greek city-states)

Notes: The examples in each category are solely illustrative (not exhaustive) nor necessarily mutually exclusive, and the comparison is relative (not absolute). As generalities, they allow exceptions. Also, it does not matter what the "base" era is in the analysis of any trinity. And in the present context, the "base" era is modernity (for instance, with its "free-spirited modernity" and the other two parts). So, for pre-modernity, the trinity takes the form of, say, "pre-free-spirited pre-modernity," together with the other two parts.
Sources: From Ch.2 of *BCIV* and also the 2 volumes of *FHC*

Table 1.35. The Trinity of Modernity (Part I)

- **Free-Spirited Modernity (Modernism) and Its Internal Split**
  - *—On the True and the Holy*
    - The freedom from the dogmas of the past to the better understanding of, and union with, the world and self (Ch.3 of *FHC*)
    - Alternative discourses: about the true (e.g., anti-science discourses) and the holy (non-mainstream theologies) (Ch.3 of *FHC*)
  - *—On the Technological and the Everyday*
    - The freedom from life harshness to the higher technophilic, consumeristic lifeform (Ch.2 of *FHC*)
    - Alternative discourses: about the everyday (e.g., transcendental mindsets) and the technological (e.g., Arcadianism) (Ch.2 of *FHC*)
  - *—On the Good and the Just*
    - The freedom from the theo-aristocratic tyranny to the moral universality for a just society (Ch.5 of *FHC*)
    - Alternative discourses: about the just (e.g., Communism, Anarchism) and the good (e.g., Nazism/Fascism, and Zarathustrianism) (Chs.5-6 of *FHC*)
  - *—On the Beautiful and the Sublime*
    - The freedom from the external distortion of aesthetic pleasure to the boundless infinity of totality in artistic autonomy (Ch.4 of *FHC*)
    - Alternative discourses: about the beautiful/sublime (e.g., kitsch and historical avant-gardism) (Ch.4 of *FHC*)

- **Capitalist Modernity (Modernization) and Its Own Discontents**
  - *—During the Industrial Revolution*
    - Ex: Marx on the institution of inequality (Ch.1 of *FHC*)
  - *—During the Modern Rational-Instrumental Epoch*
    - Ex: Weber on the politics of soft liberal institutions (Ch.5 of *FHC*)
  - *—During the Great Depression*
    - Ex: Keynes on the myth of the free market (Chs.1-3 of *FHC*)
  - *—During the Cold War*
    - Ex: Lasch on the narcissistic culture industry (Chs.2-3 of *FHC*)

(continued on next page)

Table 1.35. The Trinity of Modernity (Part II)

- **Hegemonic Modernity and Its Countervailing Forces**
  —*The Legacies of Colonialism and Imperialism*
    - Ex: European colonization of most of the modern world (Ch.1 of *FHC*)
  —*The Struggle for Decolonialization*
    - Ex: The countervailing forces of resentment, rechantment, and regionalism (Chs.1 & 6 of *FHC*)

Notes: The examples in each category are solely illustrative (not exhaustive) nor necessarily mutually exclusive, and the comparison is relative (not absolute). As generalities, they allow exceptions.
Sources: From the 2 volumes of *FHC*—and also from *FCD*

Table 1.36. The Trinity of Post-Modernity

- **Free-Spirited Postmodernity (Postmodernism) and Its Internal Split**

  —*On the True and the Holy*
  - Postmodern performative turn for knowing and its enemies (Ch.3 of *FHC*)
  - Postmodern comparative theology and its opponents (Ch.3 of *FHC*)

  —*On the Technological and the Everyday*
  - Postmodern corporate technological mindset and its adversaries (Ch.2 of *FHC*)
  - Postmodern postmaterialism and its critics (Ch.2 of *FHC*)

  —*On the Good and the Just*
  - Postmodern politics of difference and its foes (Ch.5 of *FHC*)

  —*On the Beautiful and the Sublime*
  - Postmodern deconstruction and its dissenters (Ch.4 of *FHC*)

- **Capitalist Postmodernity (Postmodernization) and Its Own Discontents**

  —*During the Post-Cold War and Beyond*
  - Ex: post-Fordism and its shortcomings (Ch.6 of *FHC*; Chs.6-7 of *FCD*)

- **Hegemonic Postmodernity and Its Countervailing Forces**

  —*The Debate on the Global Village*
  - Ex: uni-civilizationalism vs. multi-civilizationalism (Ch.6 of *FHC*)

  —*The Resistance Movement*
  - Ex: rechantment and the politics of civilizational claims (e.g., Islamic, Confucian and other ethos in relation to the Same) (Ch.6 of *FHC*; Ch.10 of *FCD*)
  - Ex: resentment and the politics of resurgence (e.g., the rising Chinese superpower, the growing EU, and other players in relation to the U.S. and her allies) (Ch.6 of *FHC*; Ch.8 of *FCD*)
  - Ex: regionalism and the politics of inequality (e.g., trans- or international blocs, the North-South divide, NGO's) (Ch.6 of *FHC*; Ch.5 of *FCD*)

Notes: The examples in each category are solely illustrative (not exhaustive) nor necessarily mutually exclusive, and the comparison is relative (not absolute). As generalities, they allow exceptions.
Sources: From *FCD* and the 2 volumes of *FHC*

Table 1.37. The Trinity of After-Postmodernity

- **Free-Spirited After-Postmodernity (After-Postmodernism) and Its Internal Split**
  —The discourse of naked contingency (Ch.10 of *FCD*; Ch.4 of *FPHC*)

- **Post-Capitalist After-Postmodernity (After-Postmodernization) and Its Own Discontents**
  —Different versions of post-capitalism and post-democracy, and their enemies (Ch.10 of *FCD*; Chs.3-4 of *FPHC*)

- **Hegemonic After-Postmodernity and Its Countervailing Forces**
  —The Cyclical Progression of Hegemony in Multiverses (Chs.9-10 of *FCD*; Ch.4 of *FPHC*)

Notes: The examples in each category are solely illustrative (not exhaustive) nor necessarily mutually exclusive, and the comparison is relative (not absolute). As generalities, they allow exceptions.
Sources: From *FCD* and also *FHC*

• Part Two •

Uselessness

CHAPTER 2
USELESSNESS AND ITS DUAL POLES

> The water we drink has to be purified, but look at the trash we feed our minds.
> —Unknown (QUO 2012)

The Valuelessness of Uselessness

The study of waste involves both uselessness and usefulness—as already indicated in *Sec. 1.3*.

The relationship between uselessness and usefulness is dialectic, since there is no uselessness without usefulness—and vice versa.

Yet, a scholarly danger here is to reduce the analysis of one to that of the other; thus, this book is to devote a whole chapter on each, with the understanding, of course, that the two are related.

The issues concerning uselessness in the study of waste are to be addressed in this chapter (Chapter Two), whereas the issues concerning usefulness in the study of waste will be analyzed in the next chapter (Chapter Three) instead.

With this dialectic treatment in mind—a good way to examine the appeal of uselessness in the study of waste and its dual poles is by way of the evaluation of the extent to which it is in fact both possible (or impossible) and desirable (or undesirable).

This can be done by way of a comprehensive analysis of uselessness in the context of the study of waste from the four perspectives of the mind, nature, society, and culture (in accordance to my sophisticated methodological holism, as explained in *Sec. 1.9*).

In other words, this chapter is organized in four main sections, in relation to (2.2) uselessness and the mind, (2.3) uselessness and nature, (2.4) uselessness and society, and (2.5) uselessness and culture—to be addressed hereafter, in that order (and summarized in *Table 3.1*).

Uselessness and the Mind

Uselessness, when examined from the perspective of the mind, can illuminate its possibility (or impossibility) and desirability (or undesirability), and this can be shown by way of two case studies, namely, (2.2.1) uselessness, biodegradable waste, and epidemics, and (2.2.2) uselessness, clinical waste, and diseases—to be addressed in what follows, respectively.

Uselessness, Biodegradable Waste, and Epidemics

A good way to evaluate the possibility (or impossibility) and desirability (or undesirability) of uselessness from the perspective of the mind is to explore uselessness, biodegradable waste, and epidemics.

The term "biodegradable waste" refers to "a type of waste, typically originating from plant or animal sources, which may also be degraded by other living organisms. Waste that cannot be broken down by other living organisms are called non-biodegradable. Biodegradable waste can be commonly found in municipal solid waste (sometimes called biodegradable municipal waste, or BMW) as green waste, food waste, paper waste, and biodegradable plastics. Other biodegradable wastes include human waste, manure, sewage, slaughterhouse waste." (WK 2012s)

Like biowaste (as already addressed in *Sec. 1.5*)—biodegradable waste can contribute to epidemics, and good examples are "the Bubonic Plague, cholera and typhoid fever, to mention a few," which "were diseases that altered the populations of Europe and influenced monarchies. They were perpetuated by filth that harbored rats, and contaminated water supply. It was not uncommon for Europeans to throw their waste and human wastes out of the window which would decompose in the street."(WK 2012a)

The Case of the Bubonic Plague

The Bubonic Plague in European history is a good historical example of the relationship between biodegradable waste and epidemics (as summarized in *Sec. 2.2*). (WK 2012l)

In general, "bubonic plague is a zoonotic disease, circulating mainly among small rodents and their fleas" that feed on wastes (like biodegradable waste and biowaste, just to cite two main examples) and is "one of three types of infections caused by *Yersinia pestis* (formerly known as *Pasteurella pestis*), which belongs to the family Enterobacteriaceae." (WK 2012l; ALL 2012)

With good reason, "the term *bubonic plague* is derived from the Greek word βουβών, meaning 'groin.' Swollen lymph nodes (buboes) especially occur in the armpit and groin in persons suffering from bubonic plague. *Bubonic plague* was often used synonymously for plague, but it does in fact refer specifically to an infection that enters through the skin and travels through the lymphatics, as is often seen in flea-borne infections." (WK 2012l)

(a) Bacteria and Infection

A bubonic plague has to do with "an infection of the lymphatic system, usually resulting from the bite of an infected flea, *Xenopsylla cheopis* (the rat flea). In very rare circumstances like the septicemic plague the disease can be transmitted by direct contact with infected tissue or exposure to a cough of another human. The fleas are often found on rodents such as rats and mice, and seek out other prey when their rodent hosts die." (WK 2012l)

Generally, "the bacteria began its life harmlessly living in the digestive tracts of mammals. The bacteria ability to propagate was dependent only upon its ability to traveling from mammal host to mammal host. The bacteria remained harmless to the flea, allowing the new host to spread the bacteria. The bacteria form aggregates in the gut of infected fleas and this results in the flea regurgitating ingested blood, which is now infected, into the bite site of a rodent or human host. Once established, bacteria rapidly spread to the lymph nodes and multiply." (WK 2012l)

(b) Signs and Symptoms

The signs and symptoms of a bubonic plague are not hard to observe, because it is "an infection of the lymph glands (lymphadenitis), which become swollen and painful and are known as buboes. After being transmitted via the bite of an infected flea, the *Y. pestis* bacteria become localized in an inflamed lymph node where they begin to replicate. Buboes associated with the bubonic plague are commonly found in the armpits, upper femoral, groin and neck region. Acral gangrene (i.e. of the fingers, toes, lips and nose), is another common symptom. The black color of the necrotized tissue is the origin of the term 'Black Death,' the bubonic plague pandemic that swept Europe in the mid-fourteenth century." (WK 2012l)

Because of "its bite-based form of infection, the bubonic plague is often the first step of a progressive series of illnesses. Bubonic plague symptoms appear suddenly, usually 2–5 days after exposure to the bacteria," and common symptoms include the following: (WK 2012l; T. Inglesby 2000)

- Ex: "acral gangrene" (i.e., of "toes, fingers, lips and tip of the nose")
- Ex: "Smooth, painful lymph gland swelling called a bubo, commonly found in the groin,…may occur in the armpits or neck, most often at the site of the initial infection (bite or scratch)."
- Ex: "bleeding out of the cochlea (the hearing organ shaped like a seashell that is found in the middle ear) begins after ~12 hours of infection."
- Ex: "chills"
- Ex: "general ill feeling (malaise)"
- Ex: "high fever (39 °Celsius; 102 °Fahrenheit)"
- Ex: "muscle Cramps"
- Ex: "seizures"

Surely, there are other symptoms too (like "heavy breathing, continuous vomiting of blood hematemesis, aching limbs, coughing, and extreme pain"), so the examples above are illustrative.

(c) Deadly Impact

What makes the bubonic plague important in history is that it is deadly in its impact on the victims.

Historically, "the first recorded epidemic ravaged the Byzantine Empire during the sixth century, and was named the Plague of Justinian after emperor Justinian I, who was infected," and it was "estimated to have killed approximately 25 million people in the empire alone." (WK 2012l; HIST 2009)

Later, "in the Early Modern period (1340-1400) Europe experienced the worst human disaster in its history when the Black Death (also known as the bubonic plague) hit in 1347, destroying a third of the population. It is commonly believed that society subsequently became more violent as the mass mortality rate cheapened life and thus increased warfare, crime, popular revolt, waves of flagellants, and persecution. The Black Death originated in or near China and spread from Italy and then throughout other European countries. Research published in 2002 suggests that it began in the spring of 1346 in the stepped region, where a plague reservoir stretches from the north-western shore of the Caspian Sea into southern Russia. The Mongols had cut off the trade route, the Silk Road, between China and Europe which halted spread of the Black Death from the east Russia towards western Europe." (WK 2012l; S. Cohn 2002)

Problems with the Study of Biodegradable Waste and Epidemics

However, for the critics, this study of biodegradable waste and epidemics has some major problems. Consider, for illustration, a few examples below (as summarized in *Sec. 2.2*).

Firstly, one main criticism of the study of biodegradable waste and epidemics is that epidemics can produce desirable effects instead.

For example, in the case of the Bubonic plague which caused "the Black Death that swept through Europe in the 14th century,...because the plague killed so many of the working population, wages rose and some historians have seen this as a turning point in European economic development." (WK 2012l; W. Bowsky 1971; A. Bridbury 1983)

Secondly, another main criticism of the study of biodegradable waste and epidemics is that epidemics can be stopped or prevented.

For instance, for the critics, "several classes of antibiotics are effective in treating bubonic plague. These include aminoglycosides such as streptomycin and gentamicin, tetracyclines (especially doxycycline), and the fluoroquinolone ciprofloxacin. Mortality associated with treated cases of bubonic plague is about 1-15%, compared to a mortality rate of 50-90% in untreated cases." (WK 2012l; S. Dufel 2011)

And thirdly, still another main criticism of the study of biodegradable waste and epidemics is that it is not clear to what extent exactly biodegradable waste (and biowaste, of that matter) have really caused epidemics in history.

For instance, epidemics can be influenced by other factors, and good examples involves "biological warfare," when, for illustration, "armies of the 14th century were recorded catapulting diseased corpses over the walls of towns and villages in order to spread the pestilence. Later, plague was used during the Second Sino-Japanese War as a bacteriological weapon by the Imperial Japanese Army. These weapons were provided by Shirō Ishii's units and used in experiments on humans before being used on the field. For example, in 1940, the Imperial Japanese Army Air Service bombed Ningbo with fleas carrying the bubonic plague. During the Khabarovsk War Crime Trials, the accused, such as Major General Kiyashi Kawashima, testified that, in 1941, some 40 members of Unit 731 air-dropped plague-contaminated fleas on Changde. These operations caused epidemic plague outbreaks." (WK 2012l; D. Barenblatt 2004)

Beyond Biodegradable Waste and Epidemics

These criticisms of the study of biodegradable waste and epidemics should not be misunderstood as a total rejection of its validity but serve to show us the two opposing sides of the debate, such that the possibility (or impossibility) and desirability (or undesirability) of uselessness (from the perspective of the mind with uselessness, biodegradable waste, and epidemics as a case study here) are not to the extent that the respective defenders would like us to believe.

Furthermore, the analysis of uselessness, biodegradable waste, and epidemics can teach us something important about the ontological principles in existential dialectics, and good examples include the formalness-informalness principle, the absoluteness-relativeness principle, the partiality-totality principle, the predictability-unpredictability principle, the explicability-inexplicability principle, the fiction-reality principle, the finiteness-transfiniteness principle, the preciseness-vagueness principle, the simpleness-complicatedness principle, the openness-hiddenness principle, the denseness-emptiness principle, the change-constancy principle, the order-chaos principle, the slowness-quickness principle, the expansion-contraction principle, the theory-praxis principle, the convention-novelty principle, the evolution-transformation principle, the symmetry-asymmetry principle, the softness-hardness principle, the seriousness-playfulness principle, the activeness-inactivess principle, the selfness-otherness principle, the regression-progression principle, the same-difference principle, the stability-reaction principle, the functionality-nonfunctionality principle, and the intentionality-nonintentionality principle.

For instance, in relation to the formalness-informalness principle, if there is formalness (e.g., the formal logical requirement of soundness and completeness in a system of ideas, as in the study of biodegradable waste and epidemics), there is informalness (e.g., the non-formal existence of unsoundness and incompleteness in the study of biodegradable waste and epidemics, as shown in the criticisms that epidemics can produce desirable effects instead, that epidemics can be stopped or prevented, that it is not clear to what extent exactly biodegradable waste and biowaste have really caused epidemics in history, etc.). And the reverse direction also holds true.

In relation to the absoluteness-relativeness principle, if there is absoluteness (e.g., the absolute view of the Imperial Japanese Army Air Service when bombing Ningbo with fleas carrying the bubonic plague in 1940), there is relativeness (e.g., what is right for the Imperial Japanese Army Air Service to bomb Ningbo with fleas carrying the bubonic plague in 1940 is not necessarily so for the Chinese victims and their family members). And the reverse direction also holds true.

In relation to the partiality-totality principle, if there is partiality (e.g., the partial view of the bubonic plague as a biological weapon by the Imperial Japanese Army Air Service when bombing Ningbo with fleas carrying the bubonic plague in 1940), there is totality (e.g., the more holistic view of the bubonic plague, such that the whole is not the sum of its parts, that is, it cannot be reduced to the sum of opposing views of the Imperial Japanese Army Air Service, the Chinese victims, etc., since there will emerge new views in the future not yet known today, just as there are already alternative views nowadays which question the claims by the Chinese Communist Party and their critics about the incident). And the reverse direction also holds true.

In relation to the predictability-unpredictability principle, if there is predictability (e.g., the predictable tendency of scientists to use antibiotics like "streptomycin," "gentamicin," and "tetracyclines" for the treatment of bubonic plague), there is unpredictability (e.g., the more difficult task to predict exactly the extent to which the use of antibiotics like "streptomycin," "gentamicin," and "tetracyclines" will be successful on a particular future occasion of plague occurrence, since they are not totally effective as "mortality associated with treated cases of bubonic plague" still happens). And the reverse direction also holds true.

In relation to the explicability-inexplicability principle, if there is explicability (e.g., the explanation by the Imperial Japanese Army Air Service to bomb Ningbo with fleas carrying the bubonic plague in 1940 because it is an effective biological weapon), there is inexplicability (e.g., the lack of sufficient explanation by the Imperial Japanese Army Air Service of why bombing Ningbo with fleas carrying the bubonic plague in 1940 is necessarily desirable, especially from the vantage point of the Chinese victims and their family members). And the reverse direction also holds true.

In relation to the fiction-reality principle, if there is fiction (e.g., the fictional aspect of the conventional view about the bubonic plague, insofar as it is regarded as undesirable, as questioned by the Imperial Japanese Army Air Service in 1940 in the decision to use it as an effective biological weapon instead), there is reality (e.g., the realistic aspect of the bubonic plague, insofar as it can have destructive impact on its victims). And the reverse direction also holds true.

In relation to the finiteness-transfiniteness principle, if there is finiteness (e.g., the finite number of the categories of epidemics which were perpuetuated by zoonotic diseases, "circulating mainly among small rodents and their fleas," like "the Bubonic Plague, cholera and typhoid fever" in history), there is transfiniteness (e.g., the transfinite number of all the individiual rats and fleas which have ever existed in history). And the reverse direction also holds true.

In relation to the preciseness-vagueness principle, if there is preciseness (e.g., the precise identification of the three epidemics like "the Bubonic Plague, cholera and typhoid fever" in the article aforecited), there is vagueness (e.g., the vagueness in the identification of the three epidemics like "the Bubonic Plague, cholera and typhoid fever" in the article aforecited, since it is not clear why there must be only three, not 5, 7, 8, 10, 12, and so on). And the reverse direction also holds true.

In relation to the simpleness-complicatedness principle, if there is simpleness (e.g., the relatively simple analysis of the bubonic plague as a desirable biological weapon by the Imperial Japanese Army Air Service in 1940), there is complicatedness (e.g., the relatively more complicated analysis of the bubonic plague as a biological weapon, by challenging the claims and assumptions, as shown in the problems pointed out by the critics as cited). And the reverse direction also holds true.

In relation to the openness-hiddenness principle, if there is openness (e.g., the open exploration of the use of the bubonic plague as a biological weapon by the Imperial Japanese Army Air Service in 1940), there is hiddenness (e.g., the hidden bias in the view of using the bubonic plague as a biological weapon by the Imperial Japanese Army Air Service in 1940, because the Chinese victims and their family members considered its use as immoral and inhumane). And the reverse direction also holds true.

In relation to the denseness-emptiness principle, if there is denseness (e.g., the relatively denser concentration of patients who died from the bubonic plague when untreated with antibiotics like "streptomycin and gentamicin, tetracyclines," and so on, as shown in the "mortality rate of 50-90% in untreated cases"), there is emptiness (e.g., the relatively less dense, or more empty, concentration of patients who died from the bubonic plague when treated with antibiotics like "streptomycin and gentamicin, tetracyclines," and so on, as shown in the mortality rate of only "about 1-15%"). And the reverse direction also holds true.

In relation to the change-constancy principle, if there is change (e.g., the ever new ways to understand bacteria and diseases, as shown in the study of biodegradable waste and epidemics), there is constancy (e.g., the ever constant existence of different problems in the ever new ways to understand bacteria and diseases, as shown in the problems with the study of biodegradable waste and epidemics pointed out by the critics). And the reverse direction also holds true.

In relation to the order-chaos principle, if there is order (e.g., the more or less orderly view of the streets in contemporary French cities, which had been installed with sewers and other sanitary systems since the 1950), there is chaos (e.g,. the more or less chaotic view of the streets in French cities in the 19th century and before, which were often filled with "great piles of refuse and excrement down the centre of the streets," and "cholera outbreaks were common and ravaged the population"). (WK 2012a) And the reverse direction also holds true.

In relation to the slowness-quickness principle, if there is slowness (e.g., the relatively slower speed of waste to become a critical issue in antiquity, "due to low population density, coupled with insignificant exploitation of natural resources"), there is quickness (e.g., the relatively quicker speed of waste to become a critical issue nowadays, "due to the increase in population and the massive migration of people to industrial towns and cities from rural areas"). (WK 2012a) And the reverse direction also holds true.

In relation to the expansion-contraction principle, if there is expansion (e.g., the relative spread of cities with sewer systems nowadays), there is contraction (e.g., the relative decrease of cities without any sewer systems nowadays, where the streets can be filled with "great piles of refuse and excrement down the centre of the streets"). And the reverse direction also holds true.

In relation to the theory-praxis principle, if there is theory (e.g., the theoretical construction of different antibacterial compounds in chemistry), there is praxis (e.g., the practical application of different antibacterial compounds in chemistry to the field of medicine, like the use of "streptomycin," "gentamicin," "tetracyclines," etc. for the treatment of bubonic plague). And the reverse direction also holds true.

In relation to the convention-novelty principle, if there is convention (e.g., the conventional wisdom of using antibiotics to fight the bubonic plague), there is novelty (e.g., the alternative novel challenge to the conventional wisdom about using antibiotics to fight the bubonic plague, by the new idea of the Imperial Japanese Army Air Service in 1940 to use the bubonic plague as a biological weapon instead). And the reverse direction also holds true.

In relation to the evolution-transformation principle, if there is evolution (e.g., the natural evolution of humans to get infected with diseases from time to time in the state of nature), there is transformation (e.g., the technical transformation of human ability to deal with infection by the invention of antibiotics like "streptomycin," "gentamicin," "tetracyclines," etc. for the treatment of bubonic plague). And the reverse direction also holds true.

In relation to the symmetry-asymmetry principle, if there is symmetry (e.g., the existence of different lifeforms like humans and bacteria), there is asymmetry (e.g., humans can kill bacteria by antibiotics more than bacteria can kill humans by infection, at least in absolute numbers). And the reverse direction also holds true.

In relation to the softness-hardness principle, if there is softness (e.g., the use of antibiotics to save human life against the bubonic plague), there is hardness (e.g., the use of the bubonic plague as a biological weapon by the Imperial Japanese Army Air Service). And the reverse direction also holds true.

In relation to the seriousness-playfulness principle, if there is seriousness (e.g., the serious business of those in history who tried to find treatments against the bubonic plague), there is playfulness (e.g., the playful part of those in history who tried to find treatments against the bubonic plague, when they played around with different solutions over the ages, like some "ethno-medical beliefs of prevention methods" by "walking around with flowers in or around their nose," carving "the symbol of the cross onto the front door of a house with the words 'Lord have mercy on us' near it," etc.). (WK 2012l; C. Mee 2011) And the reverse direction also holds true.

In relation to the activeness-inactivess principle, if there is activeness (e.g., the alertness of those who are not infected with diseases and remain healthy to actively work), there is inactiveness (e.g., the relative inactiveness of those who get sick because of infection with bacteria and therefore have to rest until recovery).

In relation to the selfness-otherness principle, if there is selfness (e.g., the tendency to be mostly centered on one's self or group, as shown in the use of the bubonic plague as a biological weapon by the Imperial Japanese Army Air Service in 1940), there is otherness (e.g., the tendency to be mostly centered on others, as shown by the legendary nurse Florence Nightingale to take care of patients). And the reverse direction also holds true.

In relation to the regression-progression principle, if there is regression (e.g., the regression made by the research on the bubonic plague, as shown in the use of the bubonic plague as a biological weapon by the Imperial Japanese Army Air Service in 1940), there is progression (e.g., the progress made by the research on the bubonic plague, as shown in the use of antibiotics to fight the bubonic plague aforecited). And the reverse direction also holds true.

In relation to the same-difference principle, if there is similarity in outcome (e.g., the contribution to the molding and control of beliefs, values, and behaviors, regardless of whether this be done by way of using antibiotics to fight the bubonic plague, or by way of carving "the symbol of the cross onto the front door of a house with the words 'Lord have mercy on us' near it," etc.), there is difference in outcome (e.g., the contribution to the molding and control of beliefs, values, and behaviors by way of using antibiotics to fight the bubonic plague for a relatively more scientific lifeworld—but the contribution to the molding and control of beliefs, values, and behaviors by way of carving "the symbol of the cross onto the front door of a house with the words 'Lord have mercy on us' near it," etc. for a relatively more sacral lifeworld). And the reverse direction also holds true.

In relation to the stability-reaction principle, if there is stability (e.g., the relatively more stable condition, in the older days, of appealing to folkloric and religious methods to fight the bubonic plague, like "walking around with flowers in or around their nose" and carving "the symbol of the cross onto the front door of a house with the words 'Lord have mercy on us' near it," etc.), there is reaction (e.g., the old-fashioned folkloric and religious methods to fight the bubonic plague then led to further changes later on, when nowadays, antibiotics are routinely used instead to fight the bubonic plague). And the reverse direction also holds true.

In relation to the functionality-nonfunctionality principle, if there is functionality (e.g., the presence of action for which something is fitted, as shown by "walking around with flowers in or around their nose," which served the ethno-medical function in the older days as a way to fight the bubonic plague), there is nonfunctionality (e.g., the relative lack of action for which something is fitted, as shown in the use of antibiotics to fight the bubonic plague nowadays, such that the ethno-medical method of "walking around with flowers in or around their nose" is no longer functional in the battle against the bubonic plague in our time). And the reverse direction also holds true.

And in relation to the intentionality-nonintentionality principle, if there is intentionality (e.g., the planning or design of something for a certain outcome, as shown in some European "armies of the 14th century" which "were recorded catapulting diseased corpses over the walls of towns and villages in order to spread the pestilence" in the history of Black Death), there is nonintentionality (e.g., the relative absence of planning or design of something for a certain outcome, such that the outcome can even be counter-intentional in being the opposite of what was originally planned, as shown in the unintended consequence of the Bubonic plague which caused "the Black Death that swept through Europe in the 14th century," in that "the plague killed so many of the working population [that] wages rose and some historians have seen this as a turning point in European economic development"). (WK 2012l) And the reverse direction also holds true.

Uselessness, Clinical Waste, and Diseases

Another way to evaluate the possibility (or impossibility) and desirability (or undesirability) of uselessness from the perspective of the mind is to explore uselessness, clinical waste, and diseases.

The term "clinical waste" (also known as "medical waste") refers to "waste products that cannot be considered general waste, produced from healthcare premises, such as hospitals, clinics, doctors offices, veterinary hospitals and labs." (WK 2012t)

Other definitions include the one used in Europe, where "wastes are defined by their European Waste Catalogue (EWC) Codes. Pagina EWC Codes are 6 digits long, with the first two digits defining the over-arching category of waste, the next two defining the sub-category, and the last two defining the precise waste stream. Clinical waste comes under the '18' codes, for example: '18 01 01' corresponds to healthcare waste (18), from humans (01), that is sharp and not infectious (01). It has many waste things thrown." (WK 2012t)

And in the United States, "Section 11002 of the MWTA, 'Listing of Medical Wastes,' provides a listing of definitions on what materials will be classified as 'medical wastes' under the act. The list includes within its definitions: 'cultures and stocks of infectious agents,' 'pathological wastes' such as body tissues, blood wastes and blood byproducts, sharps, contaminated carcasses and beddings of animals, surgery or autopsy wastes that were once in contact with infectious agents, 'laboratory wastes,' 'dialysis wastes that were in contact with the blood of patients undergoing hemodialysis,' medical equipment having come in contact with infectious agents, and further contaminated biological materials." (WK 2012u; EPA 2010)

Clinical Waste and Hazardous Diseases

One general argument against clinical waste is that "disposal of this waste is an environmental concern, as many medical wastes are classified as *infectious* or *biohazardous* and could potentially lead to the spread of infectious disease." (WK 2012t)

Good examples of infectious waste are "blood, potentially contaminated 'sharps' such as needles and scalpels, and identifiable body parts. Sharps include used needles, lancets, and other devices capable of penetrating skin. Infectious waste is often incinerated. The most common method of sterilization is an autoclave. The autoclave uses steam and pressure to sterilize the waste. Additionally, medical facilities produce a variety of waste hazardous chemicals, including radioactive materials. While such wastes are normally not infectious, they may be classified as hazardous wastes, and require proper disposal." (WK 2012t)

Main Methods of Clinical Waste Disposal

The methods of clinical waste disposal can vary from country to country, and "in the United States, there are three main methods for medical waste generators to dispose of their waste: On-site, truck service, and mail-back disposal," as shown below (and summarized in *Sec. 2.3*): (WK 2012t)

- "On-Site Treatment"—"involves the use of very expensive equipment, and is generally only used by very large hospitals and major universities who have the means to afford such equipment."
- "Truck Service"—"involves hiring of a medical waste disposal service whose employees are trained to collect and haul away medical waste in special containers (usually cardboard boxes, or reusable plastic bins) for treatment at a facility designed to handle large amounts of medical waste."
- "Mail-Back Medical Waste Disposal"—"is similar, except that the waste is shipped through the U.S. postal service instead of by private hauler."

Surely, there can be other methods, so the examples here are illustrative.

The Enforcement of Clinical Waste Disposal

But the enforcement of these methods depends on regulations.

For instance, in the United States, under a federal law known as "the Medical Waste Tracking Act of 1988," there are some clauses concerning the management, tracking, and enforcement of clinical (or medical) waste (as summarized in *Sec. 2.3*).

(a) Management and Tracking

In regard to management and tracking, "Section 11003, 'Tracking of Medical Waste,' outlined how the program should manage the transportation of waste materials. Four requirements were primarily identified," as shown below: (WK 2012u; EPA 2010)

- first, "to provide a means of monitoring 'the transportation of waste from the generator to the disposal facility' unless said waste had previously been incinerated"
- Second, "to be able to ensure the 'generator of the waste' that the waste had been 'received by the disposal facility'"
- Third, "to develop a uniform form for the tracking of materials across states and finally to develop a means to label and contain the wastes for the safety of the handlers"
- Fourth, "Section 11004 on 'Inspections' provides provisions allowing for agents of the EPA to 'enter… any establishment…where medical wastes are or have been generated' so as to conduct 'monitoring,' 'testing,' or to 'obtain samples from any person.' This process would allow for the Agency to legally enter generating facilities for the purpose of determining if infectious agents and materials were being handled as prescribed by the EPA Administrator."

More importantly, there is a provision for enforcement.

(b) Enforcement

The MWTA has a provision for enforcement, "as it allows for one of the first instances in which an agency of the federal government may prosecute those charged with violation of regulations dealing with medical wastes. Following enactment of the MWTA the EPA Administrator may be allowed to 'commence civil action in the United States district court in the district in which the violation occurred' against those being charged with the violation." (WK 2012u)

And "penalties ranged based on the level of violation, whether it was done with intention, and if such acts endangered the lives of other individuals. Minor violations of compliance orders would have resulted in a fine of $25,000 per day of 'continued noncompliance.' Criminal penalties against those knowingly and intentionally violating the regulations of the MWTA may face two years of imprisonment or a $50,000 fine while those guilty of knowingly endangering the life of another through noncompliance may face upwards of fifteen years imprisonment and a $1,000,000 fine." (WK 2012u)

Problems with the Study of Clinical Waste and Diseases

However, for the critics, this study of clinical waste and diseases has some major problems. Consider, for illustration, a few examples below (as summarized in *Sec. 2.3*).

Firstly, one main criticism of the study of clinical waste and diseases is that the governmental regulation of clinical waste disposal can be costly and thus produce incentives to violate it.

For example, in the 1980s, "the high cost (~$1500/ton) associated with the legal disposal of the waste" gave incentives to "private waste contractors to dump illegally to avoid high fees," as it was found that, in some cases, "sealed plastic garbage bags...were cut at the top, so their contents could disperse through the ocean" in illegal dumping to avoid high fees. (WK 2012u)

Secondly, another main criticism of the study of clinical waste and diseases is that there are no universal agreement about its enforcement.

For instance, for the critics, "in 1988 the Federal government passed The Medical Waste Tracking Act which set the standards for governmental regulation of medical waste. After the Act was repealed in 1991, States were given the responsibility to regulate and pass laws concerning the disposal of medical waste. All fifty states vary in their regulations from no regulations to very strict." (WK 2012t)

Thirdly, still another main criticism of the study of clinical waste and diseases is that the methods of clinical waste disposal have their own limits.

For instance, "although currently available in all 50 U.S. states, mail-back medical waste disposal is limited to very strict postal regulations (collection and shipping containers must be approved by the postal service for use) and only available by a handful of companies." (WK 2012t)

And fourthly, an additional main criticism of the study of clinical waste and diseases is that what constitutes "clinical waste" (or "medical waste") depends on different definitions and has crucial implications in practice.

For instance, for the critics, the definition of "clinical waste" in Europe and that in the U.S. (as described above) are not exactly similar, and this is important, because "defining what objects were to be classified as medical wastes was crucial to ensure that all municipalities...would be monitoring similar materials." (WK 2012u)

Beyond Clinical Waste and Diseases

These criticisms of the study on clinical waste and diseases, however, do not mean a total rejection of its validity but serve to show us the two opposing sides of the debate, such that the possibility (or impossibility) and desirability (or undesirability) of uselessness (from the perspective of the mind with uselessness, clinical waste, and diseases as a case study here) are not to the extent that the respective defenders would like us to believe.

In addition, the analysis of uselessness, clinical waste, and diseases is to show us something valuable about the ontological principles in existential dialectics, and good examples include the formalness-informalness principle, the absoluteness-relativeness principle, the partiality-totality principle, the predictability-unpredictability principle, the explicability-inexplicability principle, the fiction-reality principle, the finiteness-transfiniteness principle, the preciseness-vagueness principle, the simpleness-complicatedness principle, the openness-hiddenness principle, the denseness-emptiness principle, the change-constancy principle, the order-chaos principle, the slowness-quickness principle, the expansion-contraction principle, the theory-praxis principle, the convention-novelty principle, the evolution-transformation principle, the symmetry-asymmetry principle, the softness-hardness principle, the seriousness-playfulness principle, the activeness-inactivess principle, the selfness-otherness principle, the regression-progression principle, the same-difference principle, the stability-reaction principle, the functionality-nonfunctionality principle, and the intentionality-nonintentionality principle.

For instance, in relation to the formalness-informalness principle, if there is formalness (e.g., the formal logical requirement of soundness, consistency, and completeness in a system of ideas, as in the study on clinical waste and diseases), there is informalness (e.g., the non-formal existence of unsoundness, inconsistency, and incompleteness in the study on clinical waste and diseases, as shown in the criticisms that the governmental regulation of clinical waste disposal can be costly and thus produce incentives to violate it, that there are no universal agreement about its enforcement, that the methods of clinical waste disposal have their own limits, that what constitutes "clinical waste" [or "medical waste"] depends on different definitions and has crucial implications in practice, etc.). And the reverse direction also holds true.

In relation to the absoluteness-relativeness principle, if there is absoluteness (e.g., the absolute view of those "private waste contractors to dump illegally to avoid high fees" in the 1980s aforecited), there is relativeness (e.g., what is right for those "private waste contractors to dump illegally to avoid high fees" in the 1980s aforecited is not necessarily so for the federal government to enforce proper clinical waste disposal under the law). And the reverse direction also holds true.

In relation to the partiality-totality principle, if there is partiality (e.g., the partial view of those "private waste contractors to dump illegally to avoid high fees" in the 1980s aforecited), there is totality (e.g., the more holistic view of clinical waste disposal, such that the whole is not the sum of its parts, that is, it cannot be reduced to the sum of opposing views of those "private waste contractors" to dump illegally to avoid high fees" in the 1980s, the federal government, etc., since there will emerge new views in the future not yet known today, just as there are already criticisms against the two sides nowadays). And the reverse direction also holds true.

In relation to the predictability-unpredictability principle, if there is predictability (e.g., the predictable tendency of the federal government to enforce proper clinical waste disposal under a particular law like the MWTA), there is unpredictability (e.g., the more difficult task to predict exactly the extent to which the federal government will be successful in enforcing proper clinical waste disposal under a particular law in a particular future era, since different laws like the MWTA are passed and then repealed or revised over time in history). And the reverse direction also holds true.

In relation to the explicability-inexplicability principle, if there is explicability (e.g., the explanation by the advocates of the MWTA that proper clinical waste disposal is important to avoid "the spread of infectious disease"), there is inexplicability (e.g., the lack of sufficient explanation by the advocates of the MWTA of why proper clinical waste disposal is necessarily desirable, since, in the 1980s, "the high cost [~$1500/ton] associated with the legal disposal of the waste" gave incentives to "private waste contractors to dump illegally to avoid high fees"). And the reverse direction also holds true.

In relation to the fiction-reality principle, if there is fiction (e.g., the fictional aspect of the MWTA, insofar as "the Act was repealed in 1991," because the federal government was no longer deemed necessary to enforce propper clinical waste disposal, so the enforcement was then passed to the states instead), there is reality (e.g., the realistic aspect of the MWTA, insofar as the Act was in force between 1988 and 1991). And the reverse direction also holds true.

In relation to the finiteness-transfiniteness principle, if there is finiteness (e.g., the finite number of the categories of clincal waste, like "needles," "scalpels," "lancets," "identifiable body parts," etc.), there is transfiniteness (e.g., the transfinite number of all the individual "needles," "scalpels," "lancets," "identifiable body parts," etc. that have ever existed or produced in history). And the reverse direction also holds true.

In relation to the preciseness-vagueness principle, if there is preciseness (e.g., the precise identification of three methods of clinical waste disposal in the article aforecited), there is vagueness (e.g., the vagueness in the identification of the three methods of clinical waste disposal in the article aforecited, since it is not clear why there must be only 3, not 4, 5, 6, 7, and so on). And the reverse direction also holds true.

In relation to the simpleness-complicatedness principle, if there is simpleness (e.g., the relatively simple analysis of proper clinical waste disposal by the advocates), there is complicatedness (e.g., the relatively more complicated analysis of proper clinical waste disposal, by challenging the claims and assumptions, as shown in the problems pointed out by the critics as cited). And the reverse direction also holds true.

In relation to the openness-hiddenness principle, if there is openness (e.g., the open exploration of what constitutes clinical waste by a given definition), there is hiddenness (e.g., the hidden bias in each definition of clinical waste, because "defining what objects were to be classified as medical wastes was crucial to ensure that all municipalities…would be monitoring similar materials," so different definitions end up monitoring different materials). And the reverse direction also holds true.

In relation to the denseness-emptiness principle, if there is denseness (e.g., the relatively denser concentration of items like "needles," "scalpels," "lancets," "identifiable body parts," etc. to fall under the jursidiction of the MWTA), there is emptiness (e.g., the relatively less dense, or more empty, concentration of space debris to fall under the jursidiction of the MWTA). And the reverse direction also holds true.

In relation to the change-constancy principle, if there is change (e.g., the ever new ways to understand clinical waste, as shown in the study on clinical waste and diseases), there is constancy (e.g., the ever constant existence of different problems in the ever new ways to understand clinical waste, as shown in the problems with the study on clinical waste and diseases pointed out by the critics). And the reverse direction also holds true.

In relation to the order-chaos principle, if there is order (e.g., the more or less orderly appearance of clinical waste, if looked from the vantage point of the MWTA between 1988 and 1991, when it enforced proper clinical waste disposal), there is chaos (e.g,. the more or less chaotic appearance of clinical waste, if looked from the vantage point of "the '30-mile garbage slick' composed primarily of medical and household wastes" spreading in "numerous New Jersey and New York beaches" in 1987, before the enactment of the MWTA). And the reverse direction also holds true.

In relation to the slowness-quickness principle, if there is slowness (e.g., "hot air at 160 °C takes two hours to achieve" the same sterility for clinical waste that steam at 134 °C takes in three minutes), there is quickness (e.g., "steam at 134 °C can achieve in three minutes the same sterility [for clinical waste] that hot air at 160 °C takes two hours to achieve"). (WK 2012v) And the reverse direction also holds true.

In relation to the expansion-contraction principle, if there is expansion (e.g., the relative spread of the use of "many surgical instruments [such as forceps, needle holders, and scalpel handles]" which are "single-use…items" nowadays), there is contraction (e.g., the relative decline of the use of "surgical instruments [such as forceps, needle holders, and scalpel handles]" which are "reusable items" nowadays). (WK 2012v) And the reverse direction also holds true.

In relation to the theory-praxis principle, if there is theory (e.g., the theoretical understanding of "infectious or biohazardous" disease in medicine), there is praxis (e.g., the practical application of the knowledge of "infectious or biohazardous" disease in medicine to the field of waste management, like the enforcement by the MWTA for proper clinical waste disposal to avoid the spread of infectious disease). And the reverse direction also holds true.

In relation to the convention-novelty principle, if there is convention (e.g., the conventional wisdom of using "surgical instruments [such as forceps, needle holders, and scalpel handles]" which are "reusable" and "sterilizable" in the older days), there is novelty (e.g., the alternative novel challenge to the conventional wisdom of using "reusable" and "sterilizable" surgical instruments in the older days, by the new idea of using "surgical instruments [such as forceps, needle holders, and scalpel handles]" which are "single-use" nowadays). And the reverse direction also holds true.

In relation to the evolution-transformation principle, if there is evolution (e.g., the natural evolution of humans to throw away clinical waste like "identifiable body parts" in the state of nature in a primitive way), there is transformation (e.g., the technical transformation of human ability to dispose clinical waste by the invention of autoclaves). (WK 2012v) And the reverse direction also holds true.

In relation to the symmetry-asymmetry principle, if there is symmetry (e.g., the existence of different actors in the field of clinical waste disposal, like the federal government and "private waste contractors"), there is asymmetry (e.g., the federal government empowered by the MWTA can punish those "private waste contractors" for dumping illegally "to avoid high fees"). And the reverse direction also holds true.

In relation to the softness-hardness principle, if there is softness (e.g., the compliance of those "private waste contractors," when they dispose clinical waste in accordance to the regulations under the MWTA), there is hardness (e.g., the violation of those "private waste contractors," when they engage in dumping illegally "to avoid high fees"). And the reverse direction also holds true.

In relation to the seriousness-playfulness principle, if there is seriousness (e.g., the serious business of those trying to deal with proper clinical waste disposal), there is playfulness (e.g., the playful part of those trying to deal with proper clinical waste disposal, when they play around with different methods over time, like "on-site, truck service, and mail-back disposal," etc.). And the reverse direction also holds true.

In relation to the activeness-inactivess principle, if there is activeness (e.g., the activeness of those "private waste contractors," when they dispose clinical waste in accordance to the regulations under the MWTA and are allowed to stay in business), there is inactiveness (e.g., the relative inactiveness of those "private waste contractors," when they are found "guilty of knowingly endangering the life of another through noncompliance,…face upwards of fifteen years imprisonment," and therefore cannot work in that period because of imprisonment).

In relation to the selfness-otherness principle, if there is selfness (e.g., the tendency to be mostly centered on one's self or group, as shown in those "private waste contractors," when they engage in dumping illegally "to avoid high fees"), there is otherness (e.g., the tendency to be mostly centered on others, as shown in those few good individuals who genuinely work under the MWTA to ensure proper clinical waste disposal for the common good). And the reverse direction also holds true.

In relation to the regression-progression principle, if there is regression (e.g., the regression made by the methods of clinical waste disposal and their enforcement, as shown in the problems pointed out by the critics), there is progression (e.g., the progress made by the methods of clinical waste disposal and their enforcement, as shown in the contribution to the prevention of infectious disease by clinical waste aforecited). And the reverse direction also holds true.

In relation to the same-difference principle, if there is similarity in outcome (e.g., the contribution to the molding and control of beliefs, values, and behaviors, regardless of whether this be done by way of those "private waste contractors to dump clinical waste "to avoid high fees," or by way of the federal government under the MWTA for the enforcement of proper clinical waste disposal), there is difference in outcome (e.g., the contribution to the molding and control of beliefs, values, and behaviors by way of those "private waste contractors for a relatively more market-oriented lifeworld—but the contribution to the molding and control of beliefs, values, and behaviors by way of the federal government under the MWTA for a relatively more state-regulated lifeworld). And the reverse direction also holds true.

In relation to the stability-reaction principle, if there is stability (e.g., the relatively more stable condition of the enforcement of proper clinical waste disposal between 1988 and 1991 under the MWTA), there is reaction (e.g., the enforcement of proper clinical waste disposal between 1988 and 1991 under the MWTA then led to further changes later on, when "the Act was repealed in 1991" and "states were given the responsibility to regulate and pass laws concerning the disposal of medical waste"). And the reverse direction also holds true.

In relation to the functionality-nonfunctionality principle, if there is functionality (e.g., the presence of action for which something is fitted, as shown in the passing of the MWTA in 1988 to serve the function of enforcing proper clinical waste dispoasal by the federal government), there is nonfunctionality (e.g., the relative lack of action for which something is fitted, as shown in the repeal of the MWTA in 1991, when the federal government was no longer deemed necessary to serve the function of enforcing proper clinical waste dispoasal, so "states were given the responsibility to regulate and pass laws concerning the disposal of medical waste" instead). And the reverse direction also holds true.

And in relation to the intentionality-nonintentionality principle, if there is intentionality (e.g., the planning or design of something for a certain outcome, as shown in the enactment of governmental regulation for better clinical waste disposal), there is nonintentionality (e.g., the relative absence of planning or design of something for a certain outcome, such that the outcome can even be counter-intentional in being the opposite of what was originally planned, as shown in the unintended consequences of the governmental regulation for clinical waste disposal, in that it can be costly and thus produce incentives to violate it and that the methods of clinical waste disposal have their own limits). And the reverse direction also holds true.

Uselessness and Nature

Uselessness, when examined from the perspective of nature, can show us its possibility (or impossibility) and desirability (or undesirability), and this can be shown by way of two case studies, namely, (2.3.1) uselessness, litter, and the natural environment, and (2.3.2) uselessness, space waste, and the Earth's orbit—to be addressed in what follows, respectively.

Uselessness, Litter, and the Natural Environment

A good way to evaluate the possibility (or impossibility) and desirability (or undesirability) of uselessness from the perspective of nature is to explore uselessness, litter, and the natural environment.

The term "litter" refers to "waste products that have been disposed of improperly, without consent, in an inappropriate location....While most litter is associated with containers, wrappers and paper product; dumped items may include furniture, appliances (white goods), old electronics (e-waste), abandoned vehicles or construction materials. These categories of waste often contain hazardous materials." (WK 2012w)

History of Litter

Historically, "people have disposed of unwanted materials without fear of retribution, onto streets, roadsides, in small local dumps or often in remote locations. Prior to reforms within cities in the mid-to-late 19th century, sanitation was not a government priority. The growing piles of waste led to the spread of disease." (WK 2012w; S. Spacek 2012)

But with the growth of litter in contemporary times, for instance, "in the United States, the Solid Waste Disposal Act of 1965 was enacted. In 1976 the Federal government amended the Solid Waste Disposal Act, creating the Resource Conservation and Recovery Act (RCRA), which requires a 'cradle to grave' approach to the proper handling of potentially hazardous materials. RCRA gives authority to the Environmental Protection Agency (EPA) to regulate and enforce proper hazardous waste disposal." (WK 2012w)

Elsewhere, "many countries now have laws that require that household hazardous waste be deposited in a special location rather than sent to landfills with regular refuse. Household hazardous waste includes paints and solvents, chemicals, light bulbs, fluorescent lights, spray cans, and yard products such as fertilizers, pesticides, herbicides and insecticides. Additionally, medical waste generated at home is considered a hazardous waste and must be disposed of properly." (WK 2012w)

Causes of Litter

Littering can have different causes, both intentional and unintentional.

For instance, "in addition to intentional littering, almost half of litter on U.S. roadways is now a result of accidental or unintentional litter, usually debris that falls off of improperly secured trash, recycling collection vehicles and pickup trucks. Population levels, traffic density and proximity to waste disposal sites are factors known to correlate with higher litter rates. Government neglect, the inability of governments to remove litter in a timely manner, is also a reason why humans are tempted to litter." (WK 2012w; ABC 2007; P. Leigh 2007)

In addition, "illegally dumped hazardous waste may be affected by the costs associated with dropping materials at designated sites; some facilities charge a fee for depositing hazardous material. Access to nearby facilities that accept hazardous waste may deter use. Additionally, ignorance of the laws that regulate the proper disposal of hazardous waste may have an impact on proper disposal." (WK 2012w)

In other cases, "negligent or lenient law enforcement contributes to littering behavior. Other causes are inconvenience, entitlement and economic conditions. A survey of dumping in Pennsylvania found that the largest number of illegal dumps were in townships without municipal trash hauling. The same report also cites unavailability of curbside trash and recycling service, shortage of enforcement, and habit as possible causes." (WK 2012w; M. Argento 2010)

In a study by the Dutch organization VROM, "80 percent of the people claim that 'everybody leaves of a piece of paper, tin or something, on the street behind.' Young people from 12 to 24 years cause more litter than the average (Dutch or Belgian) person. Eighteen percent of people who regularly cause litter were 50 years of age or older. However, a 2010 survey of littering in Maine, New Hampshire and Vermont in the United States, placed litterers aged 55 and over at less than five percent. The same observational study estimated the overall average of litterers to be 78 percent male. Nevertheless, automobile drivers and recreationalists, smokers and youth are specific target groups within many campaigns conducted to keep countries free of litter. In 1999, research by Keep America Beautiful found that 75% of Americans admitted to littering the last five years, yet 99% of the same individuals admitted they enjoyed a clean environment." (WK 2012w)

Life Cycle of Litter

Whatever the causes of littering may be, “litter can remain either visible for extended periods of time before it eventually biodegrades, with some items made of condensed glass, styrofoam or plastic possibly remaining in the environment for over a million years.” (WK 2012w) Macfarlane, Christina. 2009

In a study, it was estimated that “about 18 percent of litter, usually traveling through stormwater systems, ends up in local streams, rivers, and waterways. Uncollected litter can accrete and flow into streams, local bays and estuaries. Litter in the ocean either washes up on beaches or collects in Ocean gyres such as the Great Pacific Garbage Patch. About 80 percent of marine debris comes from land-based sources.” (WK 2012w)

And “some litter that is collected can be recycled; however, degraded litter cannot be recycled and eventually degrades to sludge, often toxic. The majority of litter that is collected goes to landfills.” (WK 2012w)

Effects on the Natural Environment

Litter can have different effects on the natural environment.

For instance, litter can produce effects on wildlife, in that “animals may get trapped or poisoned with litter in their habitats. Cigarette butts and filters are a threat to wildlife and have been found in the stomachs of fish, birds and whales, who have mistaken them for food. Also animals can get trapped in the rubbish and be in serious discomfort. For example, the plastic used to hold beverage cans together can get wrapped around animals' necks and cause them to suffocate as they grow.” (WK 2012w; S. Spacek 2012)

There are other effects of litter, in that “organic litter in large amounts can cause water pollution and lead to algal blooms. Cigarettes could also start fires if they are not put out and then discarded in the envrionment.” (WK 2012w)

Solutions to Litter

What then are the solutions to litter? There have been different proposed solutions over time, and hereafter are some good examples.

(a) Litter Bins

The first example of a solution to litter has to do with “public waste containers or street bins,” which “are provided by local authorities to be used as a convenient place for the disposal and collection of litter. Increasingly both general waste and recycling options are provided. Local councils pick the waste up and take it to refuse or recycling.” (WK 2012w)

(b) Picking Up

The second example of a solution to litter has to do with "picking up," in which "volunteers, sometimes alone or coordinated through organisations, will pick up litter and dispose of it. Clean up events may be organised where participants will sometimes comb an area in a line to ensure that no litter is missed. In North America, Adopt a Highway programs are popular, where companies and organizations commit to cleaning stretches of road. Increasingly, geocaching is used." (WK 2012w)

However, "picking up litter can be hazardous. Hazards can include exposure to dangerous goods, sharps waste and pathogens. As a result, safety equipment is sometimes worn and tools such as litter grabbers (extendable arms) are used." (WK 2012w)

(c) Container Deposit Schemes

The third example of a solution to litter has to do with "container deposit schemes," which aim at "both reducing littering and also encouraging picking up through local recycling programs that offer incentives, particularly for aluminium cans, glass bottles and plastic bottles. In New York, an expanded bottle bill that included plastic water bottles increased recycling rates and generated 120 million dollars in revenue to the state General Fund from unclaimed deposits in 2010." (WK 2012w; BOT 2010)

(d) Litter Traps

The fourth example of a solution to litter has to do with "litter traps," which "can be used to capture litter as it exits stormwater drains into waterways. However, litter traps are only effective for large or floating items of litter and must be maintained." (WK 2012w)

(e) Monitoring Sites of Littering and Dumping

The fifth example of a solution to litter has to do with "monitoring sites of littering and dumping," which "use technology to monitor areas prone to dumping. In Japan, a study used Geographic Information Systems (GIS) to map areas of dumping based on site characteristics." (WK 2012w; T. Tasaki 2006)

(f) Legislation and Fines

And the sixth example of a solution to litter has to do with "legislation and fines," in that "actions resulting in fines can include on-the-spot fines for individuals administered by authorised officers in public or on public transport or littering from a vehicle, in which the vehicle owner is fined—reported by either responsible officer or third party, sometimes online." (WK 2012w; NAT 1975; EPAV 2011)

Problems with the Study of Litter and the Natural Environment

However, for the critics, this study of litter and the natural environment has some major problems. Consider, for illustration, a few examples below.

Firstly, one main criticism of the study of litter and the natural environment is that the prevention of littering is too complicated in practice.

For example, the causes of littering (as described earlier) are so numerous that to completely (or even mostly) eliminate litter is not practical at all.

Secondly, another main criticism of the study of litter and the natural environment is that many cases of littering are unintentional (which is subject to different uncertain circumstances), so it is difficult to control littering behavior.

For instance, a study found that "almost half of litter on U.S. roadways is now a result of accidental or unintentional litter, usually debris that falls off of improperly secured trash, recycling collection vehicles and pickup trucks." (WK 2012w)

Thirdly, still another main criticism of the study of litter and the natural environment is that each proposed solution has its own disadvantages.

For instance, in the case of litter bins, "there are issues with this approach. If the bins are not regularly emptied, then overfilling of bins occurs and can increase litter indirectly. Some local authorities will only take responsibility for rubbish that is placed in the bins, which means that litter remains a problem. People may blame a lack of well-placed bins for their littering. Hazardous materials may be incorrectly disposed of in the bins and they can encourage dumpster diving." (WK 2012w)

And fourthly, an additional main criticism of the study of litter and the natural environment is that there is no universal regulation of littering behavior.

For instance, different countries and localities have different customs and legislations concerning littering behavior. A good contrast is between Australia (where there is no national legislation about littering behavior) and the United States (where "all 50 states have anti-litter laws," but they do not necessarily converge). (WK 2012w)

Beyond Litter and the Natural Environment

These criticisms of the study on litter and the natural environment do not entail s a total rejection of its validity but serve to show us the two opposing sides of the debate, such that the possibility (or impossibility) and desirability (or undesirability) of uselessness (from the perspective of nature with uselessness, litter, and the natural environment as a case study here) are not to the extent that the respective defenders would like us to believe.

In addition, the analysis of uselessness, litter, and the natural environment can teach us a good lesson about the ontological principles in existential dialectics, and good examples include the formalness-informalness principle, the absoluteness-relativeness principle, the partiality-totality principle, the predictability-unpredictability principle, the explicability-inexplicability principle, the fiction-reality principle, the finiteness-transfiniteness principle, the preciseness-vagueness principle, the simpleness-complicatedness principle, the openness-hiddenness principle, the denseness-emptiness principle, the change-constancy principle, the order-chaos principle, the slowness-quickness principle, the expansion-contraction principle, the theory-praxis principle, the convention-novelty principle, the evolution-transformation principle, the symmetry-asymmetry principle, the softness-hardness principle, the seriousness-playfulness principle, the activeness-inactivess principle, the selfness-otherness principle, the regression-progression principle, the same-difference principle, the stability-reaction principle, the functionality-nonfunctionality principle, and the intentionality-nonintentionality principle.

For instance, in relation to the formalness-informalness principle, if there is formalness (e.g., the formal logical requirement of soundness and completeness in a system of ideas, as in the study on litter and the environment), there is informalness (e.g., the non-formal existence of unsoundness and incompleteness in the study on litter and the environment, as shown in the criticisms that the prevention of littering is too complicated in practice, that many cases of littering are unintentional [which is subject to different uncertain circumstances], that each proposed solution has its own disadvantages, that there is no universal regulation of littering behavior, etc.). And the reverse direction also holds true.

In relation to the absoluteness-relativeness principle, if there is absoluteness (e.g., the absolute view of those "Dutch police and local supervisors [known as *buitengewoon opsporingsambtenaar*, or BOA]," who "fine citizens for throwing away cans, bottles or wrappers onto the street"), there is relativeness (e.g., what is right for those "Dutch police and local supervisors [known as *buitengewoon opsporingsambtenaar*, or BOA]," who "fine citizens for throwing away cans, bottles or wrappers onto the street," is not necessarily so for those citizens who are fined but disagree about the punishment). (WK 2012w) And the reverse direction also holds true.

In relation to the partiality-totality principle, if there is partiality (e.g., the partial view of litter by the advocates of "Clean Up the World," which is a worldwide anti-litter campaign), there is totality (e.g., the more holistic view of litter, such that the whole is not the sum of its parts, that is, it cannot be reduced to the sum of opposing views held by the advocates of "Clean Up the World" and their critics, since there will emerge new views in the future not yet known today, just as there are already critics about both sides nowadays). And the reverse direction also holds true.

In relation to the predictability-unpredictability principle, if there is predictability (e.g., the predictable tendency of the advocates of "Clean Up the World" to campaign against litter), there is unpredictability (e.g., the more difficult task to predict exactly the extent to which the advocates of "Clean Up the World" will be successful in the campaign against litter on a particular future occasion, because many cases of littering are unintentional [which is subject to different uncertain circumstances], so it is difficult to control littering behavior). And the reverse direction also holds true.

In relation to the explicability-inexplicability principle, if there is explicability (e.g., the explanation by "Dutch police and local supervisors [known as *buitengewoon opsporingsambtenaar*, or BOA]," who "fine citizens for throwing away cans, bottles or wrappers onto the street" on the basis of anti-litter enforcement), there is inexplicability (e.g., the lack of sufficient explanation by "Dutch police and local supervisors [known as *buitengewoon opsporingsambtenaar*, or BOA]" of why fining "citizens for throwing away cans, bottles or wrappers onto the street" is necessarily desirable, because those citizens who are fined can disagree, which also explains why they litter in the first place). And the reverse direction also holds true.

In relation to the fiction-reality principle, if there is fiction (e.g., the fictional aspect of anti-litter solutions, insofar as each proposed solution has its own disadvantages and does not often work as claimed), there is reality (e.g., the realistic aspect of anti-litter solutions, insofar as each proposed solution has some impact on cleaning up litter). And the reverse direction also holds true.

In relation to the finiteness-transfiniteness principle, if there is finiteness (e.g., the finite number of the categories of litter, like "containers," "wrappers and paper product," "furniture," "appliances," "old electronics," "abandoned vehicles" and "construction materials"), there is transfiniteness (e.g., the transfinite number of all the individual pieces of "containers," "wrappers and paper product," "furniture," "appliances," "old electronics," "abandoned vehicles" and "construction materials" in history that have ever been produced). And the reverse direction also holds true.

In relation to the preciseness-vagueness principle, if there is preciseness (e.g., the precise identification of 6 solutions to litter in the article aforecited), there is vagueness (e.g., the vagueness in the identification of the 6 solutions to litter in the article aforecited, since it is not clear why there must be only 6, not 7, 8, 9, and so on). And the reverse direction also holds true.

In relation to the simpleness-complicatedness principle, if there is simpleness (e.g., the relatively simple analysis of the prevention of litter by the advocates), there is complicatedness (e.g., the relatively more complicated analysis of the prevention of litter, by challenging the claims and assumptions, as shown in the problems with complicated causes and other issues pointed out by the critics as cited). And the reverse direction also holds true.

In relation to the openness-hiddenness principle, if there is openness (e.g., the open exploration of using legislation to enforce proper litter disposal), there is hiddenness (e.g., the hidden bias in using legislation to enforce proper litter disposal, since there is no universal regulation of littering behavior and, in addition, different countries and localities have different customs and legislations concerning littering behavior). And the reverse direction also holds true.

In relation to the denseness-emptiness principle, if there is denseness (e.g., the relatively denser concentration of litter in litter traps), there is emptiness (e.g., the relatively less dense, or more empty, concentration of litter in areas benefited from regular clean-up activities). And the reverse direction also holds true.

In relation to the change-constancy principle, if there is change (e.g., the ever new ways to deal with waste, as shown in the study on litter and the environment), there is constancy (e.g., the ever constant existence of different problems in the ever new ways to deal with waste, as shown in the problems with the study on litter and the environment pointed out by the critics). And the reverse direction also holds true.

In relation to the order-chaos principle, if there is order (e.g., the more or less orderly appearance of clean-up areas in the aftermath of intensive anti-litter campaigns), there is chaos (e.g,. the more or less chaotic appearance of abandoned areas suffering from massive litter). And the reverse direction also holds true.

In relation to the slowness-quickness principle, if there is slowness (e.g., the relatively slower speed of some items of litter like "condensed glass, styrofoam or plastic" waste to biodegrade, because they can "possibly remain…in the environment for over a million years"), there is quickness (e.g., the relatively quicker speed of some biodegradable items of litter to eventually decompose). And the reverse direction also holds true.

In relation to the expansion-contraction principle, if there is expansion (e.g., the relative spread of the practice of using litter bins on the streets nowadays), there is contraction (e.g., the relative decline of the practice of throwing "waste and human wastes out of the window which would decompose in the street" nowadays, although this was common for Europeans in the older days). And the reverse direction also holds true.

In relation to the theory-praxis principle, if there is theory (e.g., the theoretical argument about the harmful effects of litter on wildlife), there is praxis (e.g., the practical application of the knowledge about the harmful effects of litter on wildlife to the field of litter treatment, like different legislations in the United States, where "all 50 states have anti-litter laws"). And the reverse direction also holds true.

In relation to the convention-novelty principle, if there is convention (e.g., the conventional practice of the Europeans in the older days when they threw "waste and human wastes out of the window which would decompose in the street"), there is novelty (e.g., the alternative novel challenge to the conventional practice of the Europeans in the older days to throw waste on the street, by the new idea of different solutions to litter nowadays, like the use of litter bins, litter traps, etc.). And the reverse direction also holds true.

In relation to the evolution-transformation principle, if there is evolution (e.g., the natural evolution of humans to throw away waste in the state of nature in a primitive way), there is transformation (e.g., the technical transformation of human ability to throw away waste by the invention of picking-up, container deposit schemes, monitering sites, etc.). And the reverse direction also holds true.

In relation to the symmetry-asymmetry principle, if there is symmetry (e.g., the existence of different actors on the street, like cops and citizens), there is asymmetry (e.g., "police and local supervisors" like those known as "*buitengewoon opsporingsambtenaar*, or BOA" can "fine citizens for throwing away cans, bottles or wrappers onto the street"). And the reverse direction also holds true.

In relation to the softness-hardness principle, if there is softness (e.g., the cooperation with the legal authorities for anti-litter enforcement), there is hardness (e.g., the punishment of those who litter on the street). And the reverse direction also holds true.

In relation to the seriousness-playfulness principle, if there is seriousness (e.g., the serious business of those who try to clean up litter), there is playfulness (e.g., the playful part of those who try to clean up litter, when they play around with different solutions over time, like the use of litter bins, litter traps, picking up, monitering sites, legislation and fines, etc.). And the reverse direction also holds true.

In relation to the activeness-inactivess principle, if there is activeness (e.g., the activeness of those animals which are fortunate enough to live in areas without much litter, so they do not suffer from any sickness or pain due to it), there is inactiveness (e.g., the relative inactiveness of those animals which "get…poisoned with litter in their habitats" and are "in serious discomfort," so they cannot move around freely).

In relation to the selfness-otherness principle, if there is selfness (e.g., the tendency to be mostly centered on one's self or group, as shown in the littering behavior of some individuals), there is otherness (e.g., the tendency to be mostly centered on others, as shown in the efforts of the government to clean up litter for the common good). And the reverse direction also holds true.

In relation to the regression-progression principle, if there is regression (e.g., the regression made by different solutions to litter, as shown in the disadvantages of each solution pointed out by the critics), there is progression (e.g., the progress made by different solutions to litter, as shown in their contribution to anti-litter efforts aforecited). And the reverse direction also holds true.

In relation to the same-difference principle, if there is similarity in outcome (e.g., the contribution to the molding and control of beliefs, values, and behaviors, regardless of whether this be done by way of anti-litter campaigns, or by way of throwing away waste on the street), there is difference in outcome (e.g., the contribution to the molding and control of beliefs, values, and behaviors by way of anti-litter campaigns for a relatively more environment-conscious lifeworld—but the contribution to the molding and control of beliefs, values, and behaviors by way of throwing away waste on the street for a relatively more environment-dominated lifeworld). And the reverse direction also holds true.

In relation to the stability-reaction principle, if there is stability (e.g., the relatively more stable condition of using litter bins to deal with litter in the modern history of anti-litter campaign), there is reaction (e.g., the practice of using litter bins to deal with litter in the modern history of anti-litter campaign then led to further changes later on, when it is discovered that "if the bins are not regularly emptied, then overfilling of bins occurs and can increase litter indirectly" and "some local authorities will only take responsibility for rubbish that is placed in the bins, which means that litter remains a problem"). And the reverse direction also holds true.

In relation to the functionality-nonfunctionality principle, if there is functionality (e.g., the presence of action for which something is fitted, as shown in the habit of some Europeans in the 19th century and before to throw waste on the street, which served the function to deal with waste at the time), there is nonfunctionality (e.g., the relative lack of action for which something is fitted, as shown in the wider use of litter bins, container deposit schemes, litter traps, etc. nowadays, so that the older habit of some Europeans to throw waste on the street is no longer functional in our time). And the reverse direction also holds true.

And in relation to the intentionality-nonintentionality principle, if there is intentionality (e.g., the planning or design of something for a certain outcome, as shown in the effort to prevent littering), there is nonintentionality (e.g., the relative absence of planning or design of something for a certain outcome, such that the outcome can even be counter-intentional in being the opposite of what was originally planned, as shown in the unintended consequences of the prevention of littering, in that the prevention of littering is too much more complicated in practice than previously thought and that many cases of littering are unintentional and thus it is difficult to control littering behavior). And the reverse direction also holds true.

Uselessness, Space Waste, and the Earth's Orbit

Another good way to evaluate the possibility (or impossibility) and desirability (or undesirability) of uselessness from the perspective of nature is to explore uselessness, space waste (or space debris), and the Earth's orbit.

The term "space waste" (also known as "space debris," "space junk," and "orbital debris") refers "the collection of objects in orbit around Earth that were created by humans but no longer serve any useful purpose. These objects consist of everything from spent rocket stages and defunct satellites to erosion, explosion and collision fragments." (WK 2012x)

In general, "the vast majority of the estimated tens of millions of pieces of space debris are small particles, less than 1 centimetre (0.39 in). These include dust from solid rocket motors, surface degradation products such as paint flakes, and coolant released by RORSAT nuclear powered satellites." (WK 2012x)

Characteristics of Space Waste

Space debris (or space waste) can be understood in terms of their major characteristics; consider, for illustration, three major ones below (as summarized in *Sec. 2.4*).

(a) Large vs. Small Space Debris

The first characteristic of space waste (or space debris) to be considered here has to do with "large vs. small" types.

In regard to large space debris, the word "large" is defined "not by its size so much as the current ability to detect objects of some lower size limit. Generally, large is taken to be 10 centimetres (3.9 in) across or larger, with typical masses on the order of 1 kilogram (2.2 lb). Logically it would follow that small debris would be anything smaller than that, but in fact the cutoff is normally 1 centimetre (0.39 in) or smaller. Debris between these two limits would normally be considered 'large' as well, but goes unmeasured due to our inability to track them." (WK 2012x; UN 1999)

And in relation to small space debris, "the great majority of debris consists of smaller objects, 1 centimetre (0.39 in) or less. The mid-2009 update to the NASA debris FAQ places the number of large debris items over 10 centimetres (3.9 in) at 19,000, between 1 and 10 centimetres (3.9 in) approximately 500,000, and that debris items smaller than 1 centimetre (0.39 in) exceeds tens of millions. In terms of mass, the vast majority of the overall weight of the debris is concentrated in larger objects, using numbers from 2000, about 1,500 objects weighing more than 100 kilograms (220 lb) each account for over 98% of the 1,900 tons of debris then known in low earth orbit." (WK 2012x; NASA 2009; J. Carroll 2002)

Fortunately, "since space debris comes from man-made objects, the total *possible* mass of debris is easy to calculate: it is the total mass of all spacecraft and rocket bodies that have reached orbit. The actual mass of debris will be necessarily less than that, as the orbits of some of these objects have since decayed. As debris mass tends to be dominated by larger objects, most of which have long ago been detected, the total mass has remained relatively constant in spite of the addition of many smaller objects. Using the figure of 8,500 known debris items from 2008, the total mass is estimated at 5,500 tonnes (5,400 long tons; 6,100 short tons)." (WK 2012x; R. McKie 2008)

(b) Debris in Low Earth Orbit

The second characteristic of space waste (or space debris) to be considered here has to do with "debris in Low Earth Orbit" (LEO).

This is important, because “every satellite, space probe and manned mission has the potential to create space debris. Any impact between two objects of sizeable mass can spall off shrapnel debris from the force of collision. Each piece of shrapnel has the potential to cause further damage, creating even more space debris. With a large enough collision (such as one between a space station and a defunct satellite), the amount of cascading debris could be enough to render Low Earth Orbit essentially unusable.” (WK 2012x; D. Kessler 2009)

Worse, this problem in LEO is “compounded by the fact that there are few 'universal orbits' that keep spacecraft in particular rings, as opposed to GEO, a single widely used orbit. The closest would be the sun-synchronous orbits that maintain a constant angle between the sun and orbital plane. But LEO satellites are in many different orbital planes providing global coverage, and the 15 orbits per day typical of LEO satellites results in frequent approaches between object pairs. Since sun-synchronous orbits are polar, the polar regions are common crossing points.” (WK 2012x; M. Ford 2009)

And after the creation of a space debris, “orbital perturbations mean that the orbital plane's direction will change over time, and thus collisions can occur from virtually any direction. Collisions thus usually occur at very high relative velocities, typically several kilometres per second. Such a collision will normally create large numbers of objects in the critical size range, as was the case in the 2009 collision.” (WK 2012x; ESA 2009)

(c) Debris at Higher Altitudes

And the third characteristic of space waste (or space debris) to be considered here has to do with “debris in higher altitudes.”

Unlike LEO, “at higher altitudes, where atmospheric drag is less significant, orbital decay takes much longer. Slight atmospheric drag, lunar perturbations, and solar radiation pressure can gradually bring debris down to lower altitudes where it decays, but at very high altitudes this can take millennia. Thus…these orbits are generally less used than LEO, and the problem onset is slower as a result….” (WK 2012x’ D. Kessler 1991)

On the bright side, “relative velocities in GEO are low, compared with those between objects in largely random low earth orbits. The impact velocities peak at about 1.5 kilometres per second (0.93 mi/s). This means that the debris field from such a collision is not the same as a LEO collision and does not pose the same sort of risks, at least over the short term. It would, however, almost certainly knock the satellite out of operation. Large-scale structures, like solar power satellites, would be almost certain to suffer major collisions over short periods of time.” (WK 2012x; M. Hechler 1981)

Sources of Space Waste

Space waste (or space debris) can be caused by different sources. Consider, for illustration, a few main examples below (as summarized in *Sec. 2.4*).

(a) Dead Spacecraft

The first source of space waste (or space debris) to be described here involves "dead spacecraft."

For instance, "in a catalog listing known launches up to July 2009, the Union of Concerned Scientists listed 902 operational satellites. This is out of a known population of 19,000 large objects and about 30,000 objects ever launched. Thus, operational satellites represent a small minority of the population of man-made objects in space. The rest are, by definition, debris." (WK 2012x; UCS 2009)

(b) Lost Equipment

The second source of space waste (or space debris) to be described here involves "lost equipment."

For instance, "according to Edward Tufte's book *Envisioning Information*, space debris objects have included a glove lost by astronaut Ed White on the first American space-walk (EVA); a camera Michael Collins lost near the spacecraft Gemini 10; garbage bags jettisoned by the Soviet cosmonauts throughout the Mir space station's 15-year life; a wrench and a toothbrush. Sunita Williams of STS-116 lost a camera during EVA. In an EVA to reinforce a torn solar panel during STS-120, a pair of pliers was lost and during STS-126, Heidemarie Stefanyshyn-Piper lost a briefcase-sized tool bag in one of the mission's EVAs." (WK 2012x; J. Smith 2007)

(c) Boosters

The third source of space waste (or space debris) to be described here involves "boosters."

In fact, "boosters that remain on orbit are a serious debris problem, and one of the major known impact events was due to an Ariane booster. During the initial attempts to characterize the space debris problem, it became evident that a good proportion of all debris was due to the breaking up of rocket stages. Although NASA and the USAF quickly made efforts to improve the survivability of their boosters, other launchers did not implement similar changes....As of 21 February 2007, over 1,000 fragments had been identified." (WK 2012x; H. Klinkrad 2006; K. Than 2007; SPACE 2007)

(d) Debris from and as a Weapon

And the fourth source of space waste (or space debris) to be described here involves "debris from and as a weapon."

For instance, "one major source of debris in the past was the testing of anti-satellite weapons carried out by both the U.S. and Soviet Union in the 1960s and '70s....The U.S.'s only active weapon, Program 437, was shut down in 1975. The U.S. re-started their ASAT programs in the 1980s with the Vought ASM-135 ASAT. A 1985 test destroyed a 1 tonne (2,200 lb) satellite orbiting at 525 kilometres (326 mi) altitude, creating thousands of pieces of space debris larger than 1 centimetre (0.39 in). Because it took place at relatively low altitude, atmospheric drag caused the vast majority of the large debris to decay from orbit within a decade." (WK 2012x; J. Schefter 1982; C. Chun 1999)

Then in 2007, China conducted its "anti-satellite missile test, both for the military implications....On 20 February 2008, the U.S. launched an SM-3 Missile from the USS Lake Erie specially to destroy a defective U.S. spy satellite thought to be carrying 1,000 pounds (450 kg) of toxic hydrazine propellant." (WK 2012x; L. David 2007)

Effects of Space Waste

Space waste (or space debris) can have serious effects both in space and on Earth. Consider, for illustration, three main examples below (as summarized in *Sec. 2.4*).

(a) Threat to Unmanned Spacecraft

The first effect of space waste (or space debris) to be introduced here concerns "threat to unmanned spacecraft."

For instance, "spacecraft in a debris field are subject to constant wear as a result of impacts with small debris. Critical areas of a spacecraft are normally protected by Whipple shields, eliminating most damage. However, low-mass impacts have a direct impact on the lifetime of a space mission, if the spacecraft is powered by solar panels. These panels are difficult to protect because their front face has to be directly exposed to the sun. As a result, they are often punctured by debris....Impacts with larger debris normally destroy the spacecraft." (WK 2012x)

(b) Threat to Manned Spacecraft

The second effect of space waste (or space debris) to be introduced here concerns "threat to manned spacecraft."

For instance, "the International Space Station (ISS) uses extensive Whipple shielding to protect itself from minor debris threats. However, large portions of the ISS cannot be protected, notably its large solar panels. In 1989 it was predicted that the International Space Station's panels would suffer about 0.23% degradation over four years, which was dealt with by overdesigning the panel by 1%....Like the Shuttle, the only protection against larger debris is avoidance. On two occasions the crew have been forced to abandon work and take refuge in the Soyuz capsule while the threat passed." (WK 2012x; K. Thoma 2006; H. Nahra 1989; BBC 2009 & 2011)

(c) Hazard on Earth

And the third effect of space waste (or space debris) to be introduced here concerns "hazard on Earth."

This problem exists, because, "although most debris will burn up in the atmosphere, larger objects can reach the ground intact and present a risk." (WK 2012x)

For instance, "in 1969, five sailors on a Japanese ship were injured by space debris, probably of Russian origin. In 1997 an Oklahoma woman named Lottie Williams was hit in the shoulder by a 10 x 13 centimetres (5.1 in) piece of blackened, woven metallic material that was later confirmed to be part of the propellant tank of a Delta II rocket which had launched a U.S. Air Force satellite in 1996." (WK 2012x; OTA 1990)

Solutions to Space Waste

Because of these serious negative effects of space debris, there are proposed solutions over time. Consider, for illustration, three main ones below (as summarized in *Sec. 2.4*).

(a) Growth Mitigation

The first solution to space waste (or space debris) to be considered here is about "growth mitigation."

For instance, "the passivation of spent upper stages by the release of residual propellants is aimed at reducing the risk of on-orbit explosions that could generate thousands of additional debris objects. The modification of the Delta boosters, at a time when the debris problem was first becoming apparent, essentially eliminated their further contribution to the problem." (WK 2012x; N. Johnson 2011; J. Schefter 1982)

(b) Self-Removal

The second solution to space waste (or space debris) to be considered here is about "self-removal."

For instance, "it is an ITU requirement that geostationary satellites be able to remove themselves to a 'graveyard orbit' at the end of their lives.... Rocket stages or satellites that retain enough propellant can power themselves into a decaying orbit. In cases when a direct (and controlled) de-orbit would require too much propellant, a satellite can be brought to an orbit where atmospheric drag would cause it to de-orbit after some years." (WK 2012x)

(c) External Removal

And the third solution to space waste (or space debris) to be considered here is about "external removal."

For instance, "a well-studied solution is to use a remotely controlled vehicle to rendezvous with debris, capture it, and return to a central station. The commercially developed MDA Space Infrastructure Servicing vehicle is a refuelling depot and service spacecraft for communication satellites in geosynchronous orbit, slated for launch in 2015." (WK 2012x; E. Carlson 1990; CNW 2011)

There are also alternative proposals, ranging "from foamy ball of aerogel or spray of water, inflatable balloons, electrodynamic tethers, boom electroadhesion, or dedicated 'interceptor satellites.'" (WK 2012x; D. Michaels 2009; GAC 2010; STAR 2012; J. Foust 2011; J. Palmer 2011)

Problems with the Study of Space Waste

However, for the critics, this study of space waste (or space debris) and the Earth's orbit has some major problems. Consider, for illustration, a few examples below (as summarized in *Sec. 2.4*).

Firstly, one main criticism of the study of space waste (or space debris) and the Earth's orbit is that the measurement of space debris is not an accurate business in the first place, since the exact nature of the problem is still not very well known.

For example, "GEO orbit is too distant to make accurate measurements of the existing debris field for objects under 1 metre (3 ft 3 in), so the precise nature of the existing problem is not well known." (WK 2012x)

Secondly, another main criticism of the study of space waste (or space debris) and the Earth's orbit is that to minimize space debris by preventing space collisions is easier said than done technologically.

For instance, although "the ITU has placed increasingly strict requirements on the station-keeping ability of new satellites and demands that the owners guarantee their ability to safely move the satellites out of their orbital slots at the end of their lifetimes..., studies have suggested that even the existing ITU requirements are not enough to have a major effect on collision frequency." (WK 2012x; L. Anselmo 2000)

Thirdly, still another main criticism of the study of space waste (or space debris) and the Earth's orbit is that many of the proposed solutions are too expensive to be cost-effective.

For instance, for the critics, "as of 2006, the cost of launching any of these solutions [in terms of external removal, for example] is about the same as launching any spacecraft," and "none of the existing solutions are currently cost-effective." (WK 2012x; S. Lovgren 2006)

Fourthly, an additional main criticism of the study of space waste (or space debris) and the Earth's orbit is that there is no international treaty banning space debris.

For instance, for the critics, "there is no international treaty mandating behaviour to minimize space debris," although "the United Nations Committee on the Peaceful Uses of Outer Space (COPUOS) did publish voluntary guidelines in 2007." (WK 2012x; UN 2010)

And fifthly, another additional main criticism of the study of space waste (or space debris) and the Earth's orbit is that there are simply too many pieces of space debris to be able to cope with them all (or even most of them).

For instance, in accordance to "the mid-2009 update to the NASA debris FAQ,...the number of...debris items smaller than 1 centimetre (0.39 in) exceeds tens of millions." (WK 2012x; NASA 2009; J. Carroll 2002) Of course, the number will increase even more, if one wants to include those debris items larger than 1 centimetre (0.39 in).

Beyond Space Waste

These criticisms of the study on space waste (or space debris), of course, do not mean a total rejection of its validity but serve to show us the two opposing sides of the debate, such that the possibility (or impossibility) and desirability (or undesirability) of uselessness (from the perspective of nature with uselessness, space waste, and the Earth's orbit as a case study here) are not to the extent that the respective defenders would like us to believe.

Besides, the analysis of uselessness, space waste (or space debris), and the Earth's orbit can cast a bright light on the ontological principles in existential dialectics, and good examples include the formalness-informalness principle, the absoluteness-relativeness principle, the partiality-totality principle, the predictability-unpredictability principle, the explicability-inexplicability principle, the fiction-reality principle, the finiteness-transfiniteness principle, the preciseness-vagueness principle, the simpleness-complicatedness principle, the openness-hiddenness principle, the denseness-emptiness principle, the change-constancy principle, the order-chaos principle, the slowness-quickness principle, the expansion-contraction principle, the theory-praxis principle, the convention-novelty principle, the evolution-transformation principle, the symmetry-asymmetry principle, the softness-hardness principle, the seriousness-playfulness principle, the activeness-inactivess principle, the selfness-otherness principle, the regression-progression principle, the same-difference principle, the stability-reaction principle, the functionality-nonfunctionality principle, and the intentionality-nonintentionality principle.

For instance, in relation to the formalness-informalness principle, if there is formalness (e.g., the formal logical requirement of soundness and completeness in a system of ideas, as in the study on space debris), there is informalness (e.g., the non-formal existence of unsoundness and incompleteness in the study on space debris, as shown in the criticisms that the measurement of space debris is not an accurate business in the first place, that to minimize space debris by preventing space collisions is easier said than done technologically, that many of the proposed solutions are too expensive to be cost-effective, that there is no international treaty banning space debris, that there are simply too many pieces of space debris to be able to cope with them all or even most of them, etc.). And the reverse direction also holds true.

In relation to the absoluteness-relativeness principle, if there is absoluteness (e.g., the absolute view of the ITU, which "has placed increasingly strict requirements on the station-keeping ability of new satellites and demands that the owners guarantee their ability to safely move the satellites out of their orbital slots at the end of their lifetimes"), there is relativeness (e.g., what is right for the ITU in regard to its "strict requirements on the station-keeping ability of new satellites" is not necessarily so for the critics, since "studies have suggested that even the existing ITU requirements are not enough to have a major effect on collision frequency"). And the reverse direction also holds true.

In relation to the partiality-totality principle, if there is partiality (e.g., the partial view of the ITU in regard to its "strict requirements on the station-keeping ability of new satellites" to manage space debris), there is totality (e.g., the more holistic view of the future of space debris, such that the whole is not the sum of its parts, that is, it cannot be reduced to the sum of the opposing views of the ITU, the skeptics, and others, since there will emerge new views in the future not yet known today, just as there are already criticisms of them all nowadays). And the reverse direction also holds true.

In relation to the predictability-unpredictability principle, if there is predictability (e.g., the predictable tendency of "the United Nations Committee on the Peaceful Uses of Outer Space" or COPUOS to provide "voluntary guidelines" for space debris management in the foreseeable future), there is unpredictability (e.g., the more difficult task to predict exactly the extent to which COPUOS will be successful in managing space debris on a particular farther future occasion, since "there is no international treaty mandating behaviour to minimize space debris"). And the reverse direction also holds true.

In relation to the explicability-inexplicability principle, if there is explicability (e.g., the explanation, by the advocates of different ways of "external removal" to manage space debris, on the basis of their technological feasibility), there is inexplicability (e.g., the lack of sufficient explanation, by the advocates of different ways of "external removal" to manage space debris, of why these solutions are necessarily desirable, since for the critics, "as of 2006, the cost of launching any of these solutions [in terms of external removal, for example] is about the same as launching any spacecraft," and "none of the existing solutions are currently cost-effective"). And the reverse direction also holds true.

In relation to the fiction-reality principle, if there is fiction (e.g., the fictional aspect of the measurement of space debris, insofar as "GEO orbit is too distant to make accurate measurements of the existing debris field for objects under 1 metre [3 ft 3 in], so the precise nature of the existing problem is not well known"), there is reality (e.g., the realistic aspect of the measurement of space debris, insofar as "the mid-2009 update to the NASA debris FAQ, had estimated that "the number of...debris items smaller than 1 centimetre [0.39 in] exceeds tens of millions"). And the reverse direction also holds true.

In relation to the finiteness-transfiniteness principle, if there is finiteness (e.g., the finite number of spacecrafts which were lanuched by humans so far, as of 2012), there is transfiniteness (e.g., the transfinite number of all the individual space particles, both natural and man-made, which have ever traveled near or to the Earth in the history of the universe and could hit any spacecraft or any object in the way). (WK 2012x) And the reverse direction also holds true.

In relation to the preciseness-vagueness principle, if there is preciseness (e.g., the precise identification of 3 solutions to space debris in the article aforecited), there is vagueness (e.g., the vagueness in the identification of the 3 solutions to space debris in the article aforecited, since it is not clear why there must be only 3, not 4, 5, 6, and so on). And the reverse direction also holds true.

In relation to the simpleness-complicatedness principle, if there is simpleness (e.g., the relatively simple analysis of the solutions to space debris by the advocates), there is complicatedness (e.g., the relatively more complicated analysis of the solutions to space debris, by challenging the claims and assumptions, as shown in the problems pointed out by the critics as cited). And the reverse direction also holds true.

In relation to the openness-hiddenness principle, if there is openness (e.g., the open exploration of the so-called "Kessler Syndrome" by Donald Kessler and later other researchers, in that, "instead of the density of debris being a measure of the number of items launched into orbit, it was that number plus any new debris caused when they collided," so "if the new debris did not decay from orbit before impacting another object, the number of debris items would continue to grow even if there were no new launches"), there is hiddenness (e.g., the hidden nature of the "Kessler Syndrome," because, as explained by Kessler, "the start of a cascade would not be obvious until the situation was well advanced, which might take years"). (WK 2012x; D. Kessler 2009; J. Stupl 2010) And the reverse direction also holds true.

In relation to the denseness-emptiness principle, if there is denseness (e.g., the relatively denser concentration of "collisions…at very high relative velocities, typically several kilometres per second," in the LEO region), there is emptiness (e.g., the relatively less dense, or more empty, concentration of major collisions at higher altitudes, unlike those in LEO, and "this means that the debris field from such a collision is not the same as a LEO collision and does not pose the same sort of risks"). (WK 2012x; ESA 2009) And the reverse direction also holds true.

In relation to the change-constancy principle, if there is change (e.g., the ever new ways to use technology to handle space problems, as shown in the study on space debris), there is constancy (e.g., the ever constant existence of different problems with the ever new ways to use technology to handle space problems, as shown in the problems with the study on space debris pointed out by the critics). And the reverse direction also holds true.

In relation to the order-chaos principle, if there is order (e.g., the more or less orderly appearance of those orbits "with a low enough density" of space debris, so that "the addition of debris through impacts is slower than their rate of decay, and the problem does not become significant," so things do not look so chaotic, with so many fewer collisions), there is chaos (e.g,. the more or less chaotic appearance of space debris in those orbits characterized by a "Kessler Syndrome," in that, "instead of the density of debris being a measure of the number of items launched into orbit, it was that number plus any new debris caused when they collided," so "if the new debris did not decay from orbit before impacting another object, the number of debris items would continue to grow even if there were no new launches," and more collisions would continue). And the reverse direction also holds true.

In relation to the slowness-quickness principle, if there is slowness (e.g., "relative velocities in GEO are low, compared with those between objects in largely random low earth orbits"), there is quickness (e.g., relative velocities of "objects in largely random low earth orbits" like LEO are higher). And the reverse direction also holds true.

In relation to the expansion-contraction principle, if there is expansion (e.g., the relative expansion of "the testing of anti-satellite weapons carried out by both the U.S. and Soviet Union in the 1960s and 1970s"), there is contraction (e.g., the relative decline of ASAT testing in late 1970s, when "the U.S.'s only active weapon, Program 437, was shut down in 1975"). (WK 2012x; C. Chun 1999) And the reverse direction also holds true.

In relation to the theory-praxis principle, if there is theory (e.g., the theoretical understanding of "micro-meteorites" by Fred Whipple and Fletcher Watson for the study of "the interplanetary dust cloud" in cosmology after WWII), there is praxis (e.g., the practical application of the theoretical understanding of "micro-meteorites" in cosmology to the field of space race for the measurement of space debris). And the reverse direction also holds true.

In relation to the convention-novelty principle, if there is convention (e.g., the conventional wisdom about space debris in the older days when it was thought that "the density of debris" is "a measure of the number of items launched into orbit"), there is novelty (e.g., the alternative novel challenge to the conventional wisdom about space debris in the older days, by the new idea of the so-called "Kessler Syndrome" by Donald Kessler and later other researchers, when they argued that, "instead of the density of debris being a measure of the number of items launched into orbit, it was that number plus any new debris caused when they collided," so "if the new debris did not decay from orbit before impacting another object, the number of debris items would continue to grow even if there were no new launches"). And the reverse direction also holds true.

In relation to the evolution-transformation principle, if there is evolution (e.g., the natural evolution of humans to live on Earth in the state of nature in a primitive way), there is transformation (e.g., the technical transformation of human ability to live by the invention of spacecrafts for space exploration). And the reverse direction also holds true.

In relation to the symmetry-asymmetry principle, if there is symmetry (e.g., the existence of different countries on Earth), there is asymmetry (e.g., only some countries, like the USSR in the older days or Russia nowadays, the U.S., and now China, can send manned spacecrafts for space exploration). And the reverse direction also holds true.

In relation to the softness-hardness principle, if there is softness (e.g., the cooperation by Russia and America for the International Space Station nowadays), there is hardness (e.g., the rivalry between Russia and America in international politics nowadays). And the reverse direction also holds true.

In relation to the seriousness-playfulness principle, if there is seriousness (e.g., the serious business of those who try to propose solutions to space debris), there is playfulness (e.g., the playful part of those who try to propose solutions to space debris, when they play around with different proposals over time, like "foamy ball of aerogel or spray of water, inflatable balloons, electrodynamic tethers, boom electroadhesion, or dedicated 'interceptor satellites'"). And the reverse direction also holds true.

In relation to the activeness-inactivess principle, if there is activeness (e.g., the activeness of "the U.S. and Soviet Union in the 1960s and 1970s" when they regularly engaged in "the testing of anti-satellite weapons"), there is inactiveness (e.g., the relative inactiveness of the U.S. and Russia later on, when, "following the U.S. test in 1985, there was a de-facto moratorium on such tests"). (WK 2012x; D. Wright 2997) And the reverse direction also holds true.

In relation to the selfness-otherness principle, if there is selfness (e.g., the tendency to be mostly centered on one's self or group, as shown in the numerous tests of "anti-satellite weapons" by the U.S. and Soviet Union in the 1960s and 1970s" for global domination), there is otherness (e.g., the tendency to be mostly centered on others, as shown by those working for the Union of Concerned Scientists about the dangers of space debris, for a better human future). And the reverse direction also holds true.

In relation to the regression-progression principle, if there is regression (e.g., the regression made by the proposed solutions to space debris, as shown in the problems pointed out by the critics), there is progression (e.g., the progress made by the proposed solutions to space debris, as shown in their contribution to the reduction of space debris aforecited). And the reverse direction also holds true.

In relation to the same-difference principle, if there is similarity in outcome (e.g., the contribution to the molding and control of beliefs, values, and behaviors, regardless of whether this be done by way of hunting and gathering in the older days, or by way of space exploration nowadays), there is difference in outcome (e.g., the contribution to the molding and control of beliefs, values, and behaviors by way of hunting and gathering in the older days for a relatively more harmonious co-existence with nature—but the contribution to the molding and control of beliefs, values, and behaviors by way of space exploration nowadays for a relatively more dominative conquest of nature). And the reverse direction also holds true.

In relation to the stability-reaction principle, if there is stability (e.g., the relatively more stable condition of space exploration in this space age of ours, when major powers like the USSR/Russia, the U.S.A., and now China send spacecrafts into space), there is reaction (e.g., the launches of spacecrafts then led to further changes later on, in that, in accordance to the Kessler Syndrome, "instead of the density of debris being a measure of the number of items launched into orbit, it was that number plus any new debris caused when they collided," so "if the new debris did not decay from orbit before impacting another object, the number of debris items would continue to grow even if there were no new launches" and collisions would continue in the cascades). And the reverse direction also holds true.

In relation to the functionality-nonfunctionality principle, if there is functionality (e.g., the presence of action for which something is fitted, as shown in the rivalry between "the U.S. and Soviet Union in the 1960s and 1970s" when they regularly engaged in "the testing of anti-satellite weapons" to serve the political function of global domination), there is nonfunctionality (e.g., the relative lack of action for which something is fitted, as shown in the decision of the U.S. and Russia later on, when, "following the U.S. test in 1985, there was a de-facto moratorium on such tests," so "widespread ASAT testing had ended" and it no longer serves the function of global domination in the post-Cold War era, since the attention is now paid to other things instead). And the reverse direction also holds true.

And in relation to the intentionality-nonintentionality principle, if there is intentionality (e.g., the planning or design of something for a certain outcome, as shown in the efforts to get rid of space debris), there is nonintentionality (e.g., the relative absence of planning or design of something for a certain outcome, such that the outcome can even be counter-intentional in being the opposite of what was originally planned, as shown in the unintended consequences of the efforts to get rid of space debris, in that the measurement of space debris is not an accurate business in the first place since the exact nature of the problem is still not very well known, that to minimize space debris by preventing space collisions is easier said than done technologically, and that many of the proposed solutions are too expensive to be cost-effective). And the reverse direction also holds true.

Uselessness and Society

Uselessness, when examined from the perspective of society, can reveal its possibility (or impossibility) and desirability (or undesirability), and this can be shown by way of two case studies, namely, (2.4.1) uselessness, hazardous waste, and the Basel Convention, and (2.4.2) uselessness, waste management, and minorities—to be addressed hereafter.

Uselessess, Hazardous Waste, and the Basel Convention

A good way to evaluate the possibility (or impossibility) and desirability (or undesirability) of uselessness from the perspective of society is to explore uselessness, hazardous waste, and the Basel Convention.

The term "Basel Convention" is a shortcut for "the Basel Convention on the Control of Transboundary Movements of Hazardous Wastes and Their Disposal" and refers to "an international treaty that was designed to reduce the movements of hazardous waste between nations, and specifically to prevent transfer of hazardous waste from developed to less developed countries (LDCs)." (WK 2012y)

In addition, "the Convention is...intended to minimize the amount and toxicity of wastes generated, to ensure their environmentally sound management as closely as possible to the source of generation, and to assist LDCs in environmentally sound management of the hazardous and other wastes they generate." (WK 2012y)

Originally, "the Convention was opened for signature on 22 March 1989, and entered into force on 5 May 1992." (WK 2012y)

Causes of the Basel Convention

There are different historical reasons for the creation of the Basel Convention.

One often cited reason has to do with high disposal costs for hazardous waste in developed countries, which experienced "the tightening of environmental laws (for example, RCRA)...in the

1970s....At the same time, globalization of shipping made transboundary movement of waste more accessible, and many LDCs were desperate for foreign currency. Consequently, the trade in hazardous waste, particularly to LDCs, grew rapidly." (WK 2012y)

For instance, "one of the incidents which led to the creation of the Basel Convention was the *Khian* Sea waste disposal incident, in which a ship carrying incinerator ash from the city of Philadelphia in the United States after having dumped half of its load on a beach in Haiti, was forced away where it sailed for many months, changing its name several times. Unable to unload the cargo in any port, the crew was believed to have dumped much of it at sea." (WK 2012y)

Another example is "the 1988 Koko case in which 5 ships transported 8,000 barrels of hazardous waste from Italy to the small town of Koko in Nigeria in exchange for $100 monthly rent which was paid to a Nigerian for the use of his farmland." (WK 2012y)

Incidents like this "have been deemed 'Toxic Colonialism' by many developing countries." (WK 2012y)

To things in a proper perspective, "according to Maureen Walsh [1992], only around 4% of hazardous wastes that come from OECD countries are actually shipped across international borders. These wastes include, among others, chemical waste, radioactive waste, municipal solid waste, asbestos, incinerator ash, and old tires. Of internationally shipped waste that comes from developed countries, more than half is shipped for recovery and the remainder for final disposal." (WK 2012y)

In any event, the Basel Convention can be analyzed in terms of some major issues; for illustration, consider three major ones below.

The Issue of Definition

The first issue addressed by the Basel Convention has to do with the "definition" of hazardous waste.

In accordance to the Basel Convention, "a waste will fall under the scope of the Convention if it is within the category of wastes listed in Annex I of the Convention and it does exhibit one of the hazardous characteristics contained in Annex III. In other words it must both be listed and contain a characteristic such as being explosive, flammable, toxic, or corrosive. The other way that a waste may fall under the scope of the Convention is if it is defined as or considered to be a hazardous waste under the laws of either the exporting country, the importing country, or any of the countries of transit." (WK 2012y)

And "the definition of the term *disposal* is made in Article 2 al 4 and just refers to annex IV, which gives a list of operations which are understood as disposal or recovery. The examples of disposal are broad and include also recovery, recycling. Annex II lists other wastes such as household wastes and residue that comes from incinerating household waste." (WK 2012y)

The Issue of Obligations

The second issue addressed by the Basel Convention has to do with "obligations."

For instance, "in addition to conditions on the import and export of the above wastes, there are stringent requirements for notice, consent and tracking for movement of wastes across national boundaries. It is of note that the Convention places a general prohibition on the exportation or importation of wastes between Parties and non-Parties....Parties to the Convention must honor import bans of other Parties." (WK 2012y)

Also, "Article 4 of the Basel Convention calls for an overall reduction of waste generation. By encouraging countries to keep wastes within their boundaries and as close as possible to its source of generation, the internal pressures should provide incentives for waste reduction and pollution prevention." (WK 2012y)

And "according to Article 12, Parties are directed to adopt a protocol that establishes liability rules and procedures that are appropriate for damage that comes from the movement of hazardous waste across borders." (WK 2012y)

The Issue of Less Developed Countries

And the third issue addressed by the Basel Convention has to do with "less developed countries."

For instance, "after the initial adoption of the Convention,...[m]any nations and NGOs argued for a total ban on shipment of all hazardous waste to LDCs. In particular, the original Convention did not prohibit waste exports to any location except Antarctica but merely required a notification and consent system known as 'prior informed consent' or PIC. Further, many waste traders sought to exploit the good name of recycling and begin to justify all exports as moving to recycling destinations. Many believed a full ban was needed including exports for recycling. These concerns led to several regional waste trade bans, including the Bamako Convention." (WK 2012y)

So, by the 1990s, "lobbying at the 1995 Basel conference by LDCs, Greenpeace and key European countries such as Denmark, led to a decision to adopt the Basel Ban Amendment to the Basel Convention. Not yet in force, but considered morally binding by signatories, the Amendment prohibits the export of hazardous waste from a list of developed (mostly OECD) countries to developing countries. The Basel Ban applies to export for any reason, including recycling. An area of special concern for advocates of the Amendment was the sale of ships for salvage, shipbreaking." (WK 2012y)

Problems with the Basel Convention

However, for the critics, the Basel Convention has some major problems. Consider, for illustration, a few examples below.

Firstly, one main criticism of the Basel Convention is that what constitutes hazardous waste is subject to contention.

For example, for the critics, the definition of hazardous waste in the Basel Convention is too narrow, because it does not include some important types of hazardous waste like "radioactive waste," "wastes from the normal operation of ships," and certain "'commodities' which are not considered wastes," etc. (WK 2012y)

Secondly, another main criticism of the Basel Convention is that there are questionable rules about "obligations" in the Basel Convention.

For instance, for the critics, although the Basel Convention prohibits "the exportation or importation of wastes between Parties and non-Parties," it makes the notorious exception to this rule whenever "the waste is subject to another treaty that does not take away from the Basel Convention." (WK 2012y) This notorious exception benefits some major powers like the United States (which is "a notable non-Party to the Convention and has a number of such agreements for allowing the shipping of hazardous wastes to Basel Party countries") and the OECD (which "allows, among other things, the OECD countries to continue trading in wastes with countries like the United States that have not ratified the Basel Convention"). (WK 2012y)

Thirdly, still another main criticism of the Basel Convention is that there is no enforcement provisions in the Basel Convention.

For instance, for the critics, "the Convention states that illegal hazardous waste traffic is criminal but contains no enforcement provisions." (WK 2012y)

And fourthly, an additional main criticism of the Basel Convention is that it does not go far enough to protect less developed countries from the spread of hazardous waste.

For instance, although the call for the "Basel Ban Amendment to the Basel Convention" represents a good effort to take into account the interests of less developed countries, the amendment has not been ratified, because "the Ban Amendment was strenuously opposed by a number of industry groups as well as nations including Australia and Canada. The number of ratification for the entry-into force of the Ban Amendment is under debate: Amendments to the convention enter into force after ratification of 'three-fourths of the Parties who accepted them' [Art. 17.5]; so far, the Parties of the Basel Convention could not yet agree whether this would be three fourth of the Parties that were Party to the Basel Convention when the Ban was adopted, or three fourth of the current Parties of the Convention." (WK 2012y)

Beyond the Basel Convention

These criticisms of the Basel Convention, however, should not be misconstrued as a total rejection of its validity but serve to show us the two opposing sides of the debate, such that the possibility (or impossibility) and desirability (or undesirability) of uselessness (from the perspective of society with uselessness, hazardous waste, and the Basel Convention as a case study here) are not to the extent that the respective defenders would like us to believe.

In fact, the analysis of uselessness, hazardous waste, and the Basel Convention can teach us something important about the ontological principles in existential dialectics, and good examples include the formalness-informalness principle, the absoluteness-relativeness principle, the partiality-totality principle, the predictability-unpredictability principle, the explicability-inexplicability principle, the fiction-reality principle, the finiteness-transfiniteness principle, the preciseness-vagueness principle, the simpleness-complicatedness principle, the openness-hiddenness principle, the denseness-emptiness principle, the change-constancy principle, the order-chaos principle, the slowness-quickness principle, the expansion-contraction principle, the theory-praxis principle, the convention-novelty principle, the evolution-transformation principle, the symmetry-asymmetry principle, the softness-hardness principle, the seriousness-playfulness principle, the activeness-inactivess principle, the selfness-otherness principle, the regression-progression principle, the same-difference principle, the stability-reaction principle, the functionality-nonfunctionality principle, and the intentionality-nonintentionality principle.

For instance, in relation to the formalness-informalness principle, if there is formalness (e.g., the formal logical requirement of soundness and completeness in a system of ideas, as in the creation of the Basel Convention), there is informalness (e.g., the non-formal existence of unsoundness and incompleteness in the creation of the Basel Convention, as shown in the criticisms that what constitutes hazardous waste is subject to contention, that there are questionable rules about "obligations" in the Basel Convention, that there is no enforcement provisions in the Basel Convention, that it does not go far enough to protect less developed countries from the spread of hazardous waste, etc.). And the reverse direction also holds true.

In relation to the absoluteness-relativeness principle, if there is absoluteness (e.g., the absolute view of "Greenpeace and key European countries such as Denmark" for "the Basel Ban Amendment to the Basel Convention"), there is relativeness (e.g., what is right for "Greenpeace and key European countries such as Denmark" for "the Basel Ban Amendment to the Basel Convention" is not necessarily so for "a number of industry groups as well as nations including Australia and Canada" against the amendment). And the reverse direction also holds true.

In relation to the partiality-totality principle, if there is partiality (e.g., the partial view of "Greenpeace and key European countries such as Denmark" for "the Basel Ban Amendment to the Basel Convention"), there is totality (e.g., the more holistic view of the future of the Basel Convention, such that the whole is not the sum of its parts, that is, it cannot be reduced to the sum of the opposing views of the pros and cons of "the Basel Ban Amendment to the Basel Convention," since there will emerge new views in the future not yet known today, just as there are already criticisms against both groups nowadays). And the reverse direction also holds true.

In relation to the predictability-unpredictability principle, if there is predictability (e.g., the predictable tendency of the parties in the Basel Convention to advocate "a general prohibition on the exportation or importation of wastes between Parties and non-Parties"), there is unpredictability (e.g., the more difficult task to predict exactly the extent to which the Basel Convention will be successful in banning "the exportation or importation of wastes between Parties and non-Parties" on a particular future occasion, because it "contains no enforcement provisions"). And the reverse direction also holds true.

In relation to the explicability-inexplicability principle, if there is explicability (e.g., the explanation by those in "the 1988 Koko case," in that "5 ships transported 8,000 barrels of hazardous waste from Italy to the small town of Koko in Nigeria in exchange for $100 monthly rent which was paid to a Nigerian for the use of his farmland"), there is inexplicability (e.g., the lack of sufficient explanationby by those in "the 1988 Koko case of why the transport of 8,000 barrels of hazardous waste from Italy to the small town of Koko in Nigeria is necessarily desirable, since, for the critics, incidents like this "have been deemed 'Toxic Colonialism' by many developing countries"). And the reverse direction also holds true.

In relation to the fiction-reality principle, if there is fiction (e.g., the fictional aspect of the definition of "hazardous waste" in the Basel Convention, insofar as it is subject to contention, because it does not include some important types of hazardous waste like "radioactive waste," "wastes from the normal operation of ships," and certain "'commodities' which are not considered wastes," etc.), there is reality (e.g., the realistic aspect of the definition of "hazardous waste" in the Basel Convention, insofar as it includes many, though not all, types of "hazardous waste"). And the reverse direction also holds true.

In relation to the finiteness-transfiniteness principle, if there is finiteness (e.g., the finite number of the properties of hazardous waste, like "ignitability," "reactivity," "corrosivity," and "toxicity"), there is transfiniteness (e.g., the transfinite number of all the individual pieces of hazardous waste in history that have ever existed). (WK 2012z) And the reverse direction also holds true.

In relation to the preciseness-vagueness principle, if there is preciseness (e.g., the precise identification of 4 properties of hazardous waste in the Basel Convention, namely, "explosive, flammable, toxic, or corrosive"), there is vagueness (e.g., the vagueness in the identification of the 4 properties of hazardous waste in the Basel Convention, namely, "explosive, flammable, toxic, or corrosive," since it is not clear why there must be only 4, not 5, 6, 7, and so on, because, for the critics, there are other properties like "radioactive," etc.). And the reverse direction also holds true.

In relation to the simpleness-complicatedness principle, if there is simpleness (e.g., the relatively simple analysis of the Basel Convention by the advocates), there is complicatedness (e.g., the relatively more complicated analysis of the Basel Convention, by challenging the claims and assumptions, as shown in the problems pointed out by the critics as cited). And the reverse direction also holds true.

In relation to the openness-hiddenness principle, if there is openness (e.g., the open exploration of "a general prohibition on the exportation or importation of wastes between Parties and non-Parties" in the Basel Convention), there is hiddenness (e.g., the hidden bias in the Basel Convention, because, for the critics, it does not go far enough to protect less developed countries from the spread of hazardous wastes and often favors major powers with those notorious exceptional provisions). And the reverse direction also holds true.

In relation to the denseness-emptiness principle, if there is denseness (e.g., the relatively denser concentration of hazardous wastes from developed countries to less developed countries prior to the creation of the Basel Convention), there is emptiness (e.g., the relatively less dense, or more empty, concentration of hazardous wastes from developed countries to less developed countries, under the Basel Ban Amendment to the Basel Convention, at least on paper). And the reverse direction also holds true.

In relation to the change-constancy principle, if there is change (e.g., the ever new ways to manage waste, as shown in the creation of the Basel Convention), there is constancy (e.g., the ever constant existence of different problems with the ever new ways to manage waste, as shown in the problems with the Basel Convention pointed out by the critics). And the reverse direction also holds true.

In relation to the order-chaos principle, if there is order (e.g., the more or less orderly appearance of hazardous waste management nowadays, with the creation of the Basel Convention, which "places a general prohibition on the exportation or importation of wastes between Parties and non-Parties"), there is chaos (e.g,. the more or less chaotic appearance of hazardous waste management in the older days before the creation of the Basel Convention, as shown in "the Khian Sea waste disposal incident, in which a ship carrying incinerator ash from the city of Philadelphia in the United States after having dumped half of its load on a beach in Haiti, was forced away where it sailed for many months, changing its name several times," and "unable to unload the cargo in any port, the crew was believed to have dumped much of it at sea"). And the reverse direction also holds true.

In relation to the slowness-quickness principle, if there is slowness (e.g., the relatively slower speed of the "Basel Ban Amendment to the Basel Convention" to be ratified, if the requirement is that "amendments to the convention enter into force after ratification of 'three-fourths of the Parties who accepted them'"), there is quickness (e.g., the relatively quicker speed of the "Basel Ban Amendment to the Basel Convention" to be ratified, if the requirement is that "amendments to the convention enter into force after ratification" of only a small number of the Parties who accepted them). And the reverse direction also holds true.

In relation to the expansion-contraction principle, if there is expansion (e.g., the relative expansion of the types of hazardous waste to be included for "the prohibition on the exportation or importation of wastes between Parties and non-Parties" under the Basel Convention and the "Basel Ban Amendment nowadays), there is contraction (e.g., the relative decline or decrease of the incidents like the *Khian* Sea waste disposal incident and the 1988 Koko case nowadays, because of the Basel Convention). And the reverse direction also holds true.

In relation to the theory-praxis principle, if there is theory (e.g., the theoretical understanding of the properties of hazardous waste, like "ignitability," "reactivity," "corrosivity," and "toxicity," in chemistry), there is praxis (e.g., the practical application of the theoretical understanding of the properties of hazardous waste, like "ignitability," "reactivity," "corrosivity," and "toxicity," in chemistry to the field of hazardous waste management, like the creation of the Basel Convention). And the reverse direction also holds true.

In relation to the convention-novelty principle, if there is convention (e.g., the conventional wisdom of the Basel Convention in the older days in regard to its exclusion of radioactive waste in the original definition of hazarouds waste), there is novelty (e.g., the alternative novel challenge to the conventional wisdom about the exclusion of radioactive waste in the original definition of hazarouds waste in the Basel Convention in the older days, by the new idea of the "Basel Ban Amendment to the Basel Convention" nowadays, which seeks a broader definition). And the reverse direction also holds true.

In relation to the evolution-transformation principle, if there is evolution (e.g., the natural evolution of humans to throw away hazardous wastes in the state of nature in a primitive way), there is transformation (e.g., the technical transformation of human ability to manage hazardous wastes by the invention of legal institutions like the Basel Convention). And the reverse direction also holds true.

In relation to the symmetry-asymmetry principle, if there is symmetry (e.g., the existence of different parties in the Basel Convention), there is asymmetry (e.g., major powers can seek exceptional treatments within the Basel Convention, which then allows the United States, for instance, to ship "hazardous wastes to Basel Party countries," and "the OECD countries to continue trading in wastes with countries like the United States that have not ratified the Basel Convention"). And the reverse direction also holds true.

In relation to the softness-hardness principle, if there is softness (e.g., the support of the Basel Convention by the advocates), there is hardness (e.g., the critique of the Basel Convention, as shown in the problems pointed out by the critics as cited). And the reverse direction also holds true.

In relation to the seriousness-playfulness principle, if there is seriousness (e.g., the serious business of those involved in the Basel Convention to deal with hazardous waste management), there is playfulness (e.g., the playful part of those involved in the Basel Convention to deal with hazardous waste management, when they play around with different definitions of hazardous waste over time). And the reverse direction also holds true.

In relation to the activeness-inactivess principle, if there is activeness (e.g., the activeness of those in "the 1988 Koko case, in which 5 ships transported 8,000 barrels of hazardous waste from Italy to the small town of Koko in Nigeria in exchange for $100 monthly rent which was paid to a Nigerian for the use of his farmland"), there is inactiveness (e.g., the relative inactiveness of those who are no longer allowed to engage in hazardous waste dumping under the Basel Convention, so they have to do something else instead).

In relation to the selfness-otherness principle, if there is selfness (e.g., the tendency to be mostly centered on one's self or group, as shown in "the 1988 Koko case in which 5 ships transported 8,000 barrels of hazardous waste from Italy to the small town of Koko in Nigeria in exchange for $100 monthly rent which was paid to a Nigerian for the use of his farmland," which is later condemned as "toxic colonialism" by the critics), there is otherness (e.g., the tendency to be mostly centered on others, as shown in some environmental NGOs campaigning for the end of toxic colonialism, for the common good of humanity). And the reverse direction also holds true.

In relation to the regression-progression principle, if there is regression (e.g., the regression made by the Basel Convention, as shown in the problems pointed out by the critics), there is progression (e.g., the progress made by the Basel Convention, as shown in the contribution to a better hazardous waste management aforecited). And the reverse direction also holds true.

In relation to the same-difference principle, if there is similarity in outcome (e.g., the contribution to the molding and control of beliefs, values, and behaviors, regardless of whether this be done by way of dumping in "the 1988 Koko case, or by way of the reduction of "the movements of hazardous waste between nations" in the Basel Convention), there is difference in outcome (e.g., the contribution to the molding and control of beliefs, values, and behaviors by way of dumping for a relatively more market-oriented lifeworld—but the contribution to the molding and control of beliefs, values, and behaviors by way of the reduction of "the movements of hazardous waste between nations" in the Basel Convention for a relatively more legal-regulative lifeworld). And the reverse direction also holds true.

In relation to the stability-reaction principle, if there is stability (e.g., the relatively more stable condition of the Basel Convention in its early days), there is reaction (e.g., the Basel Convention in its early days then led to further changes later on, as shown by the later campaign for the "Basel Ban Amendment to the Basel Convention"). And the reverse direction also holds true.

In relation to the functionality-nonfunctionality principle, if there is functionality (e.g., the presence of action for which something is fitted, as shown in "the 1988 Koko case, when "5 ships transported 8,000 barrels of hazardous waste from Italy to the small town of Koko in Nigeria in exchange for $100 monthly rent which was paid to a Nigerian for the use of his farmland," which served the function of dumping for those involved at the time), there is nonfunctionality (e.g., the relative lack of action for which something is fitted, as shown in the creation of the Basel Convention, so that the practice of dumping in "the 1988 Koko case" is no longer functional in our time, as it is more appropriate nowadays to reduce "the movements of hazardous waste between nations"). And the reverse direction also holds true.

And in relation to the intentionality-nonintentionality principle, if there is intentionality (e.g., the planning or design of something for a certain outcome, as shown in the design of the Basel Convention to handle hazardous waste), there is nonintentionality (e.g., the relative absence of planning or design of something for a certain outcome, such that the outcome can even be counter-intentional in being the opposite of what was originally planned, as shown in the unintended consequences of the Basel Convention, in that what constitutes hazardous waste is subject to contention, that there are questionable rules about "obligations" in the Basel Convention, and that it does not go far enough to protect less developed countries from the spread of hazardous waste). And the reverse direction also holds true.

Uselessness, Waste Management, and Minorities

Another good way to evaluate the possibility (or impossibility) and desirability (or undesirability) of uselessness from the perspective of society is to explore uselessness, waste management, and minorities.

A major contention here is that "waste management is a significant environmental justice issue. Many of the environmental burdens...are more often borne by marginalized groups, such as racial minorities, women, and residents of developing nations. NIMBY (not-in-my-back-yard) is a popular term used to describe the opposition of residents to a proposal for a new development close to them." (WK 2012; M. Wolsink 1994)

The term "environmental justice" (above) refers to "the fair treatment and meaningful involvement of all people regardless of race, color, sex, national origin, or income with respect to the development, implementation and enforcement of environmental laws, regulations, and policies." (WK 2012o; T. Miller 2003; D. Schlosberg 2007)

History of Environmental Discrimination

Historically, especially "in the early 1980s, environmental justice emerged as a concept in the United States, fueled by a mounting disdain within African-American, Hispanic and indigenous communities that were subject to hazardous and polluting industries located predominantly in their neighborhoods. This prompted the launch of the environmental justice movement, which adopted a civil rights and social justice approach to environmental justice and grew organically from dozens, even hundreds, of local struggles, events and a variety of other social movements." (WK 2012o) CAP). 2005

Then, in 1990, "the United States Environmental Protection Agency established its Workgroup on Environmental Equity. By September 1991, the First National People of Color Environmental Leadership Summit took place, organized and attended by more than 650 grassroots and national leaders representing more than 300 environmental groups. The Second National People of Color Environmental Leadership Summit (also called Summit II) was also held in Washington DC, from October 23–26, 2002. Materials produced at the summit included a timeline for Environmental Justice milestones." (WK 2012o; S. Cutter 1995; EJRC 2012)

By 1992, "the EPA established its Office of Environmental Equity and the Work group on Environmental Equity had finished its report. Critics of the report contend that EPA did not go far enough in examining its current activities, including its own role in reinforcing environmental inequalities. Legislatively, a number of bills were introduced into Congress, including the Environmental Justice Act 1992. President Clinton signed Executive Order 12898 (federal actions to address environmental justice in minority populations and low-income populations) into law on February 11, 1994." (WK 2012o; S. Cutter 1995)

Barriers to Minority Participation

The launch of the environmental justice movement, however, confronts some major challenges over the years. Consider, for illustration, two major ones below (as summarized in *Sec. 2.5*).

(a) Initial Barriers

The first barrier to minority participation to be discussed here has to do with "initial barriers."

For instance, "when environmentalism first became popular during the first half of the 20th century, the focus was wilderness protection and wildlife preservation. These goals reflected the interests of the movement's initial supporters. The actions of many mainstream environmental organizations still reflect these early principles." (WK 2012o)

Thus, "many low-income minorities felt isolated or even negatively impacted by the movement. Supporters of the early environmental movement excluded minorities and working class people from their organizations. They regarded inner-city areas as places of degradation and ignored the environmental problems prevalent there. In addition, the NIMBY movement has transferred locally unwanted land uses (LULUs) from middle-class neighborhoods to poor communities with large minority populations. Therefore, vulnerable communities with fewer political opportunities are more often exposed to hazardous waste and toxins. This has resulted in the PIBBY principle, or at least the PIMBY (Place-in-minorities'-backyard), as supported by the United Church of Christ's study in 1987." (WK 2012o; M. Gerrard 1993; G. Roberts 1998)

(b) Cost Barriers

And the second barrier to minority participation to be discussed here has to do with "cost barriers."

For instance, "one of the major initial barriers to minority participation in Environmental Justice are the initial costs of trying to change the system and prevent companies from dumping their toxic waste and other pollutants in areas with high numbers of minorities living in them. There are massive legal fees involved in fighting for environmental justice and trying to shed environmental racism." (WK 2012o)

A good case in point is the United Kingdom, where "there is a rule that the claimant may have to cover the fees of their opponents, which further exacerbates any cost issues, especially with lower income minority groups; also, the only way for environmental justice groups to hold companies accountable for their pollution and breaking any licensing issues over waste disposal would be to sue the government for not enforcing rules. This would lead to the forbidding legal fees that most could not afford. This can be seen by that fact that out of 210 judicial review cases between 2005 and 2009, 56% did not proceed due to costs." (WK 2012o; E. Jeffries 2011)

Contributions of the Civil Rights Movement

In spite of these barriers to minority participation, the environmental justice movement has made some progress, especially with the help of the Civil Rights Movement, in three major ways (as described below and summarized in *Sec. 2.5*).

(a) Similar Goals and Tactics

Firstly, "the Environmental Justice movement and the Civil Rights Movement have many commonalities. At their core, the goals of movements are the same: 'social justice, equal protection, and an end to institutional discrimination.' By stressing the similarities of the two movements, it emphasizes that environmental equity is a right for all citizens. Because the two movements have parallel goals, it is useful to employ similar tactics that often emerge on the grassroots level. Common confrontational strategies include protests, neighborhood demonstrations, picketing, political pressure, and demonstration." (WK 2012o; G. Roberts 1998; R. Bullard 1992)

(b) Existing Organizations and Leaders

Secondly, "just as the civil rights movement of the 1960s began in the South, the modern civil rights movement and the fight for environmental equity has been largely based in the South, where environmental discrimination is most prominent. In these southern communities, black churches and other voluntary associations are used to organize resistance efforts, including research and demonstrations, such as the protest in Warren County, North Carolina. As a result of the existing community structure, many church leaders and civil rights activists, such as Reverend Benjamin Chavis Muhammad, have spearheaded the Environmental Justice movement." (WK 2012o; R. Bullard 1992)

(c) Litigation

And thirdly, "some of the most successful Environmental Justice lawsuits are based on violations of civil rights laws. The first case to use civil rights as a means to legally challenge the siting of a waste facility was in 1979. With the legal representation of Linda McKeever Bullard, the wife of Robert D. Bullard, residents of Houston's Northwood Manor opposed the decision of the city and Browning Ferris Industries to construct a solid waste facility near their mostly African-American neighborhood." (WK 2012o; J. Worsham 2009)

Also, "the Equal Protection Clause of the Fourteenth Amendment, which was used many times to defend minority rights during the 1960s, has also been used in numerous Environmental Justice cases." (WK 2012o; R. Bullard 1992)

Problems with the Environmental Justice Movement

However, for the critics, the environmental justice movement has some major problems. Consider, for illustration, a few examples below (as summarized in *Sec. 2.5*).

Firstly, one main criticism of the environmental justice movement is that what constitutes fairness or its opposite is in the eyes of the beholder.

For instance, for the critics, although "President Bill Clinton issued Executive Order 12898 on February 11, 1994 directing federal agencies to consider environmental factors that negatively affect people of color"—"these actions toward Environmental Justice did not achieve what was hoped for. By the end of Clinton's second term, resistance from congress, industry, and even within the ranks of the EPA, compounded with a systemic disconnect between the environmental protection model and the aims of Environmental Justice, rendered Clinton's initial push for in EJ largely ineffectual." (WK 2012o; K. Harn 2010) As the old saying goes, "[w]hat looks fair to the goose seems foul to the gander." (P. Samuelson 2001: 332)

Secondly, another main criticism of the environmental justice movement is that the environmental movement has been elitist in producing the opposite effects.

For instance, for the critics, the environmental movement has been plagued with "three different forms" of elitism, namely, (a) "compositional," in that "environmentalists are from the middle and upper class," (b) "ideological," in that "the reforms benefit the movement's supporters but impose costs on nonparticipants," and (c) "impactful," in that "the reforms have 'regressive social impacts.' They disproportionately benefit environmentalists and harm underrepresented populations." (WK 2012o; D. Morrison 1986)

Thirdly, still another main criticism of the environmental justice movement is that the legal process is too costly.

For instance, for the critics, "there are massive legal fees involved in fighting for environmental justice and trying to shed environmental racism." (WK 2012o)

And fourthly, an additional main criticism of the environmental justice movement is that there is no universal agreement on the definition of "environmental justice" and "environmental injustice."

For instance, for the critics, the United States Environmental Protection Agency, the United States Department of Transportation, and participants of the Central and Eastern European Workshop on Environmental Justice all have different definitions of "environmental justice." (WK 2012o) Even worse, there are "myriad definitions of Environment Justice found within the EJ movement" itself. (WK 2012o; D. Schlosberg 2007)

Beyond the Environmental Justice Movement

These criticisms of the environmental justice movement, however, do not entail a total rejection of its validity but serve to show us the two opposing sides of the debate, such that the possibility (or impossibility) and desirability (or undesirability) of uselessness (from the perspective of society with uselessness, waste management, and minorities as a case study here) are not to the extent that the respective defenders would like us to believe.

Indeed, the analysis of uselessness, waste management, and minorities can reveal some important insights of the ontological principles in existential dialectics, and good examples include the formalness-informalness principle, the absoluteness-relativeness principle, the partiality-totality principle, the predictability-unpredictability principle, the explicability-inexplicability principle, the fiction-reality principle, the finiteness-transfiniteness principle, the preciseness-vagueness principle, the simpleness-complicatedness principle, the openness-hiddenness principle, the denseness-emptiness principle, the change-constancy principle, the order-chaos principle, the slowness-quickness principle, the expansion-contraction principle, the theory-praxis principle, the convention-novelty principle, the evolution-transformation principle, the symmetry-asymmetry principle, the softness-hardness principle, the seriousness-playfulness principle, the activeness-inactivess principle, the selfness-otherness principle, the regression-progression principle, the same-difference principle, the stability-reaction principle, the functionality-nonfunctionality principle, and the intentionality-nonintentionality principle.

For instance, in relation to the formalness-informalness principle, if there is formalness (e.g., the formal logical requirement of soundness and completeness in a system of ideas, as in the launch of the environmental justice movement), there is informalness (e.g., the non-formal existence of unsoundness and incompleteness in the launch of the environmental justice movement, as shown in the criticisms that what constitutes fairness or its opposite is in the eyes of the beholder, that the environmental movement has been elitist in producing the opposite effects, that the legal process is too costly, that there is no universal agreement on the definition of "environmental justice" and "environmental injustice," etc.). And the reverse direction also holds true.

In relation to the absoluteness-relativeness principle, if there is absoluteness (e.g., the absolute view of the advocates of the Environmental Justice movement and the Civil Rights Movement for "social justice, equal protection, and an end to institutional discrimination"), there is relativeness (e.g., what is right for the advocates of the Environmental Justice movement and the Civil Rights Movement in regard to "social justice, equal protection, and an end to institutional discrimination" is not necessarily so for the opponents like some conservative industry groups, white supremacists, etc.). And the reverse direction also holds true.

In relation to the partiality-totality principle, if there is partiality (e.g., the partial view of the advocates of the Environmental Justice movement and the Civil Rights Movement for "social justice, equal protection, and an end to institutional discrimination"), there is totality (e.g., the more holistic view of the future of environmental justice, such that the whole is not the sum of its parts, that is, it cannot be reduced to the sum of opposing views of the advocates of the Environmental Justice movement and their critics, since there will emerge new views in the future not yet known today, just as there are already criticisms against them all nowadays). And the reverse direction also holds true.

In relation to the predictability-unpredictability principle, if there is predictability (e.g., the predictable tendency of the Environmental Justice movement nowadays to encourage the poor to participate), there is unpredictability (e.g., the more difficult task to predict exactly the extent to which the Environmental Justice Movement nowadays will be successful in encouraging the poor to participate in a particular future historical era, because of the problems that it has been plagued with "three different forms" of elitism, that it is not cost-effective, etc.). And the reverse direction also holds true.

In relation to the explicability-inexplicability principle, if there is explicability (e.g., the explanation, by the advocates of the Environmental Justice Movement, that it is desirable because of its campaign to end environmental discrimination), there is inexplicability (e.g., the lack of sufficient explanation, by the advocates of the Environmental Justice

Movement, of why its campaign is necessarily desirable, since, for the critics, it has been plagued with "three different forms" of elitism, with opposite effects). And the reverse direction also holds true.

In relation to the fiction-reality principle, if there is fiction (e.g., the fictional aspect of the Environmental Justice Movement, insofar as it has been plagued with "three different forms" of elitism, with opposite effects), there is reality (e.g., the realistic aspect of the Environmental Justice Movement, insofar as it has received help from the Civil Rights Movement). And the reverse direction also holds true.

In relation to the finiteness-transfiniteness principle, if there is finiteness (e.g., the finite number of "different kinds of sites," where environmental discrimination has historically occurred, like "waste disposal," "manufacturing," "energy production," "highways," "ports," and "airports"), there is transfiniteness (e.g., the transfinite number of all the individual waste items in all the sites of "waste disposal," "manufacturing," "energy production," "highways," "ports," and "airports" in history that have affected minorities). And the reverse direction also holds true.

In relation to the preciseness-vagueness principle, if there is preciseness (e.g., the precise identification of 2 barriers to minority participation in the environmental justice movement in the article aforecited), there is vagueness (e.g., the vagueness in the identification of the 2 barriers to minority participation in the environmental justice movement in the article aforecited, since it is not clear why there must be only 2, not 3, 4, 5, and so on). And the reverse direction also holds true.

In relation to the simpleness-complicatedness principle, if there is simpleness (e.g., the relatively simple analysis of environmental justice movement by the advocates), there is complicatedness (e.g., the relatively more complicated analysis of the environmental justice movement, by challenging the claims and assumptions, as shown in the problems pointed out by the critics as cited). And the reverse direction also holds true.

In relation to the openness-hiddenness principle, if there is openness (e.g., the open exploration of using the environmental justice movement to end environmental discrimination against minorities), there is hiddenness (e.g., the hidden bias in the Environmental Justice Movement, because, for the critics, it has been plagued with "three different forms" of elitism, with opposite effects). And the reverse direction also holds true.

In relation to the denseness-emptiness principle, if there is denseness (e.g., the relatively denser concentration of minorities and working class people in the environmental justice movement nowadays), there is emptiness (e.g., the relatively less dense, or more empty, concentration of minorities and working class people in the early environmental justice movement in the older days). And the reverse direction also holds true.

In relation to the change-constancy principle, if there is change (e.g., the ever new ways to fight for environmental justice, as shown in the history of the environmental justice movement), there is constancy (e.g., the ever constant existence of different problems in the ever new ways to fight for environmental justice, as shown in the problems with the history of the environmental justice movement pointed out by the critics). And the reverse direction also holds true.

In relation to the order-chaos principle, if there is order (e.g., the more or less orderly view of the environmental justice movement, if looked from the vantage point of the three clear contributions of the Civil Rights Movement to the movement), there is chaos (e.g,. the more or less chaotic view of the environmental justice movement, if looked from the myriad conflicting vantage points of its different forms of elitism by some groups against other groups, further compounded by the different conflicts between different "supporters of economic growth" and different proponents of environmental justice). And the reverse direction also holds true.

In relation to the slowness-quickness principle, if there is slowness (e.g., the relatively slower speed of the early environmental justice movement in the older days to fight for environmental justice, when "supporters of economic growth have taken advantage of environmentalists' neglect of minorities"), there is quickness (e.g., the relatively quicker speed of the environmental justice movement nowadays to fight for environmental justice, when it receives help from the Civil Rights Movement). And the reverse direction also holds true.

In relation to the expansion-contraction principle, if there is expansion (e.g., the relative spread of the environmental justice movement nowadays), there is contraction (e.g., the relative decrease of market-oriented waste management nowadays). And the reverse direction also holds true.

In relation to the theory-praxis principle, if there is theory (e.g., the theoretical indoctrination of the Civil Rights Movement in regard to "social justice, equal protection, and an end to institutional discrimination"), there is praxis (e.g., the practical application of the theoretical indoctrination of the Civil Rights Movement in regard to "social justice, equal protection, and an end to institutional discrimination" to the field of environmental justice, like its practical influence on the environmental justice movement). And the reverse direction also holds true.

In relation to the convention-novelty principle, if there is convention (e.g., the conventional wisdom of the early environmental movement to focus on "wilderness protection and wildlife preservation"), there is novelty (e.g., the alternative novel challenge to the conventional wisdom of the early environmental movement to focus on "wilderness protection and wildlife preservation, by the new idea nowadays of the environmental movement to focus more on "low-income minorities" to ban environmental discrimination). And the reverse direction also holds true.

In relation to the evolution-transformation principle, if there is evolution (e.g., the natural evolution of humans to throw away wastes in the state of nature in a primitive way), there is transformation (e.g., the technical transformation of the human treatment of wastes by the invention of the environmental justice movement to ban environmental discrimination against minorities). And the reverse direction also holds true.

In relation to the symmetry-asymmetry principle, if there is symmetry (e.g., the existence of different groups in the environmental justice movement), there is asymmetry (e.g., "minorities and working class people" are less represented than "the middle and upper class" in the environmental justice movement). And the reverse direction also holds true.

In relation to the softness-hardness principle, if there is softness (e.g., the inclusion of "the middle and upper class" in the early environmental justice movement), there is hardness (e.g., the exclusion of "minorities and working class people" in the early environmental justice movement). And the reverse direction also holds true.

In relation to the seriousness-playfulness principle, if there is seriousness (e.g., the serious business of those fighting for environmental justice), there is playfulness (e.g., the playful part of those fighting for environmental justice, when they play around with different goals and tactics over time, as shown in the controversial history of its evolution). And the reverse direction also holds true.

In relation to the activeness-inactivess principle, if there is activeness (e.g., the activeness of those from "the middle and upper class" in the early environmental justice movement), there is inactiveness (e.g., the relative inactiveness of those "minorities and working class people" in the early history of the environmental justice movement, because there were excluded).

In relation to the selfness-otherness principle, if there is selfness (e.g., the tendency to be mostly centered on one's self or group, as shown in the supporters of economic growth without caring for environmental justice in the older days), there is otherness (e.g., the tendency to be mostly centered on others, as shown in the more mature environmental justice movement nowadays, when they try to include different groups of people in the campaign, to avoid the charge of elitism which had plagued its early history). And the reverse direction also holds true.

In relation to the regression-progression principle, if there is regression (e.g., the regression made by the environmental justice movement, as shown in the problems pointed out by the critics), there is progression (e.g., the progress made by the environmental justice movement, as shown in its contribution to environmental justice aforecited). And the reverse direction also holds true.

In relation to the same-difference principle, if there is similarity in outcome (e.g., the contribution to the molding and control of beliefs, values, and behaviors, regardless of whether this be done by way of market-oriented economic growth, or by way of environmental justice), there is difference in outcome (e.g., the contribution to the molding and control of beliefs, values, and behaviors by way of market-oriented economic growth for a relatively more competitive lifeworld—but the contribution to the molding and control of beliefs, values, and behaviors by way of environmental justice for a relatively more egalitarian lifeworld). And the reverse direction also holds true.

In relation to the stability-reaction principle, if there is stability (e.g., the relatively more stable condition of the early environmental movement, when, "during the first half of the 20th century, the focus was wilderness protection and wildlife preservation"), there is reaction (e.g., the early focus of the environmental movement during the first half of the 20th century then led to further changes later on, when, nowadays, it pays more attention to the issues affecting "minorities and working class people" like the one concerning environmental discrimination). And the reverse direction also holds true.

In relation to the functionality-nonfunctionality principle, if there is functionality (e.g., the presence of action for which something is fitted, as shown in the focus on "wilderness protection and wildlife preservation" in the early environmental movement during the first half of the 20th century, because it served the function of reflecting the interests of "the middle and upper class" in the organization at the time), there is nonfunctionality (e.g., the relative lack of action for which something is fitted, as shown in the inclusion of the interests of "minorities and working class people" in the environmental movement nowadays, since the focus on wilderness protection and wildlife preservation is no longer functional for the need to address environmental discrimination against minorities at our time). And the reverse direction also holds true.

And in relation to the intentionality-nonintentionality principle, if there is intentionality (e.g., the planning or design of something for a certain outcome, as shown in the founding of the environment justice movement to protect minorities from environmental discrimination), there is nonintentionality (e.g., the relative absence of planning or design of something for a certain outcome, such that the outcome can even be counter-intentional in being the opposite of what was originally planned, as shown in the unintended consequences of the environmental justice movement, in that what constitutes fairness or its opposite is in the eyes of the beholder, that the environmental movement has been elitist in producing the opposite effects, that the legal process is too costly, etc.). And the reverse direction also holds true.

Uselessness and Culture

Uselessness, when examined from the perspective of culture, can cast an illuminating light on its possibility (or impossibility) and desirability (or undesirability), and this can be shown by way of two case studies, namely, (2.5.1) uselessness, waste ideology, and ethical consumerism, and (2.5.2) uselessness, waste reduction, and simple lifestyle—to be addressed hereafter, in that order.

Uselessness, Waste Ideology, and Ethical Consumerism

An excellent case study here to understand uselessness from the perspective of culture has to do with uselessness, waste ideology, and ethical consumerism.

As an ideology for waste reduction (and for that matter, waste reuse and recycle, as well as other related concerns like animal rights, human rights, etc.), "ethical consumerism" is also known as "ethical consumption," "ethical purchasing," "moral purchasing," "ethical sourcing," "ethical shopping," or "green consumerism"—and refers to "a type of consumer activism practiced through 'positive buying' in that ethical products are favoured." (WK 2012aa; ETH 2012)

And the term "ethical consumer" was "first popularised by the UK magazine the Ethical Consumer, first published in 1989. Ethical Consumer magazine's key innovation was to produce 'ratings tables,' inspired by the criteria-based approach of the then emerging ethical investment movement. Ethical Consumer's ratings tables awarded companies negative marks (and from 2005 overall scores) across a range of ethical and environmental categories such a 'animal rights', 'human rights' and 'pollution and toxics', empowering consumers to make ethically informed consumption choices and providing campaigners with reliable information on corporate behaviour. Such criteria-based ethical and environmental ratings have subsequently become a commonplace both in providing consumer information and in business-to-business corporate social responsibility and sustainability ratings such as those provided by Innovest, Calvert, Domini, IRRC, TIAA-CREF and KLD Analytics." (WK 2012aa)

Moral Basis

The moral basis of "ethical consumerism" can be understood in the article "The Global Markets as an Ethical System" by John McMurtry, who "argues that no purchasing decision exists that does not itself imply some moral choice, and that there is no purchasing that is not ultimately moral in nature. This mirrors older arguments, especially by the Anabaptists, e.g. Mennonites, Amish, that one must accept all personal moral and spiritual liability of all harms done at any distance in space or time to anyone by one's own choices." (WK 2012aa)

In fact, this absolute aspect of ethical consumerism is often linked to "the Judeo-Christian scriptures," which "further direct followers towards practising good stewardship of the Earth, under an obligation to a God who is believed to have created the planet for us to share with other creatures... It should be noted, however, that a very similar argument can be presented from an entirely secular humanist point of view, and there are many people who believe that it is simply better for human beings to acknowledge that the planet supports life only because of a delicate balance of many different factors." (WK 2012aa)

Thus, "sustainability is required and purchasing for vanity or status is abhorred and shunned. This theory is echoed in some modern eco-villages who adopt very similar stances, effectively blocking all goods that do not satisfy their moral criteria at the village gate, and relying on internally produced food and tools as much as possible." (WK 2012aa)

Moral Concerns

The ethical consumers, because of their moral beliefs, are highly concerned with some moral issues, as shown by "the five main areas" outlined by "Ethical Consumer, the alternative consumer organisation," and they constitute the so-called "Ethiscore" (as shown below and summarized in *Sec. 2.6*): (WK 2012aa)

- "Environment"—"Environmental Reporting, Nuclear Power, Climate Change, Pollution & Toxics, Habitats & Resources"
- "People"—"Human Rights, Workers' Rights, Supply Chain Policy, Irresponsible Marketing, Armaments"
- "Animals"—"Animal Testing, Factory Farming, Other Animal Rights"
- "Politics"—"Political Activity, Boycott Call, Genetic Engineering, Anti-Social Finance, Company Ethos"
- "Product Sustainability"—"Organic, Fairtrade, Positive Environmental Features, Other Sustainability"

Surely, there can be other examples, but these are the ones outlined by Ethical Consumer.

Moral Buying

Because of their concerns, it is not surprising that ethical consumers advocate a kind of "positive buying" which means "favoring ethical products, be they fair trade, cruelty free, organic, recycled, re-used, or produced locally. This option is arguably the most important since it directly supports progressive companies." (WK 2012aa).

Over the years, "a number of standards and labels have been introduced to induce positive buying," and good examples include the following: (WK 2012aa)

- Ex: "dolphin safe fish"
- Ex: "fairtrade"
- Ex: "FSC-certified sustainably sourced wood"
- Ex: "grass fed beef"
- Ex: "green America seal of approval"
- Ex: "Halal (religious standard)"
- Ex: "Kosher (religious standard)"
- Ex: "Made in USA"
- Ex: "MSC-certified sustainably sourced seafood"
- Ex: "organic food"
- Ex: "rainforest Alliance certified"
- Ex: "shade-grown coffee"
- Ex: "union-made"
- Ex: "vegan"

In addition, "with disclosure of ingredients, some mandatory labelling of origins of clothing or food is required in all developed nations. This practice has been extended in some developing nations, e.g., where every item carries the name, phone number and fax number of the factory where it was made so a buyer can inspect its conditions. And, more importantly, to prove that the item was not made by 'prison labor,' use of which to produce export goods is banned in most developed nations." (WK 2012aa)

As a practical example, "in 2009, Atlanta's Virginia-Highland became the first Carbon-Neutral Zone in the United States. Seventeen merchants of Atlanta's Virginia-Highland allowed their carbon footprint to be audited. Now, they are partnered with the Valley Wood Carbon Sequestration Project—thousands of acres of forest in rural Georgia—through the Chicago Climate Exchange. The businesses involved in the partnership display the Verus Carbon Neutral seal in each storefront and posted a sign prominently declaring the area's Carbon Neutral status." (WK 2012aa; K. Jay 2008; J. Auchmutey 2009)

Moral Boycott

In order to enforce ethical consumerism, the proponents engage in moral boycott against those products, corporations, and countries that do not comply with their concerns, as described below (and summarized in *Sec. 2.6*).

(a) Products

There are different reasons for ethical consumers to boycott certain products, such as the following: (WK 2012aa)

- Ex: "factory farming"
- Ex: "environmental harm"
- Ex: "excessive packaging"
- Ex: "strike action"

In other cases, ethical consumers target the corporations instead.

(b) Corporations

Hereafter are some good examples of corporations which are targeted by ethical consumers: (WK 2012aa)

- Ex: those which "are perceived to espouse unethical behavior by one of its subsidiaries"
- Ex: those "investing a portion of their profits in, for example, the arms industry"

Indeed, "such boycotts can cause great damage to reputations, not to mention loss of profits, and has, in part, led to the development of the concept of corporate social responsibility." (WK 2012aa

(c) Countries

And ehical consumers can also target certain countries violating the moral basis of positive buying, and good examples include the following: (WK 2012aa)

- Ex: "Made in Germany"
- Ex: "consumer boycotts of South Africa over apartheid," as "these boycotts were mirrored in state policy over time, and contributed to the fall of the Apartheid regime"
- Ex: "the ongoing boycott of Israeli products due to the perceived treatment of ethnic Palestinians by the government of Israel"

Surely, there can other examples, so the ones above are illustrative.

Problems with Ethical Consumerism

However, for the critics, the ideology of ethical consumerism has some major problems. Consider, for illustration, a few examples below (as summarized in *Sec. 2.6*).

Firstly, one main criticism of the ideology of ethical consumerism is that the actual effect of ethical consumerism is limited.

For example, "critics argue that the ability to affect structural change is limited in ethical consumerism. Some cite the preponderance of niche markets as the actual effect of ethical consumerism." (WK 2012aa)

Secondly, another main criticism of the ideology of ethical consumerism is that moral boycott can be based on incorrect or limited information.

For instance, for the critics, "information is limited regarding the outcomes of a given purchase, preventing consumers from making informed ethical choices." (WK 2012aa)

Thirdly, still another main criticism of the ideology of ethical consumerism is that it does not give sufficient attention to the power structure of society.

For instance, "critics have also argued that the uneven distribution of wealth prevents consumerism, ethical or otherwise, from fulfilling its democratic potential." (WK 2012aa)

Fourthly, an additional main criticism of the ideology of ethical consumerism is that it can be counter-productive, in encouraging immoral behavior instead.

For instance, for the critics, “one recent study suggests that 'Buying Green' serves as a license for unethical behavior. In their 2009 paper, 'Do Green Products Make Us Better People?,'” Nina Mazar (2010) and Chen-Bo Zhong explained that “people act less altruistically and are more likely to cheat and steal after purchasing green products as opposed to conventional products. Together, the studies show that consumption is more tightly connected to our social and ethical behaviors in directions and domains other than previously thought.” (WK 2012aa) In another study, it was found that “a polluter that buys a pollution license in effect has 'bought green,' which makes them more likely to engage in unethical behavior, leading to increased pollution. Without considering human psychology, such programs could aggravate the very problems they hope to solve.” (WK 2012aa)

Fifthly, another additional main criticism of the ideology of ethical consumerism is that ethical consumerism is a serious mistake in thinking.

For instance, for the critics, “in a 2010 newspaper article, British environmental writer and activist George Monbiot [2010] described green consumerism as 'a catastrophic mistake' on the grounds that 'it strengthens extrinsic values' (those that 'concern status and self-advancement'), thereby 'making future campaigns less likely to succeed.'” (WK 2012aa)

And sixthly, still another additional main criticism of the ideology of ethical consumerism is that there are inconsistent definitions of what constitutes something “moral” in the first place.

For instance, for the critics, “as large corporations have tried to position themselves as moral, principled or ethical organisations, the definition has become wider and means different things to different groups of people. For example, McDonald's started to sell salads, (a more healthy choice) and has a corporate social responsibility blog. Ethical Consumerism can be seen as a movement in marketing, which may or may not reflect actual changes in the practices of businesses.” (WK 2012aa)

Beyond Ethical Consumerism

These criticisms of ethical consumerism, however, do not mean a total rejection of its validity but serve to show us the two opposing sides of the debate, such that the possibility (or impossibility) and desirability (or undesirability) of uselessness (from the perspective of culture with uselessness, waste ideology, and ethical consumerism as a case study here) are not to the extent that the respective defenders would like us to believe.

Furthermore, the analysis of uselessness, waste ideology, and ethical consumerism can reveal the refreshing insights of the ontological principles in existential dialectics, and good examples include the formalness-informalness principle, the absoluteness-relativeness principle, the partiality-totality principle, the predictability-unpredictability principle, the explicability-inexplicability principle, the fiction-reality principle, the finiteness-transfiniteness principle, the preciseness-vagueness principle, the simpleness-complicatedness principle, the openness-hiddenness principle, the denseness-emptiness principle, the change-constancy principle, the order-chaos principle, the slowness-quickness principle, the expansion-contraction principle, the theory-praxis principle, the convention-novelty principle, the evolution-transformation principle, the symmetry-asymmetry principle, the softness-hardness principle, the seriousness-playfulness principle, the activeness-inactivess principle, the selfness-otherness principle, the regression-progression principle, the same-difference principle, the stability-reaction principle, the functionality-nonfunctionality principle, and the intentionality-nonintentionality principle.

For instance, in relation to the formalness-informalness principle, if there is formalness (e.g., the formal logical requirement of soundness, consistency, and completeness in a system of ideas, as in ethcal consumerism), there is informalness (e.g., the non-formal existence of unsoundness, inconsistency, and incompleteness in ethical consumerism, as shown in the criticisms that the actual effect of ethical consumerism is limited, that moral boycott can be based on incorrect or limited information, that it does not give sufficient attention to the power structure of society, that it can be counter-productive in encouraging immoral behavior instead, that ethical consumerism is a mistake in thinking, that there are inconsistent definitions of what constitutes something "moral" in the first place, etc.). And the reverse direction also holds true.

In relation to the absoluteness-relativeness principle, if there is absoluteness (e.g., the absolute view of the advocates of ethical consumerism in regard to the moral basis), there is relativeness (e.g., what is right for the advocates of ethical consumerism in regard to the moral basis is not necessarily so for the critics, who argued that moral boycott can be based on incorrect or limited information, that it can be counter-productive in encouraging immoral behavior instead, that ethical consumerism is a mistake in thinking, etc.). And the reverse direction also holds true.

In relation to the partiality-totality principle, if there is partiality (e.g., the partial view of the advocates of ethical consumerism in regard to the moral basis), there is totality (e.g., the more holistic view of the future of ethical consumerism, such that the whole is not the sum of its parts, that is, it cannot be reduced to the sum of opposing views of the advocates and the critics, since there will emerge new views in the future not yet known today, just as there already some nowadays who do not side with either of the two opposing sides). And the reverse direction also holds true.

In relation to the predictability-unpredictability principle, if there is predictability (e.g., the predictable tendency of the proponents of ethical consumerism to advocate moral or positive buying), there is unpredictability (e.g., the more difficult task to predict exactly the extent to which ethical consumerism will be effective on a particular future occasion, since, for the critics, the actual effect of ethical consumerism is limited in niche markets, that moral boycott can be based on incorrect or limited information, etc.). And the reverse direction also holds true.

In relation to the explicability-inexplicability principle, if there is explicability (e.g., the explanation, by the advocates of ethical consumerism, that it is desirable because of its positive effects on waste reduction, animal rights, etc.), there is inexplicability (e.g., the lack of sufficient explanation, by the advocates of ethical consumerism, of why it is necessarily desirable because, for the critics, it can be counter-productive in encouraging immoral behavior instead, and so on). And the reverse direction also holds true.

In relation to the fiction-reality principle, if there is fiction (e.g., the fictional aspect of ethical consumerism, insofar as it can be "a catastrophic mistake" as pointed out by George Monbiot), there is reality (e.g., the realistic aspect of ethical consumerism, insofar as it can contribute to the buying of ethical products). And the reverse direction also holds true.

In relation to the finiteness-transfiniteness principle, if there is finiteness (e.g., the finite number of the categories of moral concerns by ethical consumers, like "the environment," "people," "animals," "politics," and "product"), there is transfiniteness (e.g., the transfinite number of all the individual things in "the environment," "people," "animals," "politics," and "product" that have ever existed in history). And the reverse direction also holds true.

In relation to the preciseness-vagueness principle, if there is preciseness (e.g., the precise identification of 4 types of "products" that are targeted by ehtical consumers in the article aforecited), there is vagueness (e.g., the vagueness in the identification of the 4 types of "products" that are targeted by ehtical consumers in the article aforecited, since it is not clear why there must be only 4, not 5, 6, 7, and so on). And the reverse direction also holds true.

In relation to the simpleness-complicatedness principle, if there is simpleness (e.g., the relatively simple analysis of ethical consumerism by its advocates), there is complicatedness (e.g., the relatively more complicated analysis of ethical consumerism, by challenging the claims and assumptions, as shown in the problems pointed out by the critics as cited). And the reverse direction also holds true.

In relation to the openness-hiddenness principle, if there is openness (e.g., the open exploration, by ethical consumers, of buying "ethical products"), there is hiddenness (e.g., the hidden bias in ethical consumerism, because, for the critics, it can be counter-productive in encouraging immoral behavior instead, that ethical consumerism is a mistake in thinking, etc.). And the reverse direction also holds true.

In relation to the denseness-emptiness principle, if there is denseness (e.g., the relatively denser concentration of concerns with moral issues by ethical consumers), there is emptiness (e.g., the relatively less dense, or more empty, concentration of concerns with moral issues by market-oriented capitalists, who are not environment-friendly, for instance). And the reverse direction also holds true.

In relation to the change-constancy principle, if there is change (e.g., the ever new ways to study consumption, as shown in ethical consumerism), there is constancy (e.g., the ever constant existence of different problems in the ever new ways to study consumption, as shown in the problems with ethical consumerism pointed out by the critics). And the reverse direction also holds true.

In relation to the order-chaos principle, if there is order (e.g., the more or less orderly appearance of ethical consumerism, if looked from the vantage point of a single definition of what constitutes "moral," as in the "Judeo-Christian scriptures," which "direct followers towards practising good stewardship of the Earth, under an obligation to a God who is believed to have created the planet for us to share with other creatures"), there is chaos (e.g,. the more or less chaotic appearance of ethical consumerism, if looked from the different vantage points of competing inconsistent definitions of what constitutes something "moral" in the first place, because "the definition has become wider and means different things to different groups of people" over time). And the reverse direction also holds true.

In relation to the slowness-quickness principle, if there is slowness (e.g., the relatively slower speed of ethical consumerism to be influential in less developed/poor countries, because it has spread primarily in developed countries), there is quickness (e.g., the relatively quicker speed of the fall of the Apartheid regime, because of the contribution by "consumer boycotts of South Africa over apartheid"). (WK 2012aa) And the reverse direction also holds true.

In relation to the expansion-contraction principle, if there is expansion (e.g., the relative spread of ethical consumerism in developed countries nowadays), there is contraction (e.g., the relative decline of corporations which still engage in unethical behavior nowadays in developed countries). And the reverse direction also holds true.

In relation to the theory-praxis principle, if there is theory (e.g., the theoretical construction of the "good stewardship of the Earth" in "Judeo-Christian scriptures"), there is praxis (e.g., the practical application of the theoretical construction of the "good stewardship of the Earth" in "Judeo-Christian scriptures" to the field of ethics for ethical consumerism). And the reverse direction also holds true.

In relation to the convention-novelty principle, if there is convention (e.g., the conventional custom of some corporations which invest "a portion of their profits in, for example, the arms industry" in the older days), there is novelty (e.g., the alternative novel challenge to the conventional custom of some corporations which invest "a portion of their profits in, for example, the arms industry" in the older days, by the new idea of ethical consumerism to boycott these corporations). And the reverse direction also holds true.

In relation to the evolution-transformation principle, if there is evolution (e.g., the natural evolution of humans to throw wastes or to pollute the environment in the state of nature in a primitive way), there is transformation (e.g., the technical transformation of human practice of dealing with wastes and the environment by the invention of ethics, like ethical consumerism for a "carbon-neutral zone"). And the reverse direction also holds true.

In relation to the symmetry-asymmetry principle, if there is symmetry (e.g., the co-existence of different social groups), there is asymmetry (e.g., "critics have…argued that the uneven distribution of wealth prevents consumerism, ethical or otherwise, from fulfilling its democratic potential"). And the reverse direction also holds true.

In relation to the softness-hardness principle, if there is softness (e.g., the endorsement of ethical consumerism by its advocates), there is hardness (e.g., the rejection of ethcial consumerism by its critics). And the reverse direction also holds true.

In relation to the seriousness-playfulness principle, if there is seriousness (e.g., the serious business of the advocates of ethical consumerism to stop unethical behavior), there is playfulness (e.g., the playful part of the advocates of ethical consumerism, when they play around with different definitions of what constitutes something "moral" in the first place, as "the definition has become wider and means different things to different groups of people" over time). And the reverse direction also holds true.

In relation to the activeness-inactivess principle, if there is activeness (e.g., the activeness of those ethical consumers who buy "ethical products" which "are favored"), there is inactiveness (e.g., the relative inactiveness of those ethical consumers who refuse to buy "products" which violate their moral basis).

In relation to the selfness-otherness principle, if there is selfness (e.g., the tendency to be mostly centered on one's self or group, as shown in unethical behavior of those market-oriented capitalists who were not environment-friendly in the older days), there is otherness (e.g., the tendency to be mostly centered on others, as shown in the 2009 finding that "people act more altruistically after…exposure to green than conventional products," although there is the danger of unethical behavior too). (WK 2012aa) And the reverse direction also holds true.

In relation to the regression-progression principle, if there is regression (e.g., the regression made by ethical consumerism, as shown in the problems pointed out by the critics), there is progression (e.g., the progress made by ethical consumerism, as shown in its contribution to moral buying aforecited). And the reverse direction also holds true.

In relation to the same-difference principle, if there is similarity in outcome (e.g., the contribution to the molding and control of beliefs, values, and behaviors, regardless of whether this be done by way of ethical consumerism, or by way of market-oriented capitalism), there is difference in outcome (e.g., the contribution to the molding and control of beliefs, values, and behaviors by way of ethical consumerism for a relatively more moralistic lifeworld—but the contribution to the molding and control of beliefs, values, and behaviors by way of market-oriented capitalism for a relatively more amoral lifeworld). And the reverse direction also holds true.

In relation to the stability-reaction principle, if there is stability (e.g., the relatively more stable condition of market-oriented capitalism in the older days, when moral issues were often not explicitly considered), there is reaction (e.g., market-oriented capitalism in the older days then then to further changes later on, when moral basis is given more attention in capitalism of our time). And the reverse direction also holds true.

In relation to the functionality-nonfunctionality principle, if there is functionality (e.g., the presence of action for which something is fitted, as shown in the neglect of environmental concerns in the older days, when it served the function of market-oriented capitalism at the time), there is nonfunctionality (e.g., the relative lack of action for which something is fitted, as shown in the rise of ethical consumerism nowadays, such that market-oriented capitalism can no longer serve the function of moral purchase and environment-conscious consumption nowadays). And the reverse direction also holds true.

And in relation to the intentionality-nonintentionality principle, if there is intentionality (e.g., the planning or design of something for a certain outcome, as shown in the creation of ethical consumerism for waste reduction), there is nonintentionality (e.g., the relative absence of planning or design of something for a certain outcome, such that the outcome can even be counter-intentional in being the opposite of what was originally planned, as shown in the unintended consequences of ethical consumerism, in that the actual effect of ethical consumerism is limited, that moral boycott can be based on incorrect or limited information, that it can be counter-productive in encouraging immoral behavior instead, etc.). And the reverse direction also holds true.

Uselessness, Waste Reduction, and Simple Lifestyle

Another way to evaluate the possibility (or impossibility) and desirability (or undesirability) of uselessness from the perspective of culture concerns uselessness, waste reduction, and simple lifestyle.

One major argument here is that a simple lifestyle is conducive to waste reduction and a more sustainable environment—and thus refers to "simple living" which "encompasses a number of different voluntary practices to simplify one's lifestyle. These may include reducing one's possessions or increasing self-sufficiency, for example. Simple living may be characterized by individuals being satisfied with what they need rather than want….[A]sceticism generally promotes living simply and refraining from luxury and indulgence….Simple living is distinct from those living in forced poverty, as it is a voluntary lifestyle choice." (WK 2012q; L. Pierce 2000; M. Griffiths 2010)

In this way, "simple living can…be a reaction to materialism and conspicuous consumption" (due to its excessive waste generation), although there are other possible reasons too. (WK 2012q)

History of Simple Lifestyle

The history of lifestyle has both religious and secular roots, as explained below (and summarized in *Sec. 2.7*).

(a) Religious and Spiritual

Simple lifestyle can be religious and spiritual, because "a number of religious and spiritual traditions encourage simple living. Early examples include the Shramana traditions of Iron Age India, Gautama Buddha, biblical Nazirites (notably John the Baptist) and the Islamic prophet Muhammad. Various notable individuals have claimed that spiritual inspiration led them to a simple living lifestyle, such as Francis of Assisi, Ammon Hennacy, Leo Tolstoy, Rabindranath Tagore, Albert Schweitzer, and Mohandas Gandhi." (WK 2012q; H. Echlin 2006; N. Slocock 2004; D. Shi 2001)

In fact, "simple living has traditions that stretch back to the Orient, resonating with leaders such as Zarathustra, Buddha, Laozi, and Confucius and was heavily stressed in both Greco-Roman culture and Judeo-Christian ethics. Diogenes of Sinope was a Greek cynic who advocated extreme voluntary poverty and purportedly lived in a barrel in the agora." (WK 2012q; D. Shi 2001)

Other people include "Christian groups who have for centuries practiced lifestyles in which some forms of wealth or technology are excluded for religious or philosophical reasons. Groups include the Shakers, Mennonites, Amish, Harmony Society, and some Quakers. There is a Quaker belief called *Testimony of Simplicity* that a person ought to live her or his life simply." (WK 2012q)

And "Jean-Jacques Rousseau strongly praised the simple life in many of his writings, especially in his *Discourse on the Arts and Sciences* (1750) and *Discourse on Inequality* (1754)." (WK 2012q; P. Marshall 1996)

(b) Secular

And simple lifestyle can also be secular; for example, "epicureanism, based on the teachings of the Athens-based philosopher Epicurus, flourished from about the fourth century BC to the third century AD. Epicureanism upheld the untroubled life as the paradigm of happiness, made possible by carefully considered choices. Specifically, Epicurus pointed out that troubles entailed by maintaining an extravagant lifestyle tend to outweigh the pleasure of partaking in it. He therefore concluded that what is necessary for happiness, bodily comfort, and life itself should be maintained at minimal cost, while all things beyond what is necessary for these should either be tempered by moderation or completely avoided." (WK 2012q; M. Smith 2001)

Later, "Henry David Thoreau, a North American naturalist and author, is often considered to have made the classic secular statement advocating a life of simple and sustainable living in his book *Walden*

(1854). Thoreau conducted a two year experiment living a plain and simple life on the shores of Walden Pond." (WK 2012q)

One of Thoreau's admirer, Henry Stephens Salt in Victorian Britain "popularised the idea of 'Simplification, the saner method of living.' Other British advocates of the simple life included Edward Carpenter, William Morris, and the members of 'The Fellowship of the New Life.' C.R. Ashbee and his followers also practiced some of these ideas, thus linking simplicity with the Arts and Crafts Movement. British novelist John Cowper Powys advocated the simple life in his 1933 book *A Philosophy of Solitude*. John Middleton Murry and Max Plowman practised a simple lifestyle at their Aldephi Centre in Essex in the 1930s. Irish poet Patrick Kavanagh championed a 'right simplicity' philosophy based on ruralism in some of his work." (WK 2012q; P. Gould 1988; F. Maccarthy 1981; J. Powys 1933; D. Hardy 2000)

Another example is George Lorenzo Noyes, who "is known as the Thoreau of Maine. He lived a wilderness lifestyle, advocating through his creative work a simple life and reverence for nature. During the 1920s and 1930s, the Vanderbilt Agrarians of the Southern United States advocated a lifestyle and culture centered upon traditional and sustainable agrarian values as opposed to the progressive urban industrialism which dominated the Western world at that time." (WK 2012q)

Then, there is Thorstein Veblen, who "warned against the conspicuous consumption of the materialistic society with *The Theory of the Leisure Class* (1899); Richard Gregg coined the term 'voluntary simplicity' in *The Value of Voluntary Simplicity* (1936). From the 1920s, a number of modern authors articulated both the theory and practice of living simply, among them Gandhian Richard Gregg, economists Ralph Borsodi and Scott Nearing, anthropologist-poet Gary Snyder, and utopian fiction writer Ernest Callenbach. E. F. Schumacher argued against the notion that 'bigger is better' in *Small Is Beautiful* (1973); and Duane Elgin continued the promotion of the simple life in *Voluntary Simplicity* (1981)." (WK 2012q)

Practices of Simple Lifestyle

The practices of simple lifestyle can also take different forms. Consider, for illustration, some main examples below (as summarized in *Sec. 2.7*).

(a) Reducing Consumption, Income and Possessions

The first example of simple lifestyle to be considered here has to do with "reducing consumption, income and possessions."

For exmaple, "some people practice simple living by reducing consumption. By lowering expenditure on goods or services, it is possible to reduce income and the time spent earning money. The time saved may be used to help family or volunteering. For example during the Christmas and holiday season, some people often perform alternative giving. Others may spend the extra free time to improve their quality of life, for example pursuing creative activities such as art and crafts....Developing a detachment from money has led some individuals, such as Suelo and Mark Boyle, to live with no money at all." (WK 2012q; H. Osborne 2009; J. Salter 2010)

Alternatively, "another approach is to focus more fundamentally on the underlying motivation of buying and consuming so many resources for a good quality of life"; for instance, "the grassroots awareness campaign, National Downshifting Week (UK) (founded 1995) encourages participants to positively embrace living with less. Campaign creator, British writer and broadcaster on downshifting and sustainable living, Tracey Smith says, 'The more money you spend, the more time you have to be out there earning it and the less time you have to spend with the ones you love.' National Downshifting Week encourages participants to 'Slow Down and Green Up' and contains a list of suggestions for individuals, companies, children and schools to help adopt green or eco-friendly policies and habits, develop corporate social and environmental responsibility in the workplace, and create eco-protocols and lessons that work alongside the national curriculum, respectively." (WK 2012q)

Also, "reducing possessions or the size of home can also form part of simple living. The 100 Thing Challenge is a grassroots movement to whittle down possessions to a mere 100 items, with the aim of decluttering and simplifying people's lives. The small house movement includes individuals who chose to live in small mortgage-free low-impact dwellings, such as log cabins or beach huts." (WK 2012q; BBC 2011a)

(b) Increasing Self-Sufficiency

The second example of simple lifestyle to be considered here has to do with "increasing self-sufficiency."

For instance, for some people, "the do it yourself ethic refers to the principle of undertaking necessary tasks oneself rather than having others, who are more skilled or experienced, complete them for you"—and "one way to simplify life is to get back-to-the-land and grow your own food, as increased self-sufficiency reduces dependency on money and the economy. Tom Hodgkinson believes the key to a free and simple life is to stop consuming and start producing." (WK 2012q; T. Hodgkinson 2006)

In fact, "the idea of food miles, the number of miles a given item of food or its ingredients has travelled between the farm and the table, is used by simple living advocates to argue for locally grown food. This is now gaining mainstream acceptance, as shown by the popularity of books such as *The 100-Mile Diet*, and Barbara Kingsolver's *Animal, Vegetable, Miracle: A Year of Food Life*. In each of these cases, the authors devoted a year to reducing their carbon footprint by eating locally." (WK 2012q; K. Taylor 2007)

For those living in cities, "city dwellers can also produce fresh home grown fruit and vegetables in pot gardens or miniature indoor greenhouses. Tomatoes, lettuce, spinach, Swiss chard, peas, strawberries, and several types of herbs can all thrive in pots. Jim Merkel says that a person 'could sprout seeds. They are tasty, incredibly nutritious, and easy to grow....We grow them in wide mouthed mason jars with a square of nylon window screen screwed under a metal ring.'" (WK 2012q; J. Merkel 2003)

(c) Reconsidering Technology

The third example of simple lifestyle to be considered here has to do with "reconsidering technology."

For instance, "people who practice simple living have diverse views on the role of technology. For example, people who eschew all technology are often referred to as Luddites or neo-Luddites. Others, like Scott Nearing, are skeptical about how humanity will use new technology, citing destructive inventions such as nuclear weapons. Although simple living is often a secular pursuit, it may still involve reconsidering personal definitions of appropriate technology, as Anabaptist groups such as the Amish or Mennonites have done." (WK 2012q; K. Sale 1997; S. Nearing 2006)

Others have a different interpretation, in that they "see cutting-edge technologies as a way to make a simple lifestyle within mainstream culture easier and more sustainable. They argue that the internet can reduce an individual's carbon footprint through telecommuting and paper usage. Some have also calculated their energy consumption and have shown that one can live simply and in an emotionally satisfying way by using much less energy than is used in western countries. Technologies they may embrace include computers, photovoltaic arrays, wind and water turbines. However, others, like Evgeny Morozov, warn that tools like the internet can facilitate mass surveillance and political repression." (WK 2012q; A. Rajvanshi 2012; E. Morozov 2011)

In addition, "advertising is criticised for encouraging a consumerist mentality. Many advocates of simple living tend to agree that cutting out, or cutting down on, television viewing is a key ingredient in simple living. Some see the Internet, podcasting, community radio or pirate radio as viable alternatives." (WK 2012q)

(d) Simplifying Diet

And the fourth example of simple lifestyle to be considered here has to do with "simplifying diet."

For instance, some people adopt "a simplified diet. Diets that may simplify domestic food production and consumption include vegan diets and the Gandhi diet. In the United Kingdom, the Movement for Compassionate Living was formed by Kathleen and Jack Jannaway in 1984, to spread the vegan message and promote simple living and self-reliance as a remedy against the exploitation of humans, animals, and the Earth." (WK 2012q)

Problems with Simple Lifestyle

However, the practice of simple lifestyle has some major problems. Consider, for illustration, a few examples below (as summarized in *Sec. 2.7*).

Firstly, one main criticism of the practice of simple lifestyle is that simple living can be decadent.

For instance, for the critics like Friedrich Nietzsche, asceticism in the Judeo-Christian tradition is essentially life-denying.

Secondly, another main criticism of the practice of simple lifestyle is that simple living is not possible for everyone.

For instance, for the critics, those who are consumeristic in the contemporary world of capitalism find it impractical to live a life without material comfort and sensual pleasure, since, for them, more—not less—is better.

Thirdly, still another main criticism of the practice of simple lifestyle is that it is ideological, in that simple lifestyle has been used by different groups for their ulterior purposes.

For instance, for the critics, "simple living may be undertaken by environmentalists," like those in "Green parties," who "often advocate simple living as a consequence of their 'four pillars' or the 'Ten Key Values' of the Green Party of the United States. This includes, in policy terms, their rejection of genetic modification and nuclear power and other technologies they consider to be hazardous." (WK 2012q) And others allege a "relationship between economic growth and war, when fought for control and exploitation of natural and human resources....Avoiding the perpetuation of the resource curse is a similar objective of many simple living adherents." (WK 2012q)

Fourthly, an additional main criticism of the practice of simple lifestyle is that it is not so practical as romantically assumed by the advocates.

For instance, for the critics, in response to the call for "increasing self-sufficiency," it was found that "one argument against gardening at home is how often people begin a garden and later stop it." (WK 2012q)

And fifthly, another additional main criticism of the practice of simple lifestyle is that different arguments by different advocates in the movement can be mutually inconsistent or contradictory.

For instance, some people in the movement "eschew all technology" and "are often referred to as…neo-Luddites," who, "like Scott Nearing, are skeptical about how humanity will use new technology, citing destructive inventions such as nuclear weapons"—but others in the same movement have the opposite interpretation, in that they "see cutting-edge technologies as a way to make a simple lifestyle within mainstream culture easier and more sustainable." (WK 2012q; K. Sale 1997; S. Nearing 2006)

Beyond Simple Lifestyle

These criticisms of the practice of simple lifestyle, however, should not be misunderstood as a total rejection of its validity but serve to show us the two opposing sides of the debate, such that the possibility (or impossibility) and desirability (or undesirability) of uselessness (from the perspective of culture with uselessness, waste reduction, and simple lifestyle as a case study here) are not to the extent that the respective defenders would like us to believe.

Furthermore, the analysis of uselessness, waste reduction, and simple lifestyle is to cast a bright light on the ontological principles in existential dialectics, and good examples include the formalness-informalness principle, the absoluteness-relativeness principle, the partiality-totality principle, the predictability-unpredictability principle, the explicability-inexplicability principle, the fiction-reality principle, the finiteness-transfiniteness principle, the preciseness-vagueness principle, the simpleness-complicatedness principle, the openness-hiddenness principle, the denseness-emptiness principle, the change-constancy principle, the order-chaos principle, the slowness-quickness principle, the expansion-contraction principle, the theory-praxis principle, the convention-novelty principle, the evolution-transformation principle, the symmetry-asymmetry principle, the softness-hardness principle, the seriousness-playfulness principle, the activeness-inactivess principle, the selfness-otherness principle, the regression-progression principle, the same-difference principle, the stability-reaction principle, the functionality-nonfunctionality principle, and the intentionality-nonintentionality principle.

For instance, in relation to the formalness-informalness principle, if there is formalness (e.g., the formal logical requirement of soundness, completeness, and consistency in a system of ideas, as in the simple living movement), there is informalness (e.g., the non-formal existence of unsoundness, incompleteness, and inconsistency in the simple living movement, as shown in the criticisms that simple living can be decadent, that simple living is not possible for everyone, that it is ideological, in that simple lifestyle has been used by different groups for their ulterior purposes, that it is not so practical as romantically assumed by the advocates, that different arguments by different advocates in the movement can be mutually inconsistent or contradictory, etc.). And the reverse direction also holds true.

In relation to the absoluteness-relativeness principle, if there is absoluteness (e.g., the absolute view of those people in the simple living movement who "eschew all technology" and "are often referred to as…neo-Luddites"), there is relativeness (e.g., what is right for those people in the simple living movement who "eschew all technology" and "are often referred to as…neo-Luddites" is not necessarily so for others in the movement, who have the opposite interpretation, in that they "see cutting-edge technologies as a way to make a simple lifestyle within mainstream culture easier and more sustainable"). And the reverse direction also holds true.

In relation to the partiality-totality principle, if there is partiality (e.g., the partial view of those people in the simple living movement who "eschew all technology" and "are often referred to as…neo-Luddites"), there is totality (e.g., the more holistic view of the future of technology, such that the whole is not the sum of its parts, that is, it cannot be reduced to the sum of opposing views of the neo-Luddites and their critics, since there will emerge new views in the future not yet known today, just as there are already different views nowdays which question both sides of the debate). And the reverse direction also holds true.

In relation to the predictability-unpredictability principle, if there is predictability (e.g., the predictable tendency of those in the simple living movement to advocate simple lifestyle), there is unpredictability (e.g., the more difficult task to predict exactly the extent to which simple lifestyle will remain influential in a particular future historical era, since there are already different criticisms against it nowadays). And the reverse direction also holds true.

In relation to the explicability-inexplicability principle, if there is explicability (e.g., the explanation by those in the Judeo-Christian tradition that simple living is desirable because of its religious promise), there is inexplicability (e.g., the lack of sufficient explanation by those in the Judeo-Christian tradition of why simple living is necessarily desirable, because, for the critics like Friedrich Nietzsche, asceticism in the Judeo-Christian tradition is essentially life-denying). And the reverse direction also holds true.

In relation to the fiction-reality principle, if there is fiction (e.g., the fictional aspect of "advertising," insofar as it misleadingly glorifies or encourages "a consumerist mentality"), there is reality (e.g., the realistic aspect of "advertising," insofar as it contributes to the awareness of some products on sale). And the reverse direction also holds true.

In relation to the finiteness-transfiniteness principle, if there is finiteness (e.g., the finite number of the categories of those people who "eschew all technology" or technological products, like the 2 categories known as "Luddites" and "neo-Luddites"), there is transfiniteness (e.g., the transfinite number of all the individual pieces of technological products in history that have ever been produced). And the reverse direction also holds true.

In relation to the preciseness-vagueness principle, if there is preciseness (e.g., the precise identification of 4 major traditions of simple lifestyle in the article aforecited), there is vagueness (e.g., the vagueness in the identification of the 4 major traditions of simple lifestyle in the article aforecited, since it is not clear why there must be only 4, not 5, 6, 7, and so on). And the reverse direction also holds true.

In relation to the simpleness-complicatedness principle, if there is simpleness (e.g., the relatively simple analysis of simple living by the advocates), there is complicatedness (e.g., the relatively more complicated analysis of simple living, by challenging the claims and assumptions, as shown in the problems pointed out by the critics as cited). And the reverse direction also holds true.

In relation to the openness-hiddenness principle, if there is openness (e.g., the open exploration of simple living by those in the Judeo-Christian tradition), there is hiddenness (e.g., the hidden bias in the simple living by those in the Judeo-Christian tradition, because, for the critics like Friedrich Nietzsche, asceticism in the Judeo-Christian tradition is essentially life-denying). And the reverse direction also holds true.

In relation to the denseness-emptiness principle, if there is denseness (e.g., the relatively denser concentration of "vegan diets" in the Movement for Compassionate Living), there is emptiness (e.g., the relatively less dense, or more empty, concentration of "vegan diets" among those in mainstream society who consume animal products instead). And the reverse direction also holds true.

In relation to the change-constancy principle, if there is change (e.g., the ever new ways to deal with human needs, as shown in the simple living movement), there is constancy (e.g., the ever constant existence of different problems in the ever new ways to deal with human needs, as shown in the problems with the simple living movement pointed out by the critics). And the reverse direction also holds true.

In relation to the order-chaos principle, if there is order (e.g., the more or less orderly view of the simple living movement, if looked from the vantage point of its focus on "voluntary practices to simplify one's lifestyle"), there is chaos (e.g,. the more or less chaotic view of the simple living movement, if looked from the vantage point of the myraid reasons behind those who choose simple living, like "spirituality, health, increase in 'quality time' for family and friends, work-life balance, personal taste, frugality, or reducing personal ecological footprint and stress," as well as "anti-war movements," "conservation, degrowth, social justice, ethnic diversity, tax resistance and sustainable development," just to cite some of them). (WK 2012q; NWTRCC 2007) And the reverse direction also holds true.

In relation to the slowness-quickness principle, if there is slowness (e.g., the relatively slower speed of getting broccoli if having to grow it onself, since, on average, "broccoli takes 105 days to grow a head," as ponted out by farmer Matt Moore), there is quickness (e.g., the relatively quicker speed of getting broccoli if one buys broccoli at a supermarket and gets it instantly, without waiting for 105 days). And the reverse direction also holds true.

In relation to the expansion-contraction principle, if there is expansion (e.g., the relative rise of materialist capitalist culture in our time), there is contraction (e.g., the relative decline of Judeo-Christian asceticism in our time). And the reverse direction also holds true.

In relation to the theory-praxis principle, if there is theory (e.g., the theoretical construction of environmentalism in environmental studies), there is praxis (e.g., the practical application of environmentalism in environmental studies to the field of spiritual studies, like the simple living movement for waste reduction and a more sustainable environment). And the reverse direction also holds true.

In relation to the convention-novelty principle, if there is convention (e.g., the conventional wisdom about "consumerist mentality" in modern capitalist culture), there is novelty (e.g., the alternative novel challenge to the conventional wisdom about "consumerist mentality" in modern capitalist culture, by the new idea of a new economics based on the simple living movement for "simple prosperity" by David Wann). And the reverse direction also holds true.

In relation to the evolution-transformation principle, if there is evolution (e.g., the natural evolution of humans to live by foraging in the state of nature in a primitive way), there is transformation (e.g., the technical transformation of human ability to live by the invention of the ideology of "simple living" for a more sustainable environment). And the reverse direction also holds true.

In relation to the symmetry-asymmetry principle, if there is symmetry (e.g., the existence of different lifestyles, like simple living and conspicuous consumption), there is asymmetry (e.g., conspicuous consumption in modern capitalist mainstream culture is more dominant than simple living in the subculture of the simple living movement). And the reverse direction also holds true.

In relation to the softness-hardness principle, if there is softness (e.g., the appreciation of simple lifestyle by its advocates), there is hardness (e.g., the critique of simple lifestyle by its critics, as shown in the problems aforecited). And the reverse direction also holds true.

In relation to the seriousness-playfulness principle, if there is seriousness (e.g., the serious business of those in the simple living movement to reject conspicuous consumption in modern capitalist culture), there is playfulness (e.g., the playful part of those in the simple living movement, when they play around with differnet solutions, as different groups in the movement advocate different versions of simple living over time, like reducing consumption, reconsidering technology, increasing self-sufficiency, and simplifying diets). And the reverse direction also holds true.

In relation to the activeness-inactivess principle, if there is activeness (e.g., the activeness of those who work a lot to earn money for what they want in the modern capitalist culture), there is inactiveness (e.g., the relative inactiveness of those who are happy with what they need, not what they want, and therefore decide to work much less and earn less in the simple living movement, although they may spend more time on something else more spiritual).

In relation to the selfness-otherness principle, if there is selfness (e.g., the tendency to be mostly centered on one's self or group, as shown by those who engage in conspicuous consumption in modern capitalist culture to satisfy what they want, not what they need, as "more is better" for them), there is otherness (e.g., the tendency to be mostly centered on others, as shown by those in the simple living movement when they work less and spend "the time saved…to help…or volunteer" for those who are homeless and poor). (WK 2012q) And the reverse direction also holds true.

In relation to the regression-progression principle, if there is regression (e.g., the regression made by simple lifestyle, as shown in the problems pointed out by the critics), there is progression (e.g., the progress made by simple lifestyle, as shown in its contribution to waste reduction and a more sustainable environment aforecited). And the reverse direction also holds true.

In relation to the same-difference principle, if there is similarity in outcome (e.g., the contribution to the molding and control of beliefs, values, and behaviors, regardless of whether this be done by way of simple lifestyle in the simple living movement, or by way of conspicuous consumption in modern capitalist mainstream culture), there is difference in outcome (e.g., the contribution to the molding and control of beliefs, values, and behaviors by way of simple lifestyle for a relatively more spiritual lifeworld—but the contribution to the molding and control of beliefs, values, and behaviors by way of conspicuous consumption for a relatively more materialistic lifeworld). And the reverse direction also holds true.

In relation to the stability-reaction principle, if there is stability (e.g., the relatively more stable condition of conspicuous consumption in modern times, when the consumerist mentality in modern capitalist mainstream culture has been domiant), there is reaction (e.g., conspicuous consumption in modern times then led to further changes later on, when "a new economics movement has been building since the UN conference on the environment in 1972" for "simple prosperity" instead). (WK 2012q; UN 1972) And the reverse direction also holds true.

In relation to the functionality-nonfunctionality principle, if there is functionality (e.g., the presence of action for which something is fitted, as shown in the rise of conspicuous consumption in modern times, when it serves the function to maintain the materialistic capitalist mainstream culture), there is nonfunctionality (e.g., the relative lack of action for which something is fitted, as shown in the rise of the new economics movement" nowadays, such that conspicuous consumption can no longer satisfy those segments of the population who yearn for simple living for a more spiritual and environmental lifeworld instead). And the reverse direction also holds true.

And in relation to the intentionality-nonintentionality principle, if there is intentionality (e.g., the planning or design of something for a certain outcome, as shown in the practice of simple living for waste reduction), there is nonintentionality (e.g., the relative absence of planning or design of something for a certain outcome, such that the outcome can even be counter-intentional in being the opposite of what was originally planned, as shown in the unintended consequences of simple living, in that simple living can be decadent instead, that simple living is not possible for everyone, that simple lifestyle has been used by different groups for their ulterior purposes, that it is not so practical as romantically assumed by the advocates, etc.). And the reverse direction also holds true.

The Valuableness of Uselessness

This comprehensive analysis of the possibility (or impossibility) and desirability (or undesirability) of uselessness in the context of waste, from the perspectives of the mind, nature, society, and culture, is to show us the different ways in which uselessness is both possible and desirable, but not to the extent that the spokespersons from each side would like us to believe.

Uselessness thus has its valuelessness, just as it has its valuableness too, as the other side of the same coin.

But this is only one part of a larger story, since there is the other side of the story, which concerns usefulness, the opposite of uselessness. Since this chapter already deals with uselessness, the next chapter will therefore address the issue of usefulness.

These dual analyses should not be dismissed, because those on the side of usefulness often dismiss uselessness merely as the other side of waste without really appreciating it from the vantage point of uselessness too—and vice versa.

With this in mind, let's now turn to Chapter Three for the study of usefulness and its double sides—for which we now turn to chapter three.

Table 2.1. Uselessness and Its Dual Poles

- **Uselessness and the Mind**
  —Ex: uselessness, biodegradable waste, and epidemics
  —Ex: uselessness, clinical waste, and diseases

- **Uselessness and Nature**
  —Ex: uselessness, litter, and the natural environment
  —Ex: uselessness, space waste, and the Earth's orbit

- **Uselessness and Society**
  —Ex: uselessness, hazardous waste, and the Basel Convention
  —Ex: uselessness, waste management, and minorities

- **Uselessness and Culture**
  —Ex: uselessness, waste ideology, and ethical consumerism
  —Ex: uselessness, waste reduction, and simple lifestyle

Notes: The examples in the categories are solely illustrative (not exhaustive), and the comparison is relative (not absolute), nor are they necessarily mutually exclusive. And some can be easily re-classified elsewhere. As generalities, they allow exceptions.
Source: A summary of Ch.2 of *FPHW*

Table 2.2. Uselessness, Biodegradable Waste, and Epidemics

• **The Case of the Bubonic Plague**
  —Ex: bacteria and infection
  —Ex: signs and symptoms
  —Ex: deadly impact

• **Problems with the Study of Biodegradable Waste and**
  —Firstly, one main criticism of the study of biodegradable waste and epidemics is that epidemics can produce desirable effects instead.
  —Secondly, another main criticism of the study of biodegradable waste and epidemics is that epidemics can be stopped or prevented.
  —And thirdly, still another main criticism of the study of biodegradable waste and epidemics is that it is not clear to what extent exactly biodegradable waste (and biowaste, of that matter) have really caused epidemics in history.

Notes: The examples in the categories are solely illustrative (not exhaustive), and the comparison is relative (not absolute), nor are they necessarily mutually exclusive. And some can be easily re-classified elsewhere. As generalities, they allow exceptions.
Source: From *Sec. 2.2.1* of *FPHW*. See book for citations.

Table 2.3. Uselessness, Clinical Waste, and Diseases

• **Main Methods of Clinical Waste Disposal**
—Ex: "on-site treatment"
—Ex: "truck service"
—Ex: "mail-back medical waste disposal"

• **The Enforcement of Clinical Waste Disposal**
—Ex: management and tracking
—Ex: enforcement

• **Problems with the Study of Clinical Waste and Diseases**
—Firstly, one main criticism of the study of clinical waste and diseases is that the governmental regulation of clinical waste disposal can be costly and thus produce incentives to violate it.
—Secondly, another main criticism of the study of clinical waste and diseases is that there are no universal agreement about its enforcement.
—Thirdly, still another main criticism of the study of clinical waste and diseases is that the methods of clinical waste disposal have their own limits.
—And fourthly, an additional main criticism of the study of clinical waste and diseases is that what constitutes "clinical waste" (or "medical waste") depends on different definitions and has crucial implications in practice.

Notes: The examples in the categories are solely illustrative (not exhaustive), and the comparison is relative (not absolute), nor are they necessarily mutually exclusive. And some can be easily re-classified elsewhere. As generalities, they allow exceptions.
Source: From *Sec. 2.2.2* of *FPHW.* See book for more details and citations.

Table 2.4. Uselessness, Space Waste, and the Earth's Orbit (Part I)

- **Characteristics of Space Waste**
  —Ex: "large vs. small space debris"
  —Ex: "debris in low Earth orbit"
  —Ex: "debris at higher altitudes"

- **Sources of Space Waste**
  —Ex: "dead spacecraft"
  —Ex: "lost equipment"
  —Ex: "boosters"
  —Ex: "debris from and as a weapon"

- **Effects of Space Waste**
  —Ex: "threat to unmanned spacecraft"
  —Ex: "threat to manned spacecraft"
  —Ex: "hazard on Earth"

- **Solutions to Space Waste**
  —Ex: "growth mitigation"
  —Ex: "self-removal"
  —Ex: "external removal"

(continued on next page)

Table 2.4. Uselessness, Space Waste, and the Earth's Orbit (Part II)

- **Problems with the Study of Space Waste**
  —Firstly, one main criticism of the study of space waste (or space debris) and the Earth's orbit is that the measurement of space debris is not an accurate business in the first place, since the exact nature of the problem is still not very well known.
  —Secondly, another main criticism of the study of space waste (or space debris) and the Earth's orbit is that to minimize space debris by preventing space collisions is easier said than done technologically.
  —Thirdly, still another main criticism of the study of space waste (or space debris) and the Earth's orbit is that many of the proposed solutions are too expensive to be cost-effective.
  —Fourthly, an additional main criticism of the study of space waste (or space debris) and the Earth's orbit is that there is no international treaty banning space debris.
  —And fifthly, another additional main criticism of the study of space waste (or space debris) and the Earth's orbit is that there are simply too many pieces of space debris to be able to cope with them all (or even most of them).

Notes: The examples are solely illustrative (not exhaustive), and the comparison is relative (not absolute), nor are they necessarily mutually exclusive. And some can be easily re-classified elsewhere. As generalities, they allow exceptions.
Source: From *Sec. 2.3.2* of *FPHW*. See book for citations.

Table 2.5. Uselessness, Waste Management, and Minorities

- **Barriers to Minority Participation**
  —Ex: "initial barriers"
  —Ex: "cost barriers"

- **Contributions of the Civil Rights Movement**
  —Ex: "similar goals and tactics"
  —Ex: "existing organizations and leaders"
  —Ex: "litigation"

- **Problems with the Environmental Justice Movement**
  —Firstly, one main criticism of the environmental justice movement is that what constitutes fairness or its opposite is in the eyes of the beholder.
  —Secondly, another main criticism of the environmental justice movement is that the environmental movement has been elitist in producing the opposite effects.
  —Thirdly, still another main criticism of the environmental justice movement is that the legal process is too costly.
  —And fourthly, an additional main criticism of the environmental justice movement is that there is no universal agreement on the definition of "environmental justice" and "environmental injustice."

Notes: The examples are solely illustrative (not exhaustive), and the comparison is relative (not absolute), nor are they necessarily mutually exclusive. And some can be easily re-classified elsewhere. As generalities, they allow exceptions.
Source: From *Sec. 2.4.2* of *FPHW*. See book for citations.

Table 2.6. Uselessness, Waste Ideology, and Ethical Consumerism

---

• **Moral Concerns**
—Ex: "environment"
—Ex: "people"
—Ex: "animals"
—Ex: "politics"
—Ex: "product sustainability"

• **Moral Boycott**
—Ex: "products "
—Ex: "corporations"
—Ex: "countries"

• **Problems with Ethical Consumerism**
— Firstly, one main criticism of the ideology of ethical consumerism is that the actual effect of ethical consumerism is limited.
—Secondly, another main criticism of the ideology of ethical consumerism is that moral boycott can be based on incorrect or limited information.
—Thirdly, still another main criticism of the ideology of ethical consumerism is that it does not give sufficient attention to the power structure of society.
—Fourthly, an additional main criticism of the ideology of ethical consumerism is that it can be counter-productive, in encouraging immoral behavior instead.
—Fifthly, another additional main criticism of the ideology of ethical consumerism is that ethical consumerism is a serious mistake in thinking.
—And sixthly, still another additional main criticism of the ideology of ethical consumerism is that there are inconsistent definitions of what constitutes something "moral" in the first place.

Notes: The examples are solely illustrative (not exhaustive), and the comparison is relative (not absolute), nor are they necessarily mutually exclusive. And some can be easily re-classified elsewhere. As generalities, they allow exceptions.
Source: From *Sec. 2.5.1* of *FPHW.* See book for citations.

Table 2.7. Uselessness, Waste Reduction, and Simple Lifestyle

- **History of Simple Lifestyle**
  —Ex: "religious and spiritual"
  —Ex: "secular"

- **Practices of Simple Lifestyle**
  —Ex: "reducing consumption, income and possessions"
  —Ex: "increasing self-sufficiency"
  —Ex: "reconsidering technology"
  —Ex: "simplifying diet"

- **Problems with Simple Lifestyle**
  —Firstly, one main criticism of the practice of simple lifestyle is that simple living can be decadent.
  —Secondly, another main criticism of the practice of simple lifestyle is that simple living is not possible for everyone.
  —Thirdly, still another main criticism of the practice of simple lifestyle is that it is ideological, in that simple lifestyle has been used by different groups for their ulterior purposes.
  —Fourthly, an additional main criticism of the practice of simple lifestyle is that it is not so practical as romantically assumed by the advocates.
  —And fifthly, another additional main criticism of the practice of simple lifestyle is that different arguments by different advocates in the movement can be mutually inconsistent or contradictory.

Notes: The examples are solely illustrative (not exhaustive), and the comparison is relative (not absolute), nor are they necessarily mutually exclusive. And some can be easily re-classified elsewhere. As generalities, they allow exceptions.
Source: From *Sec. 2.5.2* of *FPHW.* See book for citations.

• PART THREE •

Usefulness

CHAPTER 3
USEFULNESS AND ITS DOUBLE SIDES

> I know a man who doesn't pay to have his trash taken out. How does he get rid of his trash? He gift-wraps it, and puts it into an unlocked car.
> —Henny Youngman (TE 2012b)

The Desirability of Usefulness

As already indicated in *Sec. 1.3* and *Sec. 2.1*—the study of waste involves both uselessness and usefulness.

The relationship between uselessness and usefulness is dialectic, since there is no uselessness without usefulness—and vice versa.

The issues concerning uselessness in the study of waste were already addressed in Chapter Two, so the issues concerning usefulness are to be analyzed in this chapter (Chapter Three).

With this dialectic treatment in mind—a good way to examine the appeal of usefulness in the study of waste and its double sides is by way of the evaluation of the extent to which it is in fact both possible (or impossible) and desirable (or undesirable).

This can be done by way of a comprehensive analysis of usefulness in the context of the study of waste from the four perspectives of the mind, nature, society, and culture (in accordance to my sophisticated methodological holism, as explained in *Sec. 1.9*).

In other words, this chapter is organized in four main sections, in relation to (3.2) usefulness and the mind, (3.3) usefulness and nature, (3.4) usefulness and society, and (3.5) usefulness and culture—to be addressed hereafter, in that order (and summarized in *Table 3.1*).

Usefulness and the Mind

Usefulness, when examined from the perspective of the mind, can reveal its possibility (or impossibility) and desirability (or undesirability), and this can be shown by way of two case studies, namely, (3.2.1) usefulness, EPR, and the waste-recycling attitude and (3.2.2) usefulness, dumpster diving, and the waste-collecting habit—to be addressed in what follows, respectively.

Usefulness, EPR, and the Waste-Recycling Attitude

A good way to evaluate the possibility (or impossibility) and desirability (or undesirability) of usefulness from the perspective of the mind is to explore usefulness, EPR, and the waste-recycling attitude.

The term "EPR" stands for "extended producer responsibility" which is a strategy designed to promote a waste-recycling attitude by integrating "environmental costs associated with goods throughout their life cycles into the market price of the products. The concept was first formally introduced in Sweden by Thomas Lindhqvist in a 1990 report to the Swedish Ministry of the Environment." (WK 2012m; OECD 2001)

In a subsequent report to the Ministry, the following definition was used: "[EPR] is an environmental protection strategy to reach an environmental objective of a decreased total environmental impact of a product, by making the manufacturer of the product responsible for the entire life-cycle of the product and especially for the take-back, recycling and final disposal of the product." (WK 2012m)

Characteristics of EPR

The main characteristic of EPR is the use of "financial incentives to encourage manufacturers to design environmentally friendly products by holding producers responsible for the costs of managing their products at end of life. This policy approach, which differs from Product Stewardship, which shares responsibility across the chain of custody of a product, attempts to relieve local governments of the costs of managing certain priority products by requiring manufacturers internalize the cost of recycling within the product price. EPR is based upon the principle that because producers (usually brand owners) have the greatest control over product design and marketing and these same companies have the greatest ability and responsibility to reduce toxicity and waste." (WK 2012m; SIERRA 2012)

In addition, "EPR may take the form of a reuse, buy-back, or recycling program. The producer may also choose to delegate this responsibility to a third party, a so-called producer responsibility organization (PRO), which is paid by the producer for used-product management. In this way, EPR shifts the responsibility for waste management from government to private industry, obliging producers, importers and/or sellers to internalise waste management costs in their product prices and ensuring the safe handling of their products." (WK 2012m; C. Hanisch 2000)

For example, PRO EUROPE (Packaging Recovery Organisation Europe), was "founded in 1995" and is "the umbrella organisation for European packaging and packaging waste recovery and recycling schemes. These product stewardship organizations, such as PRO EUROPE are intended to relieve industrial companies and commercial enterprises of their individual obligation to take back used products through the operation of an organization which fulfils these obligations on a nation-wide basis on behalf of their member companies. The aim is to ensure the recovery and recycling of packaging waste in the most economically efficient and ecologically sound manner. In many countries, this is done through the Green Dot (symbol) trademark of which PRO EUROPE is the general licensor. The 'Green Dot' has evolved into a proven concept in many countries as implementation of Producer Responsibility. In twenty-five nations companies are now using the 'Green Dot' as the financing symbol for the organisation of recovery, sorting and recycling of sales packaging." (WK 2012m)

Goals of EPR

The goals of "buy-back" (or "take-back") in EPR are mainly to encourage a waste-recycling attitude among producers, as one way to deal with "the growing problem of excessive waste," such that "manufacturers [are] responsible taking back their products from end users at the end of the products' useful life, or partially financing a collection and recycling infrastructure. These policies were adopted due to the lack of collection infrastructure for certain products that contain hazardous materials, or due to the high costs to local governments of providing such collection services. The primary goals of these take-back laws therefore are to partner with the private sector to ensure that all wastes are managed in a way that protects public health and the environment." (WK 2012m)

More specifically, there are three main goals of "take-back" in EPR as shown below (and summarized in *Sec. 3.2*): (WK 2012m)

- to "encourage companies to design products for reuse, recyclability, and materials reduction"
- to "correct...market signals to the consumer by incorporating waste management costs into the product's price"
- to "promot[e]...innovation in recycling technology"

These goals are for "creating incentives for companies to redesign their products to minimize waste management costs, by designing their products to contain safer materials (so they do not need to be managed separately) or designing products that are easier to recycle and reuse (so recycling becomes more profitable)." (WK 2012m; L. Roeder 2004)

Historically, "the earliest take-back activity began in Europe, where government-sponsored take-back initiatives arose from concerns about scarce landfill space and potentially hazardous substances in component parts. The European Union adopted a directive on Waste Electrical and Electronic Equipment (WEEE). The purpose of this directive is to prevent the production of waste electronics and also to encourage reuse and recycling of such waste. The directive requires the Member States to encourage design and production methods that take into account the future dismantling and recovery of their products. These take-back programs have been now adopted in nearly every OECD countries. In the United States, most of these policies have been implemented at the state level, due to the political impasse at the federal level." (WK 2012m)

Approaches to EPR

These goals of EPR can be implemented in terms of three major approaches, as shown below (and summarized in *Sec. 3.2*): (WK 2012m)

- "mandatory"
- "negotiated"
- "voluntary"

These different approaches are producer-centered, "because of the tendency of economic policy in market-driven economies not to interfere with consumers' preferences," so "the producer-centric representation is the dominant form of viewing the environmental impacts of industrial production: in statistics on energy, emissions, water, etc., impacts are almost always presented as attributes of industries ('on-site' or 'direct' allocation) rather than as attributes of the supply chains of products for consumers. On a smaller scale, most existing schemes for corporate sustainability reporting include only impacts that arise out of operations controlled by the reporting company, and not supply-chain impacts. According to this world view, 'upstream and downstream [environmental] impacts are…allocated to their immediate producers. The institutional setting and the different actors' spheres of influence are not reflected.'" (WK 2012m; J. Spangenberg 2002)

For instance, "the nexus created by the different views on impacts caused by industrial production is exemplified by several contributions to the discussion about producer or consumer responsibility for greenhouse gas emissions. Emissions data are reported to the IPCC as contributions of producing industries located in a particular country rather than as embodiments in products consumed by a particular population, irrespective of productive origin." (WK 2012m)

By the same logic but "at the company level, 'when adopting the concept of eco-efficiency and the scope of an environmental management system stated in for example ISO 14001, it is insufficient to merely report on the carbon dioxide emissions limited to the judicial borders of the company.' 'Companies must recognise their wider responsibility and manage the entire life-cycle of their products….Insisting on high environmental standards from suppliers and ensuring that raw materials are extracted or produced in an environmentally conscious way provides a start.' A life-cycle perspective is also taken in Extended Producer Responsibility (EPR) frameworks: 'Producers of products should bear a significant degree of responsibility (physical and/or financial) not only for the environmental impacts of their products downstream from the treatment and disposal of their product, but also for their upstream activities inherent in the selection of materials and in the design of products.' 'The major impetus for EPR came from northern European countries in the late 1980s and early 1990s, as they were facing severe landfill shortages….As a result, EPR is generally applied to post-consumer wastes which place increasing physical and financial demands on municipal waste management.'" (WK 2012m; P. Cerin 2002; P. Erin 2005)

Advantages of EPR

The advantages of EPR are obvious to see. For instance, "when producers either face a financial or physical burden of recycling their electronics after use, they may be incentivized to design more sustainable, less toxic, and easily recyclable electronics. Using fewer materials and designing products to last longer can directly reduce producers' end-of-life costs. Thus, Extended Producer Responsibility is often cited as one way to fight planned obsolescence, because it financially encourages manufacturers to design for recycling and make products last longer." (WK 2012m; EPR 2003; B. Prakash 2012; NOUR 2012)

Problems with EPR

However, for the critics, EPR has some major problems. Consider, for illustration, a few examples below (as summarized in *Sec. 3.2*).

Firstly, one main criticism of EPR is that its implementation can be limited for complex products.

For instance, "some people have concerns about extended producer responsibility programs for complex electronics that can be difficult to safely recycle, such as Lithium-ion polymer batteries." (WK 2012m)

Secondly, another main criticism of EPR is that its application is not effective, even in relevant cases.

For instance, some critics pointed out that "such laws could increase the cost of electronics because producers would be adding recycling costs into the initial price tag." (WK 2012m)

Thirdly, still another main criticism of EPR is that it focuses too much on producer responsibility, not consumer responsibility.

For instance, for the critics, "a number of studies have highlighted that final consumption and affluence, especially in the industrialised world, are the main drivers for the level and growth of environmental pressure. Even though these studies provide a clear incentive for complementing producer-focused environmental policy with some consideration for consumption-related aspects, demand-side measures to environmental problems are rarely exploited." (WK 2012m; T. Princen 1999)

Fourthly, an additional main criticism of EPR is that its measurements are not consistent over time.

For instance, for the critics, "EPR has rarely been consistently quantified. Moreover, applying conventional life cycle assessment, and assigning environmental impacts to producers and consumers can lead to double-counting. Using input-output analysis, researchers have attempted for decades to account for both producers and consumers in an economy in a consistent way." (WK 2012m)

Fifthly, another additional main criticism of EPR is that it can be counter-productive.

For instance, "organizations and researchers against EPR claim that the mandate would slow technical innovation and impede technological process," because it adds extra costs to producers. (WK 2012m)

And sixthly, still another additional main criticism of EPR is that it can be misused for something else.

For instance, "other critics are concerned that manufacturers may use takeback programs to take secondhand electronics off the reuse market, by shredding rather than reusing or repairing goods that come in for recycling." (WK 2012m; R. Rivera 2008)

Beyond EPR

These criticisms of EPR, however, should not be misconstrued as a total rejection of its validity but serve to show us the two opposing sides of the debate, such that the possibility (or impossibility) and desirability (or undesirability) of usefulness (from the perspective of the mind with usefulness, EPR, and the waste-recycling attitude as a case study here) are not to the extent that the respective defenders would like us to believe.

Furthermore, the analysis of usefulness, EPR, and the waste-recycling attitude can teach us a valuable lesson about the ontological principles in existential dialectics, and good examples include the formalness-informalness principle, the absoluteness-relativeness principle, the partiality-totality principle, the predictability-unpredictability principle, the explicability-inexplicability principle, the fiction-reality principle, the finiteness-transfiniteness principle, the preciseness-vagueness principle, the simpleness-complicatedness principle, the openness-hiddenness principle, the denseness-emptiness principle, the change-constancy principle, the order-chaos principle, the slowness-quickness principle, the expansion-contraction principle, the theory-praxis principle, the convention-novelty principle, the evolution-transformation principle, the symmetry-asymmetry principle, the softness-hardness principle, the seriousness-playfulness principle, the activeness-inactivess principle, the selfness-otherness principle, the regression-progression principle, the same-difference principle, the stability-reaction principle, the functionality-nonfunctionality principle, and the intentionality-nonintentionality principle.

For instance, in relation to the formalness-informalness principle, if there is formalness (e.g., the formal logical requirement of soundness, completeness, and consistency in a system of ideas, as in EPR), there is informalness (e.g., the non-formal existence of unsoundness, incompleteness, and inconsistency in EPR, as shown in the criticisms that its implementation can be limited for complex electronic products, that its application is not effective, that it focuses too much on producer responsibility, that its measurements are not consistent over time, that it can be counter-productive, that it can be misused for something else, etc.). And the reverse direction also holds true.

In relation to the absoluteness-relativeness principle, if there is absoluteness (e.g., the absolute view of EPR in terms of focusing on producer responsibility), there is relativeness (e.g., what is right for those in EPR to focus on producer responsibility is not necessarily so for the critics, who ask for a focus on consumer responsibility too). And the reverse direction also holds true.

In relation to the partiality-totality principle, if there is partiality (e.g., the partial view of EPR in terms of its focus on the producer side for waste recycling), there is totality (e.g., the more holistic view of waste recycling, such that the whole is not the sum of its parts, that is, it cannot be reduced to the sum of opposing views of those for EPR and their critics, since there will emerge new views in the future not yet known today, just as there are already criticisms against them all nowadays). And the reverse direction also holds true.

In relation to the predictability-unpredictability principle, if there is predictability (e.g., the predictable tendency of those for EPR to advocate producer responsibility), there is unpredictability (e.g., the more difficult task to predict exactly the extent to which EPR will be successful on a particular future occasion, because of the numerous problems pointed out by the critics aforecited). And the reverse direction also holds true.

In relation to the explicability-inexplicability principle, if there is explicability (e.g., the explanation by the proponents of EPR that it is desirable because of its contribution to waste recycling), there is inexplicability (e.g., the lack of sufficient explanation by the proponents of EPR of why it is necessarily desirable, because "other critics are concerned that manufacturers may use takeback programs to take secondhand electronics off the reuse market, by shredding rather than reusing or repairing goods that come in for recycling," together with other problems aforecited). And the reverse direction also holds true.

In relation to the fiction-reality principle, if there is fiction (e.g., the fictional aspect of EPR, insofar as it "has rarely been consistently quantified"), there is reality (e.g., the realistic aspect of EPR, insofar as it contributes to the awareness of producer responsibility). And the reverse direction also holds true.

In relation to the finiteness-transfiniteness principle, if there is finiteness (e.g., the finite number of the categories of waste to fall under the jurisdiction of EPR, like "industrial waste," "electronic waste," etc.), there is transfiniteness (e.g., the transfinite number of all the individual pieces of products in history that have ever existed and can be classified under EPR). And the reverse direction also holds true.

In relation to the preciseness-vagueness principle, if there is preciseness (e.g., the precise identification of 3 approaches to EPR, like "mandatory," "negotiated," and "voluntary," in the article aforecited), there is vagueness (e.g., the vagueness in the identification of the 3 approaches to EPR, like "mandatory," "negotiated," and "voluntary," in the article aforecited, since it is not clear why there must be only 3, not 4, 5, 6, and so on). And the reverse direction also holds true.

In relation to the simpleness-complicatedness principle, if there is simpleness (e.g., the relatively simple analysis of EPR by the proponents), there is complicatedness (e.g., the relatively more complicated analysis of EPR, by challenging the claims and assumptions, as shown in the problems pointed out by the critics as cited). And the reverse direction also holds true.

In relation to the openness-hiddenness principle, if there is openness (e.g., the open exploration of producer responsibility by the advocates of EPR), there is hiddenness (e.g., the hidden bias in EPR, because it does not focus on consumer responsibility). And the reverse direction also holds true.

In relation to the denseness-emptiness principle, if there is denseness (e.g., the relatively denser concentration of concerns, by the proponents of EPR, with producer responsibility in EPR), there is emptiness (e.g., the relatively less dense, or more empty, concentration of concerns, by the proponents of EPR, with the subsequent consequences that "producers would be adding recycling costs into the initial price tag" or that "manufacturers may use takeback programs to take secondhand electronics off the reuse market, by shredding rather than reusing or repairing goods that come in for recycling," and other problems aforecited). And the reverse direction also holds true.

In relation to the change-constancy principle, if there is change (e.g., the ever new ways to deal with waste, as shown in EPR), there is constancy (e.g., the ever constant existence of different problems with the ever new ways to deal with waste, as shown in the problems with EPR pointed out by the critics). And the reverse direction also holds true.

In relation to the order-chaos principle, if there is order (e.g., the more or less orderly reality of waste recycling if looked from the vantage point of the the advocates of EPR to solely focus on producer responsibility), there is chaos (e.g,. the more or less chaotic reality of waste recycling if looked from the multiple vantage points of the critics of EPR, as shown in the numerous problems pointed out by the critics, like the examples that its measurements are not consistent over time, that it can be counter-productive, that it can be misused for something else, etc.). And the reverse direction also holds true.

In relation to the slowness-quickness principle, if there is slowness (e.g., the relatively slower speed of companies to engage in technical innovations if regulated by EPR, according to "organizations and researchers against EPR," as they claim that "the mandate would slow technical innovation and impede technological process," because it adds extra costs to producers and contributes to other problems), there is quickness (e.g., the relatively quicker speed of companies to engage in technical innovations if not burdened by EPR at all, because "organizations and researchers against EPR" claim that "the mandate [of EPR] would slow technical innovation and impede technological process," because it adds extra costs to producers and contributes to other problems). And the reverse direction also holds true.

In relation to the expansion-contraction principle, if there is expansion (e.g., the relative expansion of "more sustainable, less toxic, and easily recyclable" products nowadays among those who comply with EPR), there is contraction (e.g., the relative decline of electronics nowadays which cannot be safely recycled, such as the old-fashioned "Lithium-ion polymer batteries" and others). And the reverse direction also holds true.

In relation to the theory-praxis principle, if there is theory (e.g., the theoretical construction of "sustainability" in environmental studies), there is praxis (e.g., the practical application of the knowledge of "sustainability" in environmental studies to the field of economics, like the call for EPR in manufacturing). And the reverse direction also holds true.

In relation to the convention-novelty principle, if there is convention (e.g., the conventional wisdom about making "complex electronics that can be difficult to safely recycle, such as Lithium-ion polymer batteries" in the older days), there is novelty (e.g., the alternative novel challenge to the conventional wisdom about making "complex electronics that can be difficult to safely recycle," by the new idea of EPR, which suggests that "it financially encourages manufacturers to design for recycling and make products last longer"). (WK 2012m) And the reverse direction also holds true.

In relation to the evolution-transformation principle, if there is evolution (e.g., the natural evolution of humans to throw away wastes in the state of nature in a primitive way), there is transformation (e.g., the technical transformation of human ability to handle waste by the invention of EPR for a systematic "integration of environmental costs associated with goods throughout their life cycles into the market price of the products"). And the reverse direction also holds true.

In relation to the symmetry-asymmetry principle, if there is symmetry (e.g., the existence of different forms of responsibility, like producer-based and consumer-based), there is asymmetry (e.g., producer responsibility is more dominant than consumer responsibility under the policy of EPR nowadays). And the reverse direction also holds true.

In relation to the softness-hardness principle, if there is softness (e.g., the praise of EPR by its proponents), there is hardness (e.g., the critique of EPR by its critics, as shown in the problems as cited). And the reverse direction also holds true.

In relation to the seriousness-playfulness principle, if there is seriousness (e.g., the serious business of the proponents of EPR about producer responsibility), there is playfulness (e.g., the playful part of the proponents of EPR about producer responsibility, when they play aroumd with different forms of enforcement over time, like "reuse," "buy-back," "recycling," or even the delegation of this responsibility to a third party known as "producer responsibility organization" instead). And the reverse direction also holds true.

In relation to the activeness-inactivess principle, if there is activeness (e.g., the activeness of those companies which are able to "design for recycling and make products last longer," in compliance with EPR), there is inactiveness (e.g., the relative inactiveness of those companies which cannot comply with EPR, because it is too "expensive if the product contains hazardous materials and does not have a scrap value, such as with CRT televisions"—although they can shift their activities to something else instead). (WK 2012m) And the reverse direction also holds true.

In relation to the selfness-otherness principle, if there is selfness (e.g., the tendency to be mostly centered on one's self or group, as shown by those campanies in the older days which produced less sustainable, more toxic, and hardly recyclable electronics), there is otherness (e.g., the tendency to be mostly centered on others, as shown by the proponents of EPR who ask for a new policy of making products which are "more sustainable, less toxic, and easily recyclable electronics" for the common good of society). And the reverse direction also holds true.

In relation to the regression-progression principle, if there is regression (e.g., the regression made by EPR, as shown in the problems pointed out by the critics), there is progression (e.g., the progress made by EPR, as shown in its contribution to waste recycling aforecited). And the reverse direction also holds true.

In relation to the same-difference principle, if there is similarity in outcome (e.g., the contribution to the molding and control of beliefs, values, and behaviors, regardless of whether this be done by way of EPR, or by way of market-oriented responsibility), there is difference in outcome (e.g., the contribution to the molding and control of beliefs, values, and behaviors by way of EPR for a relatively more environment-friendly worldview—but the contribution to the molding and control of beliefs, values, and behaviors by way of market-oriented responsibility for a relatively more environment-dominated worldview). And the reverse direction also holds true.

In relation to the stability-reaction principle, if there is stability (e.g., the relatively more stable condition of EPR, when it popularly required producer responsibility in recent times), there is reaction (e.g., the implementation of EPR then led to further changes over time, as "producers would be adding recycling costs into the initial price tag"). And the reverse direction also holds true.

In relation to the functionality-nonfunctionality principle, if there is functionality (e.g., the presence of action for which something is fitted, as shown by those campanies in the older days which produced less sustainable, more toxic, and hardly recyclable electronics, as this served the function of manufacturing at the time), there is nonfunctionality (e.g., the relative lack of action for which something is fitted, as shown in the call for EPR nowadays, so the old-fashioined way of manufacturing can no longer serve the function of manufacturing in our time, which must be "more sustainable, less toxic, and easily recyclable electronics"). And the reverse direction also holds true.

And in relation to the intentionality-nonintentionality principle, if there is intentionality (e.g., the planning or design of something for a certain outcome, as shown in the design of EPR for more producer responsibility), there is nonintentionality (e.g., the relative absence of planning or design of something for a certain outcome, such that the outcome can even be counter-intentional in being the opposite of what was originally planned, as shown in the consequences that the application of EPR is not effective, that it can be counter-productive, and that it can be misused for something else). And the reverse direction also holds true.

Usefulness, Dumpster Diving, and the Waste-Collecting Habit

Another way to evaluate the possibility (or impossibility) and desirability (or undesirability) of usefulness from the perspective of the mind is to explore usefulness, dumpster diving, and the waste-collecting habit.

The term "dumpster diving" refers to "the practice of sifting through commercial or residential trash to find items that have been discarded by their owners, but that may prove useful to the dumpster diver. Dumpster Diving is also viewed as an effective urban foraging technique. Dumpster divers will forage dumpsters for items such as clothing, furniture, food, and other items of the like deemed in good working conditions." (WK 2012bb; M. Lewycka 2009; WST 2011)

Names for Dumpster Diving

But "dumpster diving" also has different names over time. For instance, it is known as "dumpster skipping (British English)"—and "the dumpster diving term originates from the best-known manufacturer of commercial trash bins, Dempster, who use the trade name 'Dumpster' for their bins, and the fanciful image of someone leaping head first into a dumpster as if it were a swimming pool. In practice, the size and design of most dumpsters makes it possible to retrieve many items from the outside of dumpsters without having to 'dive' into them." (WK 2012bb; E. McKean 2005)

In addition, "the practice of dumpster diving can additionally be referred to as bin-diving, containering, D-mart, dumpstering, tatting, or 'recycled' food." (WK 2012bb; A. Renton 2007; DW 2009; N. D'Andrea 2009; D. Colon 2008)

Also, "the term binner is often used to describe individuals who collect recyclable materials for their deposit value." (WK 2012bb)

Reasons for Dumpster Diving

Those who develop the habit of waste-collecting or become dumpster divers have numerous reasons for doing so. Consider, for illustration, some main reasons below: (WK 2012bb; TRASH 2012; E. Jackson 2010; A. Sachs 2004; C. Allison 2007)

- "In Vancouver, binners or bottle collectors search garbage cans and dumpsters for recyclable materials that can be redeemed for their deposit value. On average, these binners earn about $40 per day for several garbage bags full of discarded containers."
- "The karung guni, Zabbaleen, the rag and bone man, waste picker, junk man or bin hoker are people who make their living by sorting and trading trash. A similar process known as gleaning was practiced in rural areas and some ancient agricultural societies, where the residue from farmers' fields was collected."

- "Some dumpster divers, who self-identify as freegans, aim to reduce their ecological footprint by living exclusively from dumpster dived goods."
- "Artists often utilize discarded materials retrieved from trash receptacles to create works of found art or assemblage."
- "Students have been known to undergo dumpster diving to either use salvaged high tech items in technical projects or, to simply indulge in their curiosity for unusual items."
- "Dumpster diving can additionally be used in support of academic research. It serves as the main tool for garbologists, who study the sociology and archeology of trash in modern life. Others, because of their profession, may practice dumpster diving as a tool for private investigations to seek information and material for official purposes."
- "By reusing resources destined for the landfill, dumpster diving becomes an environmentalist endeavor (and is thus practiced by many pro-green communities). The wastefulness of consumer society and throw-away culture compels some individuals to rescue usable items (for example, computers) from destruction and divert them to those who can make use of the item in question."
- "Irregular, blemished, or damaged items that are still otherwise functional are regularly thrown away. Discarded food that might have slight imperfections, that is near its expiration date, or that is simply being replaced by newer stock is often thrown away despite being still edible. Many retailers are reluctant to sell this stock at reduced prices due to the belief that people will buy it instead of the higher priced newer stock, that extra handling time is required, and that there are liability risks."

Of course, there can be other examples, so the ones here are illustrative.

Items of Dumpster Diving

Just as there are different reasons for dumpster diving—there are likewise different items of dumpster diving, which can come in all shapes and sizes, as shown below: (WK 2012bb; J. Harrison 2004)

- "Dumpster diving is practiced differently in developed countries than in developing countries. In many developing countries, food is rarely thrown away unless it is rotten as food is scarce in comparison to developed nations. In countries like the United States where 40–50% of food is wasted, the trash contains a lot more food to gather."

- "In many countries, charities collect excess food from supermarkets and restaurants and distribute it to impoverished neighborhoods. Dumpster divers, Karung guni, Zabaleen, and rag and bone men in these countries may concentrate on looking for usable items or scrap materials to sell rather than food items."
- "In the United States, Canada, and Europe, some bakeries, grocery stores, or restaurants will routinely donate food according to a Good Samaritan Food Donation Act, but more often, because of health laws or company policy, they are required to discard food items by the expiration date, because of overstock, being overly ripened, spoiled, cosmetically imperfect, or blemished."
- "Offices, factories, department stores, and other commercial establishments may equally throw out nonperishable items that are irregular, were returned, have minor damages, or are replaced by newer inventory. Most items tend to be in such a state of disrepair or cosmetically flawed that they will require some work by the dumpster diver to make the items functionally usable. For this reason, factory workers will at times intentionally destroy their items prior to being discarded to prevent them from being reused or resold."
- "As proof to publishing houses of unsold merchandise, booksellers will routinely remove the front covers of printed materials to render them destroyed prior to tossing the remains in the dumpster. Though readable, many damaged publications have disclaimers and legal notices against their existence or sale."
- "Some consumer electronics are dumped because of their rapid depreciation, obsolescence, cost to repair, or expense to upgrade. Owners of functional computers may find it easier to dump them rather than donate because many nonprofit organizations and schools are unable, or unwilling, to work with used equipment. Some organizations like Geeks Into The Streets, reBOOT, Free Geek and Computerbank try to refurbish old computers for charity or educational use."
- "Sometimes dumpsters may contain recyclable metals and materials that can be reused or sold to recycling plants and scrap yards. The most common recyclable metals found are steel and then aluminum."
- "Residential buildings can additionally be a good source of clothing, furniture, appliances, and other housewares."

Again, there can be other examples, so the ones here are illustrative.

Permissability of Dumpster Diving

Dumpster diving is permitted (or legal) under certain circumstances, depending on local laws, although there are exceptions. Consider, for illustration, some examples below: (WK 2012bb)

- "Dumpster diving per se is often legal when not specifically prohibited by law. Abandonment of property is another principle of law that applies to recovering materials via dumpster diving."
- "Police searches of dumpsters as well as similar methods are also generally not considered violations; evidence seized in this way has been permitted in many criminal trials. The doctrine is not as well established in regards to civil litigation."
- "Companies run by private investigators specializing in dumpster diving have emerged as a result of the need for discreet, undetected retrieval of documents and evidence for civil and criminal trials. Private investigators have also written books on 'P.I. technique' in which dumpster diving or its equivalent 'wastebasket recovery' figures prominently."
- "In Italy, a law issued in 2000 declared dumpster diving to be legal."

Surely, there can be other examples, so the ones here are illustrative.

Problems with Dumpster Diving

However, for the critics, dumpster diving has some major problems. Consider, for illustration, a few examples below.

Firstly, one main criticism of dumpster diving is that it can be illegal.

For example, "opening a sealed bin to remove food is also illegal in Germany and, if caught, the dumpster diver could be charged with theft….This option is often used by supermarkets, which in many cases seal their bins or keep them in enclosed areas." (DW. 2009) In other cases, "there are…concerns around the legality of taking items that may still technically belong to the person who threw them away (or to the waste management operator), and whether the taking of some items like discarded documents is a violation of privacy." (WK 2012bb)

Secondly, another main criticism of dumpster diving is that the practice of dumpster diving is not well respected by those in mainstream society.

For instance, there is the old saying that "one man's rubbish is another man's treasure." (DW 2009) But this saying can be reversed by arguing that what is "treasure" for dumpster divers is often regarded as "rubbish" by those in mainstream society. For this reason, dumpster divers are often looked down upon by those in mainstream society.

Thirdly, still another main criticism of dumpster diving is that it is unhealthy.

For instance, "arguments against dumpster diving often focus on the health and cleanliness implications of people rummaging in trash. This exposes the dumpster divers to potential health risks, and, especially if the dumpster diver does not return the non-usable items to their previous location, may leave trash scattered around." (WK 2012bb)

Fourthly, an additional main criticism of dumpster diving is that it is dangerous.

For instance, for the critics, "divers can also be seriously injured or killed by garbage collection vehicles," in an accidental way. (WK 2012bb) And "faulty electrical and mechanical goods can be hazardous and dangerous." (WK 2012e)

And fifthly, another additional main criticism of dumpster diving is that the items in dumpsters are of inferior quality.

For instance, for the critics, a discarded item "is no longer in the same condition as it was" when it was first producedd and "may have faults which are not apparent even if examined." (WK 2012e)

Beyond Dumpster Diving

These criticisms of dumpster diving, however, do not mean a total rejection of its validity but serve to show us the two opposing sides of the debate, such that the possibility (or impossibility) and desirability (or undesirability) of usefulness (from the perspective of the mind with usefulness, dumpster diving, and the waste-collecting habit as a case study here) are not to the extent that the respective defenders would like us to believe.

Furthermore, the analysis of usefulness, dumpster diving, and the waste-collecting habit reveals the insightful relevance of the ontological principles in existential dialectics, and good examples include the formalness-informalness principle, the absoluteness-relativeness principle, the partiality-totality principle, the predictability-unpredictability principle, the explicability-inexplicability principle, the fiction-reality principle, the finiteness-transfiniteness principle, the preciseness-vagueness principle, the simpleness-complicatedness principle, the openness-hiddenness principle, the denseness-emptiness principle, the change-constancy principle, the order-chaos principle, the slowness-quickness principle, the expansion-contraction principle, the theory-praxis principle, the convention-novelty principle, the evolution-transformation principle, the symmetry-asymmetry principle, the softness-hardness principle, the seriousness-playfulness principle, the activeness-inactivess principle, the selfness-otherness principle, the regression-progression principle, the same-difference principle, the stability-reaction principle, the functionality-nonfunctionality principle, and the intentionality-nonintentionality principle.

For instance, in relation to the formalness-informalness principle, if there is formalness (e.g., the formal logical requirement of soundness and completeness in a system of ideas, as in the practice of dumpster diving), there is informalness (e.g., the non-formal existence of unsoundness and incompleteness in the practice of dumpster diving, as shown in the criticisms that it can be illegal, that the practice of dumpster diving is not well respected by those in mainstream society, that it is unhealthy, that it is dangerous, that the items in dumpsters are of inferior quality, etc.). And the reverse direction also holds true.

In relation to the absoluteness-relativeness principle, if there is absoluteness (e.g., the absolute view of "a law issued in 2000" in Italy which "declared dumpster diving to be legal"), there is relativeness (e.g., what is right for "a law issued in 2000" in Italy which "declared dumpster diving to be legal" is not necessarily so for the law in Germany where "a dumpster's contents are regarded as the property of the dumpster's owner," so "taking items from a dumpster is viewed as theft"). And the reverse direction also holds true.

In relation to the partiality-totality principle, if there is partiality (e.g., the partial view of "a law issued in 2000" in Italy which "declared dumpster diving to be legal"), there is totality (e.g., the more holistic view of the permissability of dumpster diving, such that the whole is not the sum of its parts, that is, it cannot be reduced to the sum of opposing views under the law in Italy, Germany, etc., since there will emerge new views in the future not yet known today, just as there are already other views which are critical to them all nowadays). And the reverse direction also holds true.

In relation to the predictability-unpredictability principle, if there is predictability (e.g., the predictable tendency of dumpster divers to "occasionally get in trouble for trespassing" if "dumpsters are…located on private premises"), there is unpredictability (e.g., the more difficult task to predict exactly the extent to which dumpster divers "get in trouble for trespassing" on a particular future occasion, because "the law is enforced with varying degrees of rigor" or can even be ignored by police, depending on different circumstances). (WK 2012bb) And the reverse direction also holds true.

In relation to the explicability-inexplicability principle, if there is explicability (e.g., the explanation, by dumpster divers, that dumpster diving is desirable because of the value of waste resuse), there is inexplicability (e.g., the lack of sufficient explanation by dumpster divers of why dumpster diving is necessarily desirable, because, for the critics, it is unhealthy and dangerous, or the items in dumpsters are of inferior quality, etc.). And the reverse direction also holds true.

In relation to the fiction-reality principle, if there is fiction (e.g., the fictional aspect of dumpster diving, insofar as it is unhealthy and dangerous, the items in dumpsters are of inferior quality, etc.), there is reality (e.g., the realistic aspect of dumpster diving, insofar as it can save by reusing the discarded items). And the reverse direction also holds true.

In relation to the finiteness-transfiniteness principle, if there is finiteness (e.g., the finite number of the categories of discarded items in dumpsters, like "business waste," "chemical waste" "clinical waste," "commercial waste," etc.), there is transfiniteness (e.g., the transfinite number of all the individual items of wastes in history that can be found in dumpsters). And the reverse direction also holds true.

In relation to the preciseness-vagueness principle, if there is preciseness (e.g., the precise identification of 8 reasons for dumpster diving in the article aforecited), there is vagueness (e.g., the vagueness in the identification of the 8 reasons for dumpster diving in the article aforecited, since it is not clear why there must be only 8, not 9, 10, 11, and so on). And the reverse direction also holds true.

In relation to the simpleness-complicatedness principle, if there is simpleness (e.g., the relatively simple analysis of dumpster diving by the advocates), there is complicatedness (e.g., the relatively more complicated analysis of dumpster diving, by challenging the claims and assumptions, as shown in the problems pointed out by the critics as cited). And the reverse direction also holds true.

In relation to the openness-hiddenness principle, if there is openness (e.g., the open exploration of collecting discarded items in dumpsters for reuse), there is hiddenness (e.g., the hidden nature of the discarded items in dumpsters, because they "may have faults which are not apparent even if examined," or "furniture may have not easily seen bedbugs which may cause an infestation," or "faulty electrical and mechanical goods can be hazardous and dangerous," etc.). (WK 2012e) And the reverse direction also holds true.

In relation to the denseness-emptiness principle, if there is denseness (e.g., the relatively denser concentration of discarded items in dumpsters), there is emptiness (e.g., the relatively less dense, or more empty, concentration of discarded items in refrigerators). And the reverse direction also holds true.

In relation to the change-constancy principle, if there is change (e.g., the ever new ways to handle waste, as shown in dumpster diving), there is constancy (e.g., the ever constant existence of different problems in the ever new ways to handle waste, as shown in the problems with dumpster diving pointed out by the critics). And the reverse direction also holds true.

In relation to the order-chaos principle, if there is order (e.g., the more or less orderly appearance of the permissability of dumpster diving if looked only from the vantage point of "a law issued in 2000" in Italy which "declared dumpster diving to be legal"), there is chaos (e.g,. the more or less chaotic appearance of the permissability of dumpster diving if looked from the myriad often conflicting vantage points of numerous laws in different countries and localities about dumpster diving). And the reverse direction also holds true.

In relation to the slowness-quickness principle, if there is slowness (e.g., the relatively slower speed of a discarded item left on private premises to be the target of collection by dumpster divers), there is quickness (e.g., the relatively quicker speed of a discarded item left in public dumpsters to be the target of collection by dumpster divers). And the reverse direction also holds true.

In relation to the expansion-contraction principle, if there is expansion (e.g., the relative expansion of dumpsters nowadays in developed countries, since "waste management became a critical issue," due to "the increase in population and the massive migration of people to industrial towns and cities from rural areas"), there is contraction (e.g., the relative decline of people nowadays in developed countries to simply "throw their…wastes out of the window which would decompose in the street," as was the case for the Europeans in the older days). (WK 2012a) And the reverse direction also holds true.

In relation to the theory-praxis principle, if there is theory (e.g., the theoretical construction of "sustainability" in environmental studies), there is praxis (e.g., the practical application of the knowledge of "sustainability" in environmental studies to the field of waste management, like "dumpster diving" which "becomes an environmentalist endeavor…and is…practiced by many pro-green communities"). (WK 2012bb) And the reverse direction also holds true.

In relation to the convention-novelty principle, if there is convention (e.g., the conventional wisdom that "most people who resort to dumpster-diving are forced to do so out of economic necessity"), there is novelty (e.g., the alternative novel challenge to the conventional wisdom that "most people who resort to dumpster-diving are forced to do so out of economic necessity," by the new idea nowadays that discarded items in dumpster can be used for police investigation, the creation of art works, academic research, etc.). And the reverse direction also holds true.

In relation to the evolution-transformation principle, if there is evolution (e.g., the natural evolution of humans to throw away wastes in the state of nature in a primitive way), there is transformation (e.g., the technical transformation of human ability to deal with wastes by the invention of dumpsters nowadays). And the reverse direction also holds true.

In relation to the symmetry-asymmetry principle, if there is symmetry (e.g., the existence of different practices of dumpster diving in different countries), there is asymmetry (e.g., "in many developing countries, food is rarely thrown away unless it is rotten as food is scarce in comparison to developed nations"—but in developed countries like "the United States where 40–50% of food is wasted, the trash contains a lot more food to gather"). (WK 2012bb; J. Harrison 2004) And the reverse direction also holds true.

In relation to the softness-hardness principle, if there is softness (e.g., the appreciation of dumpster diving, as shown in the reuse of "resources destined for the landfill" by dumpster divers), there is hardness (e.g., "in Germany, a dumpster's contents are regarded as the property of the dumpster's owner," and "therefore, taking items from a dumpster is viewed as theft"). And the reverse direction also holds true.

In relation to the seriousness-playfulness principle, if there is seriousness (e.g., the serious business of those who engage in dumpster diving), there is playfulness (e.g., the playful part of those who engage in dumpster diving, when they play around with discaded items for different purposes over time, like the creation of art works, criminal investigation, waste reuse, academic research, etc.). And the reverse direction also holds true.

In relation to the activeness-inactivess principle, if there is activeness (e.g., the activeness of those dumpster divers in Italy who are allowed to collect discarded items in dumpsters, because of "a law issued in 2000" which "declared dumpster diving to be legal"), there is inactiveness (e.g., the relative inactiveness of those dumpster divers in Germany who are not allowed to collect discarded items because "a dumpster's contents are regarded as the property of the dumpster's owner" and "therefore, taking items from a dumpster is viewed as theft" under the law, although they can do something else).

In relation to the selfness-otherness principle, if there is selfness (e.g., the tendency to be mostly centered on one's self or group, as shown by those "people who resort to dumpster-diving" because they "are forced to do so out of economic necessity"), there is otherness (e.g., the tendency to be mostly centered on others, as shown in those academic researchers who point out the issues of health risks and physical hazards in the habit of dumpster diving, for the enhancement of social welfare). And the reverse direction also holds true.

In relation to the regression-progression principle, if there is regression (e.g., the regression made by dumpster diving, as shown in the problems pointed out by the critics), there is progression (e.g., the progress made by dumpster diving, as shown in its contribution to waste reuse and other uses aforecited). And the reverse direction also holds true.

In relation to the same-difference principle, if there is similarity in outcome (e.g., the contribution to the molding and control of beliefs, values, and behaviors, regardless of whether this be done by way of the habit of dumpster diving by poor people, or by way of the throw-away practice of rich people), there is difference in outcome (e.g., the contribution to the molding and control of beliefs, values, and behaviors by way of the habit of dumpster diving by poor people for a relatively more impoverished lifeworld—but the contribution to the molding and control of beliefs, values, and behaviors by way of the throw-away practice of rich people for a relatively more affluent lifeworld). And the reverse direction also holds true.

In relation to the stability-reaction principle, if there is stability (e.g., the relatively more stable condition of restricting the practice of dumpster diving in Belgium in the older days, as shown in the arrest of "a Belgian dumpster diver and eco-activist nicknamed Ollie" on February 25th, 2009, for "taking food from a dumpster at an AD Delhaize market in Bruges"), there is reaction (e.g., the arrest of "a Belgian dumpster diver and eco-activist nicknamed Ollie" in 2009 then led to further changes later on, when his subsequent trial "evoked protests in Belgium against restrictions from taking discarded food items"). (WK 2012bb) And the reverse direction also holds true.

In relation to the functionality-nonfunctionality principle, if there is functionality (e.g., the presence of action for which something is fitted, as shown in the practice of throwing away things by those in consumer society in the older days, as it served the function of the "throw-away culture" at the time), there is nonfunctionality (e.g., the relative lack of action for which something is fitted, as shown in dumpster diving nowadays, which "becomes an environmentalist endeavor…and is thus practiced by many pro-green communities" as a way to protest "the wastefulness of consumer society," so the practice of throwing away things in the older days is no longer functional in our time for the new mindset for a more sustainable environment). And the reverse direction also holds true.

And in relation to the intentionality-nonintentionality principle, if there is intentionality (e.g., the planning or design of something for a certain outcome, as shown in the idea of dumpster diving for waste reuse), there is nonintentionality (e.g., the relative absence of planning or design of something for a certain outcome, such that the outcome can even be counter-intentional in being the opposite of what was originally planned, as shown in the unintended consequence that it can cost the dumpster diver even more in the end, because it can be illegal, unhealthy, dangerous, etc.). And the reverse direction also holds true.

Usefulness and Nature

Usefulness, when examined from the perspective of nature, can show us its possibility (or impossibility) and desirability (or undesirability), and this can be shown by way of two case studies, namely, (3.3.1) usefulness, waste-to-energy, and sustainability, and (3.3.2) usefulness, waste, and animal products—to be addressed in what follows, respectively.

Usefulness, Waste-to-Energy, and Sustainability

A good way to evaluate the possibility (or impossibility) and desirability (or undesirability) of usefulness from the perspective of nature is to explore usefulness, waste-to-energy, and sustainability.

The term "waste-to-energy" is also known as "energy-from-waste (EfW)" and refers "the process of creating energy in the form of electricity or heat from the incineration of waste source" and "most WtE processes produce electricity directly through combustion, or produce a combustible fuel commodity, such as methane, methanol, ethanol or synthetic fuels." (WK 2012h)

In general, "WtE is a form of energy recovery." (WK 2012h)

Resource Recovery Methods

Because WtE is a form of "energy recovery," the latter term refers to "the selective extraction of disposed materials for a specific next use, such as recycling, composting or energy generation. The aim of the resource recovery is to extract the maximum practical benefits from products, delay the consumption of virgin natural resources, and to generate the minimum amount of waste." (WK 2012cc)

For illustration, consider two examples of resource recovery methods below (as summarized in *Sec. 3.3*).

(a) Biological Processing

The first resource recovery method to be considered here has to do with "biological processing."

As the name suggests, "biological processing" targets organic wastes. For instance, "disposed materials that are organic in nature, such as plant material, food scraps, and paper products, can be recycled using biological composting and digestion processes to decompose the organic matter. The resulting organic material is then recycled as mulch or compost for agricultural or landscaping purposes. In addition, waste gas from the process (such as methane) can be captured and used for generating electricity and heat (CHP/cogeneration) maximising efficiencies. The intention of biological processing is to control and accelerate the natural process of decomposition of organic matter." (WK 2012cc)

Of coure, there are different forms of biological processing, such as the "large variety of composting and digestion methods and technologies varying in complexity from simple home compost heaps, to small town scale batch digesters, industrial-scale enclosed-vessel digestion of mixed domestic waste....Methods of biological decomposition are differentiated as being aerobic or anaerobic methods, though hybrids of the two methods also exist." (WK 2012cc)

Actual examples of "resource recovery through composting can be found in San Francisco, and the Green Bin Program in Toronto, Canada, where Source Separated Organics (such as kitchen scraps and plant cuttings) are collected in a dedicated container and then composted." (WK 2012cc)

(b) Source-Separated Curbside Collection

And the second resource recovery method to be considered here has to do with "source-separated curbside collection."

For instance, "in Australia, source-separated curbside collection is one method of resource recovery. Every urban domestic household is provided with three bins: one for recycling, another for general waste and another for garden materials—this bin is provided by the municipality if requested. Also, many households have compost bins; but this is not provided by the municipality. To encourage recycling, municipalities provide large recycle bins, which are larger than general waste bins." (WK 2012cc)

Also, "municipal, commercial and industrial, construction and demolition debris is dumped at landfills and some is recycled. Household disposal materials are segregated: recyclables sorted and made into new products, and unusable material is dumped in landfill areas. According to the ABS, the recycling rate is high and is 'increasing, with 99% of households reporting that they had recycled or reused within the past year (2003 survey), up from 85% in 1992'. This suggests that Australians are in favour of reduced or no landfilling and the recovery of materials through recycling. In 2002–03 '30% of materials from municipalities, 45% from commercial and industrial generators and 57% from construction and demolition debris' was recycled. Energy is produced is part of resource recovery as well: some landfill gas is captured for fuel or electricity generation, although this is considered the last resort, as the point of resource recovery is avoidance of landfill disposal altogether." (WK 2012cc)

The Goal of Environmental Sustainability

The main goal of resource recovery is to ensure environmental sustainability, especially when "companies are encouraged to improve their environmental efficiencies each year. One way to do this is by changing a company from a system of managing wastes to a resource recovery system (such as recycling: glass, food waste, paper and cardboard, plastic bottles etc.)." (WK 2012cc)

With this goal in mind, "the Talloires Declaration is a declaration for sustainability concerned about the unprecedented scale and speed of environmental pollution and degradation, and the depletion of natural resources. Local, regional, and global air pollution; accumulation and distribution of toxic wastes; destruction and depletion of forests, soil, and water; depletion of the ozone layer and emission of 'green house' gases threaten the survival of humans and thousands of other living species, the integrity of the earth and its biodiversity, the security of nations, and the heritage of future generations. Several universities have implemented the Talloires Declaration by establishing environmental management and resource recovery programs. University and vocational education are promoted by various organizations, e.g. WAMITAB and Chartered Institution of Wastes Management. Many supermarkets encourage customers to use their reverse vending machines to deposit used purchased containers and receive a refund from the recycling fees. Brands that manufacture such machines include Tomra and Envipco." (WK 2012cc)

In 2010, for instance, "CNBC aired the documentary *Trash Inc: The Secret Life of Garbage* about waste, what happens to it when it's 'thrown away,' and its impact on the world." (WK 2012cc; S. Carpenter 2010)

The Case of Waste-to-Energy

An excellent illustration of resource recovery for environmental sustainability is "waste-to-energy (WtE)" in action (as summarized in *Sec. 3.3*).

(a) WtE Methods

Waste-to-energy can be achieved by different methods.

A common method is "incineration," which is "the most common WtE implementation. All new WtE plants in OECD countries must meet strict emission standards, including those on nitrogen oxides (NO), sulphur dioxide (SO2), heavy metals and dioxins. Hence, modern incineration plants are vastly different from the old types, some of which neither recovered energy nor materials. Modern incinerators reduce the volume of the original waste by 95-96 percent, depending upon composition and degree of recovery of materials such as metals from the ash for recycling." (WK 2012h)

In general, "incinerators have electric efficiencies on the order of 14-28%. The rest of the energy can be utilized for e.g. district heating, but is otherwise lost as waste heat." (WK 2012h; DREDL 2009) In fact, "the method of using incineration to convert municipal solid waste (MSW) to energy is a relatively old method of waste-to-energy production. Incineration generally entails burning an RDF to boil water which powers steam generators that make electric energy to be used in our homes and businesses." (WK 2012h)

Besides incineration—"there are a number of other new and emerging technologies that are able to produce energy from waste and other fuels without direct combustion. Many of these technologies have the potential to produce more electric power from the same amount of fuel than would be possible by direct combustion," and hereafter are some examples: (WK 2012h)

- "Thermal Technologies"
  - –Ex: "gasification (produces combustible gas, hydrogen, synthetic fuels)"
  - –Ex: "thermal depolymerization (produces synthetic crude oil, which can be further refined"
  - –Ex: "pyrolysis (produces combustible tar/biooil and chars)"

–Ex: "plasma arc gasification PGP or plasma gasification process (produces rich syngas including hydrogen and carbon monoxide usable for fuel cells or generating electricity to drive the plasma arch, usable vitrified silicate and metal ingots, salt and sulphur)"

- "Non-Thermal Technologies"
  –Ex: "anaerobic digestion (biogas rich in methane)"
  –Ex: "fermentation production (examples are ethanol, lactic acid, hydrogen)"
  –Ex: "mechanical biological treatment (MBT)"

Surely, there can be other examples, so the ones above are illustrative.

(b) Global WtE Developments

WtE becomes more popular in our time; for instance, "during the 2001-2007 period, the WTE capacity increased by about four million metric tons per annum. Japan and China built several plants that were based on direct smelting or on fluidized bed combustion of solid waste. In China there are about 50 WTE plants. Japan is the largest user in thermal treatment of MSW in the world with 40 million tons. Some of the newest plants use stoker technology and others use the advanced oxygen enrichment technology. There are also over one hundred thermal treatment plants using relatively novel processes such as direct smelting, the Ebara fluidization process and the Thermo- select -JFE gasification and melting technology process. In Patras, Greece, a Greek company just finished testing a system that shows potential. It generates 25kwatts of electricity and 25kwatts of heat from waste water. In India its first energy bio-science center was developed to reduce the country's green house gases and its dependency on fossil fuel." (WK 2012h; COL 2008)

Problems with Waste-to Energy

However, for the critics, waste-to-energy has some major problems. Consider, for illustration, a few examples below (as summarized in *Sec. 3.3*).

Firstly, one main criticism of waste-to-energy is that it can be polluting.

For example, for the critics, "concerns regarding the operation of incinerators include fine particulate, heavy metals, trace dioxin and acid gas emissions, even though these emissions are relatively low from modern incinerators. Other concerns include toxic fly ash and incinerator bottom ash (IBA) management." (WK 2012h; DREDL 2009)

Secondly, another main criticism of waste-to-energy is that it can use up valuable resources.

For instance, for the critics, "discussions regarding waste resource ethics include the opinion that incinerators destroy valuable resources." (WK 2012h; DREDL 2009)

Thirdly, still another main criticism of waste-to-energy is that incineration competes with other waste-treatment activities.

For instance, for the critics, there is "the fear that they [incinerators] may reduce the incentives for recycling and waste minimization activities." (WK 2012h; DREDL 2009)

Fourthly, an additional main criticism of waste-to-energy is that each WtE method has its own costs and benefits (or advantages and disadvantages).

For instance, "esterification can also be done using waste to energy technologies, and the result of this process is biodiesel. The cost effectiveness of esterification will depend on the feedstock being used, and all the other relevant factors such as transportation distance, amount of oil present in the feedstock, and others." (WK 2012h; BIO 2012)

And fifthly, another additional main criticism of waste-to-energy is that energy efficiency has its limits.

For instance, in thermodynamics, "thermal efficiency" is limited by the fact that "real engines have many departures from ideal behavior that waste energy, reducing actual efficiencies far below the theoretical values" (which are possible only in ideal situations), and good examples of these less-than-ideal situations include: (WK 2012dd)

- Ex: "friction of moving parts"
- Ex: "inefficient combustion"
- Ex: "heat loss from the combustion chamber"
- Ex: "aerodynamic drag of air moving through the engine"
- Ex: "energy used by auxiliary equipment like oil and water pumps"
- Ex: "inefficient compressors and turbines"
- Ex: "imperfect valve timing"
- Ex: "engines must be optimized for other goals besides efficiency," like "low pollution" (or "low emissions"), "adequate acceleration, fast starting, light weight, low noise, etc.," so these additional requirements further reduce efficiency.

Surely, there can be other examples, so the ones here are illustrative.

Beyond Waste-to-Energy

These criticisms of waste-to-energy, however, do not suggest a total rejection of its validity but serve to show us the two opposing sides of the debate, such that the possibility (or impossibility) and desirability (or undesirability) of usefulness (from the perspective of nature with usefulness, waste-to-energy, and sustainability as a case study here) are not to the extent that the respective defenders would like us to believe.

Besides, the analysis of usefulness, waste-to-energy, and sustainability constitutes a good didactic about the ontological principles in existential dialectics, and good examples include the formalness-informalness principle, the absoluteness-relativeness principle, the partiality-totality principle, the predictability-unpredictability principle, the explicability-inexplicability principle, the fiction-reality principle, the finiteness-transfiniteness principle, the preciseness-vagueness principle, the simpleness-complicatedness principle, the openness-hiddenness principle, the denseness-emptiness principle, the change-constancy principle, the order-chaos principle, the slowness-quickness principle, the expansion-contraction principle, the theory-praxis principle, the convention-novelty principle, the evolution-transformation principle, the symmetry-asymmetry principle, the softness-hardness principle, the seriousness-playfulness principle, the activeness-inactivess principle, the selfness-otherness principle, the regression-progression principle, the same-difference principle, the stability-reaction principle, the functionality-nonfunctionality principle, and the intentionality-nonintentionality principle.

For instance, in relation to the formalness-informalness principle, if there is formalness (e.g., the formal logical requirement of soundness and completeness in a system of ideas, as in the idea of WtE), there is informalness (e.g., the non-formal existence of unsoundness and incompleteness in the idea of WtE, as shown in the criticisms that it can be polluting, that it can use up valuable resources, that incineration competes with other waste-treating activities, that each WtE method has its own costs and benefits, that energy efficiency has its limits, etc.). And the reverse direction also holds true.

In relation to the absoluteness-relativeness principle, if there is absoluteness (e.g., the absolute view of those who favor incineration), there is relativeness (e.g., what is right for those who favor incineration is not necessarily so for the critics who are concerned that "incinerators destroy valuable resources" and that "they may reduce the incentives for recycling and waste minimization activities"). And the reverse direction also holds true.

In relation to the partiality-totality principle, if there is partiality (e.g., the partial view of the advocates of each WtE method), there is totality (e.g., the more holistic view of WtE, such that the whole is not the sum of its parts, that is, it cannot be reduced to the sum of opposing views of the different advocates of different WtE methods and their critics, since there will emerge new views in the future not yet known today, just as there are already criticisms against each of them nowadays). And the reverse direction also holds true.

In relation to the predictability-unpredictability principle, if there is predictability (e.g., the predictable tendency of "incineration" to be "the most common WtE implementation" in the foreseeable future), there is unpredictability (e.g., the more difficult task to predict exactly the extent to which "incineration" will remain "the most common WtE implementation" in a particular more distant future historical era, since there are already competing technologies and waste-treatment alternatives nowadays). And the reverse direction also holds true.

In relation to the explicability-inexplicability principle, if there is explicability (e.g., the explanation by the advocates of incineration that it is desirable because of its energy creation), there is inexplicability (e.g., the lack of sufficient explanation by the advocates of incineration of why it is necessarily desirable, since, for the critics, "incinerators destroy valuable resources" and "they may reduce the incentives for recycling and waste minimization activities," besides other problems). And the reverse direction also holds true.

In relation to the fiction-reality principle, if there is fiction (e.g., the fictional aspect of incineration as a safe WtE method, insofar as its proponents tried to downplay the problem "associated with incinerating MSW to make electrical energy," because of "the potential for pollutants to enter the atmosphere with the flue gases from the boiler," and "these pollutants can be acidic and in the 1980s were reported to cause environmental damage by turning rain into acid rain"), there is reality (e.g., the realistic aspect of incineration as a safe WtE method, insofar as it has contributed to the creation of electrical energy). And the reverse direction also holds true.

In relation to the finiteness-transfiniteness principle, if there is finiteness (e.g., the finite number of the categories of pollutants by incinerators, like "fine particulate, heavy metals, trace dioxin and acid gas emissions"), there is transfiniteness (e.g., the transfinite number of all the individual pieces of "fine particulate, heavy metals, trace dioxin and acid gas emissions," and so on in history that have ever been in the environment). And the reverse direction also holds true.

In relation to the preciseness-vagueness principle, if there is preciseness (e.g., the precise identification of 8 WtE methods, including incineration, in the article aforecited), there is vagueness (e.g., the vagueness in the identification of the 8 WtE methods, including incineration, in the article aforecited, since it is not clear why there must be only 8, not 9, 10, 11, and so on). And the reverse direction also holds true.

In relation to the simpleness-complicatedness principle, if there is simpleness (e.g., the relatively simple analysis of each of the WtE methods by its advocates), there is complicatedness (e.g., the relatively more complicated analysis of WtE methods, by challenging the claims and assumptions, as shown in the problems pointed out by the critics as cited). And the reverse direction also holds true.

In relation to the openness-hiddenness principle, if there is openness (e.g., the open exploration of using incinerators to convert waste into energy), there is hiddenness (e.g., the hidden bias in the use of incinerators to convert waste into energy," because of "the potential for pollutants to enter the atmosphere with the flue gases from the boiler," and "these pollutants can be acidic and in the 1980s were reported to cause environmental damage by turning rain into acid rain," together with other problems over the years). And the reverse direction also holds true.

In relation to the denseness-emptiness principle, if there is denseness (e.g., the relatively denser concentration of using incinerators to convert waste into energy nowadays), there is emptiness (e.g., the relatively less dense, or more empty, concentration of using alternative WtE methods other than incineration nowadays). And the reverse direction also holds true.

In relation to the change-constancy principle, if there is change (e.g., the ever new ways to deal with wastes, as shown in the use of WtE methods), there is constancy (e.g., the ever constant existence of different problems in the ever new ways to deal with wastes, as shown in the problems with WtE methods pointed out by the critics). And the reverse direction also holds true.

In relation to the order-chaos principle, if there is order (e.g., the more or less orderly appearance of WtE methods, if looked only from the vantage point of incineration as the most common implementation), there is chaos (e.g,. the more or less chaotic appearance of WtE methods, if looked from the multiple vantage points of different competing, or even conflicting, WtE methods). And the reverse direction also holds true.

In relation to the slowness-quickness principle, if there is slowness (e.g., the relatively slower speed of "municipal solid wastes" to decompose naturally), there is quickness (e.g., the relatively quicker speed of "municipal solid wastes" to be dealt with by incinerators instead). And the reverse direction also holds true.

In relation to the expansion-contraction principle, if there is expansion (e.g., the relative expansion of the use of some new types of incinerators nowadays, which "reduce the volume of the original waste by 95-96%"), there is contraction (e.g., the relative decline nowadays of the use of the older types of "incineration plants, which neither recovered energy nor materials"). (WK 2012h) And the reverse direction also holds true.

In relation to the theory-praxis principle, if there is theory (e.g., the theoretical construction of "thermal efficiency" in thermodynamics), there is praxis (e.g., the practical application of "thermal efficiency" in thermodynamics to the field of waste treatment for WtE, but "thermal efficiency" is limited in real life, because "real engines have many departures from ideal behavior that waste energy, reducing actual efficiencies far below the theoretical values" which are possible only in ideal situations). And the reverse direction also holds true.

In relation to the convention-novelty principle, if there is convention (e.g., the conventional wisdom of using incineration for WtE), there is novelty (e.g., the alternative novel challenge to the conventional wisdom of using incineration for WtE, by the new idea of using alternative technologies for WtE, like "thermal depolymerization" and "plasma arc gasification"). And the reverse direction also holds true.

In relation to the evolution-transformation principle, if there is evolution (e.g., the natural evolution of organic wastes to decompose naturally in the state of nature), there is transformation (e.g., the technical transformation of wastes by the man-made invention of WtE methods to decompose wastes for energy creation). And the reverse direction also holds true.

In relation to the symmetry-asymmetry principle, if there is symmetry (e.g., the existence of different WtE methods), there is asymmetry (e.g., "incineration…is the most common WtE implementation"). And the reverse direction also holds true.

In relation to the softness-hardness principle, if there is softness (e.g., the appreciation of incineration by its proponents), there is hardness (e.g., the critique of incineration, as shown in the problems pointed out by the critics as cited). And the reverse direction also holds true.

In relation to the seriousness-playfulness principle, if there is seriousness (e.g., the serious business of those who treat wastes by WtE), there is playfulness (e.g., the playful part of those who treat wastes by WtE, when they play around with different options over time, as shown in the list of WtE technologies aforecited). And the reverse direction also holds true.

In relation to the activeness-inactivess principle, if there is activeness (e.g., the activeness of those wastes which can be recycled or converted to energy), there is inactiveness (e.g., the relative inactiveness of those wastes which cannot be reused, recycled, or converted to energy, since "unusable material" is often "dumped in landfill areas" instead).

In relation to the selfness-otherness principle, if there is selfness (e.g., the tendency to be mostly centered on one's self or group, as shown by those firms which found it profitable to use the older types of incinerators, even when they released pollutants), there is otherness (e.g., the tendency to be mostly centered on others, as shown by those in academic research studies nowadays who reveal the polluting effects of the older types of incinerators, for an advancement of the understanding of environmental sustainability). And the reverse direction also holds true.

In relation to the regression-progression principle, if there is regression (e.g., the regression made by WtE methods, as shown in the problems pointed out by the critics), there is progression (e.g., the progress made by WtE methods, as shown in their contribution to converting wastes into energy aforecited). And the reverse direction also holds true.

In relation to the same-difference principle, if there is similarity in outcome (e.g., the contribution to the molding and control of beliefs, values, and behaviors, regardless of whether this be done by way of letting organic wastes to decompose naturally, or by way of using incinerators to convert wastes into energy), there is difference in outcome (e.g., the contribution to the molding and control of beliefs, values, and behaviors by way of letting organic wastes to decompose naturally for a relatively more naturalistic lifeworld—but the contribution to the molding and control of beliefs, values, and behaviors by way of using incinerators to convert wastes into energy for a relatively more technophilic lifeworld). And the reverse direction also holds true.

In relation to the stability-reaction principle, if there is stability (e.g., the relatively more stable practice of using the old types of incinerators in the older days, although "they released acidic pollutants and at times "neither recovered energy nor materials"), there is reaction (e.g., the use of the old types of incinerators then led to further changes later on, when nowadays the industry depends on more environment-friendly incinerators, which use "lime scrubbers and electro-static precipitators on smokestacks," so that "by passing the smoke through the lime scrubbers, any acids that may be in the smoke are neutralized which prevents the acid from reaching the atmosphere and hurting our environment"). (WK 2012h) And the reverse direction also holds true.

In relation to the functionality-nonfunctionality principle, if there is functionality (e.g., the presence of action for which something is fitted, as shown in the use of the old types of incinerators in the older days, even though they were polluting the environment, because they served the function of being a technological solution for WtE at the time), there is nonfunctionality (e.g., the relative lack of action for which something is fitted, as shown in the use of more environment-friendly incinerators nowadays, which no longer pollute as much as the older types, which are no longer functional as a technological solution for WtE at our time, because of higher standards and expectations). And the reverse direction also holds true.

And in relation to the intentionality-nonintentionality principle, if there is intentionality (e.g., the planning or design of something for a certain outcome, as shown in the design of WtE for energy recovery), there is nonintentionality (e.g., the relative absence of planning or design of something for a certain outcome, such that the outcome can even be counter-intentional in being the opposite of what was originally planned, as shown in the unintended consequences that the design of WtE can end up destroying valuable resources, that it can actually be polluting, and that its efficiency is essentially limited in practice). And the reverse direction also holds true.

Usefulness, Waste, and Animal Products

Another good way to evaluate the possibility (or impossibility) and desirability (or undesirability) of usefulness from the perspective of nature is to explore usefulness, waste, and animal products.

The term "animal product" refers "any material derived from the body of a non-human animal. Examples are fat, flesh, blood, milk, eggs, and lesser known products such as isinglass and rennet." (WK 2012ee; N. Unklesbay 1992)

A related (though not identical) term is "animal by-product," which refers to "animal carcasses and parts of carcasses from slaughterhouses, animal shelters, zoos and veterinarians, and products of animal origin not intended for human consumption, including catering waste (all waste food from restaurants, catering facilities, central kitchens, slaughterhouses and household kitchens). These products go through a process known as 'rendering' to be made into human and non-human foodstuffs, fats, and other material that can be sold to make commercial products such as cosmetics, paint, cleaners, polishes, glue, soap and ink." (WK 2012ee)

However, these two terms are not "applied to products made from fossilized or decomposed animals, such as petroleum, which is formed from the ancient remains of marine animals. Crops grown in soil fertilized with animal remains are rarely characterized as animal products." (WK 2012ee)

Use of Meat by-Products

One way to use animal products or animal-by-products is to make good use of "meat by-products," especially for animal food (like pet food).

"Meat by products" here are "parts of slaughtered animals that don't include meat. These include lungs, spleen, kidneys, brain, liver, blood, bone, and stomach and intestines freed of their contents. It does not include hair, horns, teeth, or hooves." (WK 2012ee)

In accordance to the Association of American Feed Control Officials (AAFCO), "meat by-products" are "the non-rendered, clean parts, other than meat, derived from slaughtered mammals. It includes, but is not limited to, lungs, spleen, kidneys, brain, livers, blood, bone, partially defatted low temperature fatty tissue, and stomachs and intestines freed of their contents. It does not include hair, horns, teeth and hoofs. It shall be suitable for use in animal food." (WK 2012ee)

However, there are some new types of pet food which do not include meat by-products, as they claim "the use of human-grade meat sources only, with no animal meat by-products." (WK 2012ee)

But "AAFCO has declared that using the term 'human grade' is 'false and misleading,' to quote AAFCO in a letter sent to pet food manufacturers in (March 2004)"; for instance, in "Section IV–Pet Food Label Claims," AAFCO stated that "claims that a product contains or is made from ingredients that are 'human grade', 'human quality,' 'people foods', 'ingredients you (the purchaser) would eat,' 'food(s) that you (the purchaser) would feed your family' or similar claims are false and misleading…." (WK 2012ee)

Use of Slaughter Waste

Another way to use animal products or animal-by-products is to make good use of "carcasses," especially the "solid ouputs" (like "compost" and "digestate") from "composting and biogas plants (anaerobic digesters)," as "a fertilizer and soil amendment." (WK 2012ff)

A good example is "slaughterhouse waste," which refers to "animal body parts cut off in the preparation of carcasses for use as food. This waste can come from several sources, including slaughterhouses, restaurants, stores and farms….In the UK, slaughterhouse waste is classed as category 3 risk waste in the Animal By-Products Regulations, with the exception of condemned meat which is classed as category 2 risk." (WK 2012ee)

These "Animal By-Products Regulations (No 1774/2002) (SI 1482/2003)" permit "the treatment of some animal by-products in composting and biogas plants (anaerobic digesters)" and "describe…procedures required to allow solid outputs (compost, digestate) from composting plants and anaerobic digesters onto land in the United Kingdom." (WK 2012ff)

Types of Food

In general, animal products (and animal by-products) can be used for different types of food, as shown below: (WK 2012ee)

- Ex: "blood, especially in the form of blood sausage"
- Ex: "bone, including bone char, bone meal, etc."
- Ex: "carmine also known as cochineal (food dye)"
- Ex: "casein (found in milk and cheese)"
- Ex: "civet oil (food flavoring additive)"
- Ex: "dairy products (e.g., milk, cheese, yogurt, etc.)"
- Ex: "eggs"
- Ex: "honey"
- Ex: "honeydew (secretion)"
- Ex: "lard"
- Ex: "meat (including fish, poultry, and game)"
- Ex: "rennet (commonly used in the production of cheese)"
- Ex: "swiftlet's nest (made of saliva)"
- Ex: "whey (found in cheese and added to many other products)"
- Ex: "broths and stocks are often created with animal fat, bone, and connective tissue"
- Ex: "L-cysteine from human hair and pig bristles (used in the production of biscuits and bread)"

Surely, there can be other examples, so the ones above are illustrative.

Types of Non-Foodstuff

Besides foodstuff, animal products (and animal by-products) can also be used for different types of non-foodstuff, as shown below: (WK 2012ee)

- Ex: "animal fiber"
- Ex: "beeswax"
- Ex: "ejaculate (used in artificial insemination)"
- Ex: "feathers"
- Ex: "fur"
- Ex: "gallstones (from livestock for Traditional Chinese Medicine)"
- Ex: "horn, including antlers, ivory, tusks, etc."
- Ex: "ivory"
- Ex: "leather"
- Ex: "manure"
- Ex: "pearl or mother of pearl"
- Ex: "silk"
- Ex: "sponges"
- Ex: "tortoiseshell"
- Ex: "urine"
- Ex: "venom (used to produce human and veterinary antivenin)"
- Ex: "wool"
- Ex: "blood and some blood substitutes (blood used for transfusions is always human in origin, though some blood substitutes are made from animal sources. Many diagnostic laboratory tests use animal or human sourced reagents)"
- Ex: "casein (used in plastics, clothing, cosmetics, adhesives and paint)"
- Ex: "castoreum (secretion of the beaver used in perfumes and possibly in food flavoring)"

Again, there can be other examples, so the ones above are illustrative.

Problems with Animal Products

However, for the critics, animal products (and animal by-products) have some major problems. Consider, for illustration, a few examples below.

Firstly, one main criticism of animal products (and animal by-products) is that the use of animal products face competition from alternative sources.

For example, in the case of animal by-products, their sale by the meat industry faces economic competition from "industries selling sources of vegetable protein." (WK 2012ee; H. Ockerman 2000)

Secondly, another main criticism of animal products (and animal by-products) is that their use is prohibited by certain groups for different moral, religious, and secular reasons.

For instance, "several human diets prohibit the inclusion of some non-human animal products, including vegetarian, kosher, and halaal. Other diets, such as veganism and the raw vegan diet, exclude any material of non-human animal origin." (WK 2012ee; J. Stepaniak 2000)

And thirdly, still another main criticism of animal products (and animal by-products) is that their use is also not recommended for health reason.

For instance, nowadays, "a new category of pet food typically marketed as holistic, wellness, organic, ultra healthy, and/or simply premium pet food often emphasizes the use of human-grade meat sources only, with no animal meat by-products." (WK 2012ee)

Beyond Animal Products

These criticisms of animal products (and animal by-products), however, do not imply a total rejection of their validity but serve to show us the two opposing sides of the debate, such that the possibility (or impossibility) and desirability (or undesirability) of usefulness (from the perspective of nature with usefulness, waste, and animal products as a case study here) are not to the extent that the respective defenders would like us to believe.

In addition, the analysis of usefulness, waste, and animal products reveals the important insights about the ontological principles in existential dialectics, and good examples include the formalness-informalness principle, the absoluteness-relativeness principle, the partiality-totality principle, the predictability-unpredictability principle, the explicability-inexplicability principle, the fiction-reality principle, the finiteness-transfiniteness principle, the preciseness-vagueness principle, the simpleness-complicatedness principle, the openness-hiddenness principle, the denseness-emptiness principle, the change-constancy principle, the order-chaos principle, the slowness-quickness principle, the expansion-contraction principle, the theory-praxis principle, the convention-novelty principle, the evolution-transformation principle, the symmetry-asymmetry principle, the softness-hardness principle, the seriousness-playfulness principle, the activeness-inactivess principle, the selfness-otherness principle, the regression-progression principle, the same-difference principle, the stability-reaction principle, the functionality-nonfunctionality principle, and the intentionality-nonintentionality principle.

For instance, in relation to the formalness-informalness principle, if there is formalness (e.g., the formal logical requirement of soundness and completeness in a system of ideas, as in the use of animal products and animal by-products), there is informalness (e.g., the non-formal existence of unsoundness and incompleteness in the use of animal products and animal by-products, as shown in the criticisms that the use of animal products face competition from alternative sources, that their use is prohibited by certain groups for different moral and religious reasons, that their use is also not recommended for health reason, etc.). And the reverse direction also holds true.

In relation to the absoluteness-relativeness principle, if there is absoluteness (e.g., the absolute view of certain groups for vegetarian, kosher, and halaal diets against the inclusion of some non-human animal products), there is relativeness (e.g., what is right for certain groups like those for vegetarian, kosher, and halaal diets in regard to their prohibition of the inclusion of some non-human animal products is not necessarily so for others who consumer foods based on animal products and animal by-products). And the reverse direction also holds true.

In relation to the partiality-totality principle, if there is partiality (e.g., the partial view of certain groups for vegetarian, kosher, and halaal diets against the inclusion of some non-human animal products), there is totality (e.g., the more holistic view of the future of non-human animal products, such that the whole is not the sum of its parts, that is, it cannot be reduced to the sum of opposing views of certain groups for vegetarian, kosher, and halaal diets and those for the inclusion of some non-human animal products, since there will emerge new views in the future not yet known today, just as there are already those nowadays who do not side with either side completely). And the reverse direction also holds true.

In relation to the predictability-unpredictability principle, if there is predictability (e.g., the predictable tendency of AAFCO to claim that "using the term 'human grade' is 'false and misleading'" in the foreseeable future), there is unpredictability (e.g., the more difficult task to predict exactly the extent to which the term "human grade" will not be used "on websites or other advertising," because "the AAFCO only governs the pet food packaging and has no authority over websites or advertising," although things may change later). And the reverse direction also holds true.

In relation to the explicability-inexplicability principle, if there is explicability (e.g., the explanation by the advocates of animal products that they are desirable for food consumption), there is inexplicability (e.g., the lack of sufficient explanation by the advocates of animal products of why they are necessarily desirable, because, for the critics, their use is prohibited by certain groups for different moral and religious reasons, like those for vegetarian, kosher, and halaal diets only). And the reverse direction also holds true.

In relation to the fiction-reality principle, if there is fiction (e.g., the fictional aspect of certain new types of pet food which claim "the use of human-grade meat sources only, with no animal meat by-products," because "AAFCO has declared that using the term 'human grade' is 'false and misleading'"), there is reality (e.g., the realistic aspect of certain new types of pet food, insofar as they are useful for the advocates of pet food consumption). And the reverse direction also holds true.

In relation to the finiteness-transfiniteness principle, if there is finiteness (e.g., the finite number of the categories of foodstuffs and non-foodstuffs from animal products and animal by-products as listed earlier), there is transfiniteness (e.g., the transfinite number of all the individual pieces of foodstuffs and non-food stuffs from animal products and animal by-products in history that have ever existed). And the reverse direction also holds true.

In relation to the preciseness-vagueness principle, if there is preciseness (e.g., the precise identification of 2 major uses of animal products and animal by-products, namely, pet food and soil amendment, in the article aforecited), there is vagueness (e.g., the vagueness in the identification of the 2 major uses of animal products and animal by-products, namely, pet food and soil amendment, in the article aforecited, since it is not clear why there must be only 2, not 3, 4, 5, and so on). And the reverse direction also holds true.

In relation to the simpleness-complicatedness principle, if there is simpleness (e.g., the relatively simple analysis of animal products and animal by-products to be used for foodstuffs and non-foodstuffs by the advocates), there is complicatedness (e.g., the relatively more complicated analysis of animal products and animal by-products to be used for foodstuffs and non-foodstuffs, by challenging the claims and assumptions, as shown in the problems pointed out by the critics as cited). And the reverse direction also holds true.

In relation to the openness-hiddenness principle, if there is openness (e.g., the open exploration of selling animal products and animal by-products by the meat industry), there is hiddenness (e.g., the hidden bias in the sale of animal products and animal by-products by the meat industry, because their sale by the meat industry gives it an advantage "to compete economically with industries selling sources of vegetable protein"). And the reverse direction also holds true.

In relation to the denseness-emptiness principle, if there is denseness (e.g., the relatively denser concentration of foodstuffs from animal products among those in mainstream society who do not mind to consume non-human animal diets), there is emptiness (e.g., the relatively less dense, or more empty, concentration of foodstuffs from animal products among those for vegetarian, kosher, and halaal diets only). And the reverse direction also holds true.

In relation to the change-constancy principle, if there is change (e.g., the ever new ways to deal with waste, as shown in the use of animal products and animal by-products), there is constancy (e.g., the ever constant existence of different problems in the ever new ways to deal waste, as shown in the problems with the use of animal products and animal by-products pointed out by the critics aforecited). And the reverse direction also holds true.

In relation to the order-chaos principle, if there is order (e.g., the more or less orderly appearance of animal products and animal by-products, if looked only from the vantage point of their advocates for food consumption), there is chaos (e.g,. the more or less chaotic appearance of animal products and animal by-products, if looked from the multiple often conflicting vantage points of different groups in regard to the pros and cons of using them for food consumption, as well as the different uses for foodstuffs and non-foodstuffs aforediscussed). And the reverse direction also holds true.

In relation to the slowness-quickness principle, if there is slowness (e.g., the relatively slower speed of animal by-products to be used for foodstuffs because they have to go through the process of "rendering," which is time-consuming), there is quickness (e.g., the relatively quicker speed of animal products to be used for foodstuffs because they do not need to go through the process of "rendering" as much). And the reverse direction also holds true.

In relation to the expansion-contraction principle, if there is expansion (e.g., the relative spread of the use of food packaging nowadays, as shown by the AAFCO regulations in regard to pet food packaging), there is contraction (e.g., the relative decline of the practice of producing products without a lot of detailed info about them in antiquity). And the reverse direction also holds true.

In relation to the theory-praxis principle, if there is theory (e.g., the theoretical understanding of "heating" and "drying" in physics and chemistry), there is praxis (e.g., the practical application of the understanding of "heating" and "drying" in physics and chemistry to the field of meat industry, like the process of "rendering"). (WK 2012gg) And the reverse direction also holds true.

In relation to the convention-novelty principle, if there is convention (e.g., the conventional wisdom of using animal products and animal by-products as pet food), there is novelty (e.g., the alternative novel challenge to the conventional wisdom of using animal products and animal by-products as pet food, by the new idea of making a new category of pet food which "emphasizes the use of human-grade meat sources only, with no animal meat by-products," although AAFCO found this claim misleading). And the reverse direction also holds true.

In relation to the evolution-transformation principle, if there is evolution (e.g., the natural evolution of humans to consume animal products and animal by-products in the state of nature in a primitive way, like foraging), there is transformation (e.g., the technical transformation of human ability to deal with animal products and animal by-products by the invention of "rendering"). (WK 2012gg) And the reverse direction also holds true.

In relation to the symmetry-asymmetry principle, if there is symmetry (e.g., the existence of different intelligent life forms, like animals and humans), there is asymmetry (e.g., humans can use animal products and animal by-products for foodstuffs and non-foodstuffs in a way that animals cannot do the same to humans, at least not to the same extent). And the reverse direction also holds true.

In relation to the softness-hardness principle, if there is softness (e.g., the appreciation of animal products and animal by-products by the advocates of their use for foodstuffs and non-foodstuffs), there is hardness (e.g., the critique of the use of animal products and animal by-products for foodstuffs and non-foodstuffs, as shown in the problems pointed out by the critics as cited). And the reverse direction also holds true.

In relation to the seriousness-playfulness principle, if there is seriousness (e.g., the serious business of those who use animal products and animal by-products), there is playfulness (e.g., the playful part of those who use animal products and animal by-products, when they play around with numerous ways to use them for foodstuffs and non-foodstuffs over time, as shown earlier). And the reverse direction also holds true.

In relation to the activeness-inactivess principle, if there is activeness (e.g., the activeness of those who can convert certain animal products and animal by-products into foodstuffs and non-foodstufs), there is inactiveness (e.g., the relative inactiveness of those who cannot convert certain animal products and animal by-products into foodstuffs and non-foodstufs, although they can do something else altogether).

In relation to the selfness-otherness principle, if there is selfness (e.g., the tendency to be mostly centered on one's self or group, as shown in the sale of animal products and animal by-products by the meat industry, which "allows the meat industry to compete economically with industries selling sources of vegetable protein"), there is otherness (e.g., the tendency to be mostly centered on others, as shown in the work by academic researchers to show the problems with the use of animal products and their critics, for a better understanding of the complicated issues facing the meat industry, that is, for the advancement of knowledge). And the reverse direction also holds true.

In relation to the regression-progression principle, if there is regression (e.g., the regression made by the use of animal products and animal by-products, as shown in the problems pointed out by the critics), there is progression (e.g., the progress made by the use of animal products and animal by-products, as shown in their contribution to the creation of foodstuffs and non-foodstuffs aforecited). And the reverse direction also holds true.

In relation to the same-difference principle, if there is similarity in outcome (e.g., the contribution to the molding and control of beliefs, values, and behaviors, regardless of whether this be done by way of the vegetarian, kosher, and halaal diets, or by way of mainstream non-human animal diets), there is difference in outcome (e.g., the contribution to the molding and control of beliefs, values, and behaviors by way of the vegetarian, kosher, and halaal diets for a relatively more spiritual lifeworld—but the contribution to the molding and control of beliefs, values, and behaviors by way of the mainstream non-human animal diets for a relatively more secular lifeworld). And the reverse direction also holds true.

In relation to the stability-reaction principle, if there is stability (e.g., the relatively more stable condition of the practice of using animal by-products for pet food in the older days), there is reaction (e.g., the practice of using animal by-products for pet food in the older days then led to further changes later on, when nowadays, there are "a new category of pet food," which "often emphasizes the use of human-grade meat sources only, with no animal meat by-products"). And the reverse direction also holds true.

In relation to the functionality-nonfunctionality principle, if there is functionality (e.g., the presence of action for which something is fitted, as shown in the use of animal products for foodstuffs, since this serves the vital function of food consumption for those involved), there is nonfunctionality (e.g., the relative lack of action for which something is fitted, as shown in the prohibition of using "some non-human animal products" by those for vegetarian, kosher, and halaal diets, so that these non-human animal products can no longer serve the function of food consumption for these other groups, because of their different standards and expectations, mostly for spiritual and religious reasons). And the reverse direction also holds true.

And in relation to the intentionality-nonintentionality principle, if there is intentionality (e.g., the planning or design of something for a certain outcome, as shown in the design of animal products and animal by-products to make better use of animal wastes), there is nonintentionality (e.g., the relative absence of planning or design of something for a certain outcome, such that the outcome can even be counter-intentional in being the opposite of what was originally planned, as shown in the facts that the use of animal products and animal by-products is actually prohibited by certain groups for different moral and religious reasons and that their use is also not recommended for health reason, for instance). And the reverse direction also holds true.

Usefulness and Society

Usefulness, when examined from the perspective of society, can reveal to us its possibility (or impossibility) and desirability (or undesirability), and this can be shown by way of two case studies, namely, (3.4.1) usefulness, waste recycle, and the cost-benefit analysis, and (3.4.2) usefulness, the waste converter, and technology—to be addressed hereafter.

Usefulness, Waste Recycle, and the Cost-Benefit Analysis

A good way to evaluate the possibility (or impossibility) and desirability (or undesirability) of usefulness from the perspective of society is to explore usefulness, waste recycle, and the cost-benefit analysis.

The term "recycle" refers to the "processing" of "used materials (waste) into new products to prevent waste of potentially useful materials, reduce the consumption of fresh raw materials, reduce energy usage, reduce air pollution (from incineration) and water pollution (from landfilling) by reducing the need for 'conventional' waste disposal, and lower greenhouse gas emissions as compared to virgin production." (WK 2012g)

However, the business of waste recycle is not without controversy, because there is the cost-benefit debate over whether there are more costs than benefits (or more benefits than costs) in waste recycle—or simply "whether recycling is economically efficient." (WK 2012g)

Consider, for illustration, some main areas of disagreement between the proponents and opponents of waste recycle below (as summarized in *Sec. 3.4*).

1st Disagreement: Recycling vs. Other Methods

The first disagreement between the proponents and opponents of waste recycle has to do with recycling vs. other methods.

On the positive side, "municipalities often see fiscal benefits from implementing recycling programs, largely due to the reduced landfill costs. A study conducted by the Technical University of Denmark found that in 83 percent of cases, recycling is the most efficient method to dispose of household waste." (WK 2012g; D. Lavee 2007; ECO 2007 & 2007a)

And on the negative side, "a 2004 assessment by the Danish Environmental Assessment Institute concluded that incineration was the most effective method for disposing of drink containers, even aluminium ones." (WK 2012g; D. Vigso 2004)

2nd Disagreement: Economic vs. Fiscal Measurement

The second disagreement between the proponents and opponents of waste recycle has to do with economic vs. fiscal measurement.

Fiscal measurement does not reveal "negative externalities," like pollution, but economic measurement does. In other words, "economic analysis of recycling includes what economists call externalities, which are unpriced costs and benefits that accrue to individuals outside of private transactions." (WK 2012g)

Good examples of non-fiscal benefits are "decreased air pollution and greenhouse gases from incineration, reduced hazardous waste leaching from landfills, reduced energy consumption, and reduced waste and resource consumption, which leads to a reduction in environmentally damaging mining and timber activity. About 4000 minerals are known, of these only a few hundred minerals in the world are relatively common. At current rates, current known reserves of phosphorus will be depleted in the next 50 to 100 years. Without mechanisms such as taxes or subsidies to internalize externalities, businesses will ignore them despite the costs imposed on society." (WK 2012g; UML 2012; SCIA 2009; REAS 2010)

In fact, "to make such non-fiscal benefits economically relevant, advocates have pushed for legislative action to increase the demand for recycled materials. The United States Environmental Protection Agency (EPA) has concluded in favor of recycling, saying that recycling efforts reduced the country's carbon emissions by a net 49 million metric tonnes in 2005. In the United Kingdom, the Waste and Resources Action Programme stated that Great Britain's recycling efforts reduce CO2 emissions by 10–15 million tonnes a year. Recycling is more efficient in densely populated areas, as there are economies of scale involved." (WK 2012g; LEAG 1993; ECO 2007)

3rd Disagreement: The Amount of Energy Saved

The third disagreement between the proponents and opponents of waste recycle has to do with whether the amount of energy saved in recycling is more than the amount of energy spent in the process.

In general, "the amount of energy saved through recycling depends upon the material being recycled and the type of energy accounting that is used. Emergy (spelled with an M) analysis, for example, budgets for the amount of energy of one kind (exergy) that is required to make or transform things into another kind of product or service." (WK 2012g)

For instance, "one emergy life-cycle analysis revealed that fly ash, aluminum, recycled concrete aggregate, recycled plastic and steel yield higher efficiency ratios, whereas the recycling of lumber generates the lowest recycle benefit ratio. Hence, the specific nature of the recycling process, the methods used to analyse the process, and the products involved effects the energy savings budgets." (WK 2012g; M. Brown 2003)

To make the matter worse, "it is difficult to determine the exact amount of energy consumed or produced in waste disposal processes. How much energy is used in recycling depends largely on the type of material being recycled and the process used to do so." (WK 2012g)

4th Disagreement: The Amount of Money Saved

The fourth disagreement between the proponents and opponents of waste recycle has to do with the amount of money saved.

In general, "the amount of money actually saved through recycling depends on the efficiency of the recycling program used to do it. The Institute for Local Self-Reliance argues that the cost of recycling depends on various factors around a community that recycles, such as landfill fees and the amount of disposal that the community recycles. It states that communities start to save money when they treat recycling as a replacement for their traditional waste system rather than an add-on to it and by 'redesigning their collection schedules and/or trucks.'" (WK 2012g)

However, in other cases, "the cost of recyclable materials also exceeds the cost of raw materials. Virgin plastic resin costs 40 percent less than recycled resin. Additionally, a United States Environmental Protection Agency (EPA) study that tracked the price of clear glass from July 15 to August 2, 1991, found that the average cost per ton ranged from $40 to $60, while a USGS report shows that the cost per ton of raw silica sand from years 1993 to 1997 fell between $17.33 and $18.10." (WK 2012g)

In 1996, John Tierney (1996) argued that "it costs more money to recycle the trash of New York City than it does to dispose of it in a landfill. Tierney argued that the recycling process employs people to do the additional waste disposal, sorting, inspecting, and many fees are often charged because the processing costs used to make the end product are often more than the profit from its sale. Tierney also referenced a study conducted by the Solid Waste Association of North America (SWANA) that found in the six communities involved in the study, 'all but one of the curbside recycling programs, and all the composting operations and waste-to-energy incinerators, increased the cost of waste disposal.'" (WK 2012g; A. Dornfeld 2008)

However, the downside of this is that, as pointed out earlier, "comparing the market cost of recyclable material to the cost of new raw materials ignores economic externalities—the costs that are currently not counted by the market." (WK 2012g)

5th Disagreement: Working Conditions

The fifth disagreement between the proponents and opponents of waste recycle has to do with working conditions.

For instance, "critics...argue that while recycling may create jobs, they are often jobs with low wages and terrible working conditions. These jobs are sometimes considered to be make-work jobs that don't produce as much as the cost of wages to pay for those jobs. In areas without many environmental regulations and/or worker protections, jobs involved in recycling such as ship breaking can result in deplorable conditions for both workers and the surrounding communities." (WK 2012g; HEART 2002)

For instance, "the recycling of waste electrical and electronic equipment in India and China generates a significant amount of pollution. Informal recycling in an underground economy of these countries has generated an environmental and health disaster. High levels of lead (Pb), polybrominated diphenylethers (PBDEs), polychlorinated dioxins and furans, as well as polybrominated dioxins and furans (PCDD/Fs and PBDD/Fs) concentrated in the air, bottom ash, dust, soil, water and sediments in areas surrounding recycling sites." (WK 2012g; A. Sepúlveda 2010)

6th Disagreement: Environmental Impact

The sixth disagreement between the proponents and opponents of waste recycle has to do with environmental impact.

For instance, "economist Steven Landsburg [1993], author of a paper entitled 'Why I Am Not an Environmentalist,' has claimed that paper recycling actually reduces tree populations. He argues that because paper companies have incentives to replenish the forests they own, large demands for paper lead to large forests. Conversely, reduced demand for paper leads to fewer 'farmed' forests. Similar arguments were expressed in a 1995 article for The Free Market." (WK 2012g)

However, on the other side of the debate, "many environmentalists point out…that 'farmed' forests are inferior to virgin forests in several ways. Farmed forests are not able to fix the soil as quickly as virgin forests, causing widespread soil erosion and often requiring large amounts of fertilizer to maintain while containing little tree and wild-life biodiversity compared to virgin forests. Also, the new trees planted are not as big as the trees that were cut down, and the argument that there will be 'more trees' is not compelling to forestry advocates when they are counting saplings." (WK 2012g; C. Baird 1995)

7th Disagreement: Income Loss and Social Costs

The seventh disagreement between the proponents and opponents of waste recycle has to do with income loss and social costs.

For instance, "in some prosperous and many less prosperous countries in the world, the traditional job of recycling is performed by the entrepreneurial poor such as the karung guni, Zabaleen, the rag-and-bone man, waste picker, and junk man. With the creation of large recycling organizations that may be profitable, either by law or economies of scale, the poor are more likely to be driven out of the recycling and the remanufacturing market." (WK 2012g; NRDC 1996)

Thus, "to compensate for this loss of income to the poor, a society may need to create additional forms of societal programs to help support the poor. Like the parable of the broken window, there is a net loss to the poor and possibly the whole of a society to make recycling artificially profitable through law. However, as seen in Brazil and Argentina, waste pickers/informal recyclers are able to work alongside governments, in (semi)funded cooperatives, allowing informal recycling to be legitimized as a paying government job." (WK 2012g; PBS 2010; M. Medina 2000)

This then suggests that, "because the social support of a country is likely less than the loss of income to the poor doing recycling, there is a greater chance that the poor will come in conflict with the large recycling organizations." (WK 2012g; R. Gonzales 2008)

In fact, "one labor-intensive underused waste is electronic and computer waste. Because this waste may still be functional and wanted mostly by the poor, the poor may sell or use it at a greater efficiency than large recyclers." (WK 2012g)

8th Disagreement: Public Participation in Recycling Programmes

And the eighth disagreement between the proponents and opponents of waste recycle has to do with "public participation in recycling programmes."

For instance, "between 1960 and 2000, the world production of plastic resins increased 25-fold, while recovery of the material remained below 5%"—and "many studies have addressed recycling behaviour and strategies to encourage community involvement in recycling programmes." (WK 2012g; C. Moore 2008)

But, for the critics, "recycling behaviour is not natural because it requires a focus and appreciation for long term planning, whereas humans have evolved to be sensitive to short term survival goals; and that to overcome this innate predisposition, the best solution would be to use social pressure to compel participation in recycling programmes." (WK 2012g; T. Schackelford 2006)

Yet, "recent studies have concluded that social pressure is unviable in this context. One reason for this is that social pressure functions well in small group sizes of 50 to 150 indiviudals (common to nomadic hunter-gatherer peoples) but not in communities numbering in the millions, as we see today." (WK 2012g; M. Pratarelli 2010)

But on the positive side, "in a study done by social psychologist Shawn Burn [2006], it was found that personal contact with individuals within a neighborhood is the most effective way to increase recycling within a community. In his study, he had 10 block leaders talk to their neighbors and convince them to recycle. A comparison group was sent fliers promoting recycling. It was found that the neighbors that were personally contacted by their block leaders recycled much more than the group without personal contact. As a result of this study, Shawn Burn believes that personal contact within a small group of people is an important factor in encouraging recycling. Another study done by Stuart Oskamp [1995] examines the effect of neighbors and friends on recycling. It was found in his studies that people who had friends and neighbors that recycled were much more likely to also recycle than those who didn't have friends and neighbors that recycled." (WK 2012g)

Beyond the Cost-Benefit Analysis of Waste Recycle

The introduction to these disagreements on the cost-benefit analysis of waste recycle, however, does not presume a total rejection of its validity but shows us the two opposing sides of the debate, such that the possibility (or impossibility) and desirability (or undesirability) of usefulness (from the perspective of society with usefulness, waste recycle, and the cost-benefit analysis as a case study here) are not to the extent that the respective defenders would like us to believe.

In fact, the analysis of usefulness, waste recycle, and the cost-benefit analysis serves as a good exercise about the ontological principles in existential dialectics, and good examples include the formalness-informalness principle, the absoluteness-relativeness principle, the partiality-totality principle, the predictability-unpredictability principle, the explicability-inexplicability principle, the fiction-reality principle, the finiteness-transfiniteness principle, the preciseness-vagueness principle, the simpleness-complicatedness principle, the openness-hiddenness principle, the denseness-emptiness principle, the change-constancy principle, the order-chaos principle, the slowness-quickness principle, the expansion-contraction principle, the theory-praxis principle, the convention-novelty principle, the evolution-transformation principle, the symmetry-asymmetry principle, the softness-hardness principle, the seriousness-playfulness principle, the activeness-inactivess principle, the selfness-otherness principle, the regression-progression principle, the same-difference principle, the stability-reaction principle, the functionality-nonfunctionality principle, and the intentionality-nonintentionality principle.

For instance, in relation to the formalness-informalness principle, if there is formalness (e.g., the formal logical requirement of soundness, consistency, and completeness in a system of ideas, as in the debate on the cost-benefit analysis of waste recycle), there is informalness (e.g., the non-formal existence of unsoundness, inconsistency, and incompleteness in the debate on the cost-benefit analysis of waste recycle, as shown in the criticisms by the critics in regard to the 8 issues afore-discussed). And the reverse direction also holds true.

In relation to the absoluteness-relativeness principle, if there is absoluteness (e.g., the absolute view of those "businesses" which "ignore them [negative externalities] despite the costs imposed on society"), there is relativeness (e.g., what is good for those "businesses" which "ignore them [negative externalities] despite the costs imposed on society" is not necessarily so for the critics, who, "to make such non-fiscal benefits economically relevant,...have pushed for legislative action to increase the demand for recycled materials"). And the reverse direction also holds true.

In relation to the partiality-totality principle, if there is partiality (e.g., the partial view of those "businesses" which "ignore them [negative externalities] despite the costs imposed on society"), there is totality (e.g., the more holistic view of waste recycle, such that the whole is not the sum of its parts, that is, it cannot be reduced to the sum of opposing views of those businesses who ignore negative externalities and their critics for waste recycle, since there will emerge new views in the future not yet known today, just as there are already other views which do not side with the opposing sides nowadays). And the reverse direction also holds true.

In relation to the predictability-unpredictability principle, if there is predictability (e.g., the predictable tendency of those in "the study conducted by the Technical University of Denmark" to argue that "in 83 percent of cases, recycling is the most efficient method to dispose of household waste"), there is unpredictability (e.g., the more difficult task to predict exactly the extent to which recycling is the most efficient method on a particular future occasion, since, in another study in 2004 by "the Danish Environmental Assessment Institute," which "concluded that incineration was the most effective method for disposing of drink containers, even aluminium ones" instead). And the reverse direction also holds true.

In relation to the explicability-inexplicability principle, if there is explicability (e.g., the explanation by the advocates of waste recycle that it is desirable for environmental sustainability), there is inexplicability (e.g., the lack of sufficient explanation by the advocates of waste recycle of why it is necessarily desirable, since, for the critics, "there is a net loss to the poor and possibly the whole of a society," because "traditional job of recycling is performed by the entrepreneurial poor such as the karung guni, Zabaleen, the rag-and-bone man, waste picker, and junk man," so "with the creation of large recycling organizations that may be profitable..., the poor are more likely to be driven out of the recycling and the remanufacturing market"). And the reverse direction also holds true.

In relation to the fiction-reality principle, if there is fiction (e.g., the fictional aspect of the fiscal measurement of waste recycle, insofar as it ignores "negative externalities" like pollution), there is reality (e.g., the realistic aspect of the fiscal measurement of waste recycle, insofar as it includes the costs and benefits of waste recycle for an individual business, but not society). And the reverse direction also holds true.

In relation to the finiteness-transfiniteness principle, if there is finiteness (e.g., the finite number of the categories of "highly dispersed wastes" to be recycled, like "chlorinated organic hydrocarbons that have bioaccumulated in animal and human tissues across the globe, the copper dispersed in fungicides, the lead in widely applied paints, or the zinc oxides present in the finely dispersed rubber powder that is abraded from automobile tires"), there is transfiniteness (e.g., the transfinite number of all the individual pieces of "highly dispersed wastes" in history that have ever existed, like "chlorinated organic hydrocarbons that have bioaccumulated in animal and human tissues across the globe, the copper dispersed in fungicides, the lead in widely applied paints, or the zinc oxides present in the finely dispersed rubber powder that is abraded from automobile tires"). (WK 2012g; M. Huesemann 2003) And the reverse direction also holds true.

In relation to the preciseness-vagueness principle, if there is preciseness (e.g., the precise identification of 8 major disagreements on the cost-benefit analysis of waste recycle in the article aforecited), there is vagueness (e.g., the vagueness in the identification of the 8 major disagreements on the cost-benefit analysis of waste recycle in the article aforecited, since it is not clear why there must be only 8, not 10, 13, 15, and so on). And the reverse direction also holds true.

In relation to the simpleness-complicatedness principle, if there is simpleness (e.g., the relatively simple analysis of waste recycle with the tool of cost-benefit analysis by the advocates), there is complicatedness (e.g., the relatively more complicated analysis of waste recycle with the tool of cost-benefit analysis, by challenging the claims and assumptions, as shown in the problems pointed out by the critics as cited). And the reverse direction also holds true.

In relation to the openness-hiddenness principle, if there is openness (e.g., the open exploration of the costs and benefits of waste recycle for an individual business, by way of fiscal measurement), there is hiddenness (e.g., the hidden bias in fiscal measurement of waste recycle, because it ignores negative externalities). And the reverse direction also holds true.

In relation to the denseness-emptiness principle, if there is denseness (e.g., the relatively denser concentration of "the traditional job of recycling" among "the entrepreneurial poor such as the karung guni, Zabaleen, the rag-and-bone man, waste picker, and junk man"), there is emptiness (e.g., the relatively less dense, or more empty, concentration of "the traditional job of recycling" among the rich). And the reverse direction also holds true.

In relation to the change-constancy principle, if there is change (e.g., the ever new ways to deal with waste, as shown in the cost-benefit analysis of waste recycle), there is constancy (e.g., the ever constant existence of different problems in the ever new ways to deal with waste, as shown in the problems with the cost-benefit analysis of waste recycle pointed out by the critics). And the reverse direction also holds true.

In relation to the order-chaos principle, if there is order (e.g., the more or less orderly appearance of wastes, if looked from the vantage point of a landfill tightly packed with wastes in a small area), there is chaos (e.g,. the more or less chaotic appearance of wastes, if looked from the vantage point of "highly dispersed wastes" like "chlorinated organic hydrocarbons that have bioaccumulated in animal and human tissues across the globe," etc.). And the reverse direction also holds true.

In relation to the slowness-quickness principle, if there is slowness (e.g., the relatively slower speed of wastes to be recycled if they spread out all over the globe in "highly dispersed" and "diluted" forms), there is quickness (e.g., the relatively quicker speed of wastes to be recycled if they concentrate in a landfill). And the reverse direction also holds true.

In relation to the expansion-contraction principle, if there is expansion (e.g., the relative spread of of "large recycling organizations" to do waste recycle in developed countries nowadays), there is contraction (e.g., the relative decline of "the entrepreneurial poor such as the karung guni, Zabaleen, the rag-and-bone man, waste picker, and junk man" to do waste recycle in developed countries nowadays).

In relation to the theory-praxis principle, if there is theory (e.g., the theoretical understanding of "environmental design" to resolve wastes, like waste recycle), there is praxis (e.g., what is conceptual and theoretical is not necessarily practical, because "complete recycling is impracticable as highly dispersed wastes become so diluted that the energy needed for their recovery becomes increasingly excessive). (WK 2012g; M. Huesemann 2003) And the reverse direction also holds true.

In relation to the convention-novelty principle, if there is convention (e.g., the conventional wisdom about waste recycle to be good for the environment), there is novelty (e.g., the alternative novel challenge to the conventional wisdom about waste recycle to be good for the environment, by the new idea of Steven Landsburg, who argued instead that "paper recycling actually reduces tree populations" and is thus bad for the environment). And the reverse direction also holds true.

In relation to the evolution-transformation principle, if there is evolution (e.g., the natural evolution of humans, in that "humans have evolved to be sensitive to short term survival goals"), there is transformation (e.g., the technical transformation of human ability to deal with survival goals by the invention of "recycling behaviour," which "is not natural because it requires a focus and appreciation for long term planning"). (WK 2012g) And the reverse direction also holds true.

In relation to the symmetry-asymmetry principle, if there is symmetry (e.g., the existence of different social classes), there is asymmetry (e.g., the poor like "the karung guni, Zabaleen, the rag-and-bone man, waste picker, and junk man" do more waste recycle on the streets than the rich). And the reverse direction also holds true.

In relation to the softness-hardness principle, if there is softness (e.g., the appreciation of the cost-benefit analysis of waste recycle by the advocates), there is hardness (e.g., the critique of the cost-benefit analysis of waste recycle by the advocates, as shown in the problems pointed out by the critics as cited). And the reverse direction also holds true.

In relation to the seriousness-playfulness principle, if there is seriousness (e.g., the serious business of those who criticize the cost-benefit analysis of waste recycle), there is playfulness (e.g., the playful part of those who criticize the cost-benefit analysis of waste recycle, when they play around with different ways to undermine the cost-benefit analysis over time, as shown in the 6 disagreements over the years aforecited). And the reverse direction also holds true.

In relation to the activeness-inactivess principle, if there is activeness (e.g., the activeness of those in "large recycling organizations" to do waste recyle in developed countries nowadays), there is inactiveness (e.g., the relative inactiveness of those like "the karung guni, Zabaleen, the rag-and-bone man, waste picker, and junk man" who are "driven out of the recycling and the remanufacturing market" in developed countries nowadays and cannot do waste-recycling much). And the reverse direction also holds true.

In relation to the selfness-otherness principle, if there is selfness (e.g., the tendency to be mostly centered on one's self or group, as shown in those "businesses" which "ignore...[negative externalities] despite the costs imposed on society"), there is otherness (e.g., the tendency to be mostly centered on others, as shown by those economists who include "negative externalities" to society in the "economic analysis of recycling," so as to give a better understanding of the economic measurement of waste recycle, for the enhancement of social welfare). And the reverse direction also holds true.

In relation to the regression-progression principle, if there is regression (e.g., the regression made by the cost-benefit analysis of waste recycle, as shown in the problems pointed out by the critics), there is progression (e.g., the progress made by the cost-benefit analysis of waste recycle, as shown in its contribution to a better understanding of efficiency aforecited). And the reverse direction also holds true.

In relation to the same-difference principle, if there is similarity in outcome (e.g., the contribution to the molding and control of beliefs, values, and behaviors, regardless of whether this be done by way of the fiscal analysis of waste recycle, or by way of the economic analysis of waste recycle), there is difference in outcome (e.g., the contribution to the molding and control of beliefs, values, and behaviors by way of the fiscal analysis of waste recycle for a relatively more market-oriented economic life—but the contribution to the molding and control of beliefs, values, and behaviors by way of the economic analysis of waste recycle for a relatively more government-regulated economic life). And the reverse direction also holds true.

In relation to the stability-reaction principle, if there is stability (e.g., the relatively more stable condition of "the traditional job of recycling," which "is performed by the entrepreneurial poor such as the karung guni, Zabaleen, the rag-and-bone man, waste picker, and junk man" in the older days), there is reaction (e.g., "the traditional job of recycling...performed by the entrepreneurial poor" then led to further changes later on, when "large recycling organizations" are doing a lot of waste recycle in developed countries nowadays). And the reverse direction also holds true.

In relation to the functionality-nonfunctionality principle, if there is functionality (e.g., the presence of action for which something is fitted, as shown in "the traditional job of recycling...performed by the entrepreneurial poor such as the karung guni, Zabaleen, the rag-and-bone man, waste picker, and junk man" in many underdeveloped or poor countries, as they serve the essential function of waste recycle), there is nonfunctionality (e.g., the relative lack of action for which something is fitted, as shown in "the creation of large recycling organizations that may be profitable, either by law or economies of scale," in such a way that "the poor are…driven out of the recycling and the remanufacturing market," so that these poor people can no longer serve the essential function of waste recycle in many rich, developed countries nowadays, as they could in the older days). And the reverse direction also holds true.

And in relation to the intentionality-nonintentionality principle, if there is intentionality (e.g., the planning or design of something for a certain outcome, as shown in the design of waste recycle for a more positive environmental impact), there is nonintentionality (e.g., the relative absence of planning or design of something for a certain outcome, such that the outcome can even be counter-intentional in being the opposite of what was originally planned, as shown by the work by Steven Landsburg who showed that, for instance, "paper recycling actually reduces tree populations" and is thus bad for the environment instead). And the reverse direction also holds true.

Usefulness, the Waste Converter, and Technology

Another good way to evaluate the possibility (or impossibility) and desirability (or undesirability) of usefulness from the perspective of society is to explore usefulness, waste converter, and technology.

A major argument here is that, with the help of technology, "discards are separated into materials that may be incorporated into new products. This is different from Reuse in that energy is used to change the physical properties of the material." (WK 2012d)

A good example is the invention of the "waste converter," which is "a machine used for the treatment and recycling of solid and liquid refuse material. A converter is a self-contained system capable of performing the following functions: pasteurization of organic waste; sterilization of pathogenic or biohazard waste; grinding and pulverization of refuse into unrecognizable output; trash compaction; dehydration." (WK 2012p)

Waste Converter vs. Autoclave

However, a "waste converter" should not be confused with an "autoclave"—although "the converter is an evolution of the autoclave, invented by Sir Charles Chamberland in 1879, but differs from a waste autoclave in several key characteristics." (WK 2012p)

For instance, "while the autoclave relies on high temperature and pressure to achieve moist heat sterilization of waste, a converter operates in the atmospheric pressure range. Superheating conditions and steam generation are achieved by variable pressure control, which cycles between ambient and negative pressure within the sterilization cell. The advantage of this updated approach is a safer and less complicated operation that does not require a pressure vessel." (WK 2012p)

Also, "while autoclaves require external water input, modern converters utilize the moisture content already present in the conversion cell to generate steam sterilization conditions. Any water that is introduced into the process can be recycled in a closed-loop system as opposed to being dumped as run-off sewage. In general, the converter is a simplified, cleaner, and more efficient update to Sir Charles's invention." (WK 2012p)

Applications of the Waste Converter

The waste converter has been applied in different ways. Consider, for illustration, a few examples below.

(a) In Waste Treatment Centers

In waste treatment centers, "application of the converter is common," where "large machines process waste on an industrial scale. MSW (Municipal solid waste) or infectious waste, depending on the type of plant, is sterilized and converted into a sterilized organic and inorganic, innocuous end-product. Machines used in such large-scale applications process between 1,000 and 4,000 kg of waste per hour. At the end of each cycle, lasting as little as half-hour in Converters (that are capable of grinding), the pulverized, sanitized, and dehydrated product is off-loaded and segregated for other uses. Some of the product is routed for use in pulp production, composting, or refuse-derived fuel." (WK 2012p)

(b) In Hospitals and Clinics

In hospitals, "modern converters…allow…for the immediate treatment of potentially infected hazardous waste at its source. Hospitals and clinics equipped to have a zero hazardous waste footprint operate by having a converter placed on every floor where single use sanitary items such as needles, scalpels, bandages, and blood bags are immediately converted into innocuous product. In addition to the marked improvement in sanitation, on-site treatment of hazardous waste allows operational cost savings for these facilities. The government of Tuscany, Italy for example calculated an annual figure of 8 Million Euro that was saved by turning to on-site treatment of medical hospital waste." (WK 2012p)

(c) In Supermarkets

In supermarkets and food producers, instead of dumping "unused food waste in municipal landfills at a rate that is alarming many conservationists," there is "a use for converter technology. By processing unused and decomposing food matter together with packaging and other refuse on site, supermarkets have achieved improvements in terms of waste disposal costs. This is in addition to improvements in public perception, which had been seriously critical of the amount of waste sent to landfills by food stores. In the UK alone 6.7 million metric tons of food waste goes into landfills each year, resulting in 8 million metric tons of CO_2 being emitted." (WK 2012p)

(d) In Farms and Slaughterhouses

In farms and slaughterhouses, waste converters are used for "on-site waste conversion. Larger installations especially, where garbage hauling is a major and expensive operation, currently have economic and legislative incentives to move towards operating own converters. Recent drives toward environmentally conscious or 'green' technologies have even provided government budgets for such installations." (WK 2012p)

(e) On Ships

On ships like "naval vessels, cruise liners and off-shore installations such as gas-drilling rigs and oil platforms," there is "another logical application of converter technology. Due to the extended isolation periods of sea-going vessels and off-shore platforms there is an issue of how to store and dispose of refuse in an efficient and sanitary way. World-wide legislation on sea dumping is strict and does not allow, under stringent penalties, any ships or sea vessels to dump waste, gray water, or even ballast water that has been collected in a remote geographic location due to the danger of biological contamination. Ship-generated waste is either held and disposed of in port waste disposal facilities, or can be converted directly on the vessel for easier storage and at times (depending on waste composition) for additional fuel." (WK 2012p)

Operation of the Waste Converter

The operation of a waste converter can be understood in terms of "a typical treatment cycle," which "will begin with the loading of unsorted waste material and end in the offload of a dry powder (product), which now possesses new characteristics and that the input material did not." (WK 2012p)

Nowadays, "converters are fully automatic and will finish the computer controlled cycle autonomously, unless a failure occurs. The precise conditions needed to achieve pasteurization and sterilization are controlled by a programmable logic controller (PLC) in millisecond intervals. This level of control in modern units has allowed for the simplification of autoclave technology to the point where heavy and potentially dangerous pressure vessels are not needed, and sterilization conditions are reached by depressing the conversion cell and continuing to evaporate moisture from the product under negative pressure. The result is a statistically safer and more reliable machine, which is also smaller and lighter than older autoclaves." (WK 2012p)

In general, "the conversion cycle consists of several sequential steps or phases," as shown below: (WK 2012p)

- "The waste is first ground and pulverized to an unrecognizable mixture by a combination of fixed and actuated hardened steel blades."
- "The mixture is then heated through the injection of steam and also by the heat generated by frictional forces of the grinding phase."
- "The exact temperature required to pasteurize, and in the subsequent phase to sterilize the waste, is maintained for a time that allows for a 18 log 10 reduction in microorganisms. In order to eliminate the required amount of microorganisms required by government regulations, a complete saturation of waste matter with superheated steam is required for a minimum amount of time, also regulated by environmental agencies. The modern converter achieves saturation within 10–15 minutes due to the high degree of pulverization preceding the sterilization phase, whereas older models required up to several hours to saturate and sterilize the same load."
- "The cycle ends in a cooling phase, during which product continues to be dehydrated. Upon reaching temperature at which product is safe to handle, near ambient temperature, the cycle automatically shuts down."
- "The end product is expelled into a tray that can then be hauled off for storage. The entire process and statistics are recorded and stored in computer memory for record keeping."

Surely, there may be other phases, so the ones above are illustrative.

Environmental Impact of the Waste Converter

The waste converter is regarded as "one of the 'green technologies' available today for waste treatment," because of its "clear and definite positive environmental impact stemming from the use of waste conversion into biofuel, building material, and soil compost." (WK 2012p)

For instnace, "in 2009, ever-increasing numbers of waste exporters around the world are finding it difficult to find buyers for their cargo. Increasing numbers of local and national governments are also turning to recycling and conversion technology to relieve the pressure on already-full or overfilled landfills. Waste conversion, augmented by traditional recycling methods, now allows nearly 99% of all MSW to be reused in some way, thus sharply reducing the demand on landfills." (WK 2012p)

By contrast, "in 1980, only about 10% on municipal solid waste was recycled, and the product consisted largely of paper and glass recycled material. With the widespread use of autoclaves, that percentage climbed to a significant 45% by the year 2000, when composting and energy recovery became more common. With the evolution of the autoclave and the arrival of converters new uses of converted product are emerging on an ongoing basis." (WK 2012p)

Good examples of "some application only recently implemented are: Composting and combining with farm fertilizer, building material such as concrete additive, gasification fuel, and furnace/boiler pellet additive. The latter two are energy recovery options that only became acceptable on a large scale once the product was demonstrated to burn cleanly and within EPA emission regulations. A notable fact about energy recovery is that even though some waste was already reused in this way as early as 1980, the current generation of converters produces Biomass Fuel that burns exponentially cleaner than the incinerated waste of those times." (WK 2012p)

Waste converters are more "environmentally conscious," because they are "based on the lessons learned from older technologies. The new machines run on much less power than the large pressure vessel type autoclaves, and can be plugged into 400V power supplies or run off of a small motor as stand-alone units. As a result of this, a new degree of portability became possible, and now facilities such as hospitals are placing dish-washer sized units in every department. One important advent that allowed for a leaner, green technology is the ability of modern converters to transfer mechanical energy and friction force on the waste mass into heat energy that is used in the pasteurization and sterilization processes." (WK 2012p)

Problems with the Use of Waste Converter

However, for the critics, the use of waste converter has some major problems. Consider, for illustration, a few examples below.

Firstly, one main criticism of the use of waste converter is that its application is limited.

For instance, for the critics, "much of the difficulty inherent in recycling comes from the fact that most products are not designed with recycling in mind." (WK 2012g)

Secondly, another main criticism of the use of waste converter is that there are competing technologies.

For instance, for the critics, "waste converters do have indirect competition since there are many modes of waste disposal available in the market," and good examples are "plasma-arc gasification," "micro-wave and other irradiation sterilization," and new versions of "autoclaves." (WK 2012p)

And thirdly, still another main criticism of the use of waste converter is that it is not necessarily safer.

For instance, for the critics, some contemporary autoclaves have been so improved that they "employ the same method of sterilization as converters, but do so using larger and more complex equipment that also has higher safety… requirement…." (WK 2012p)

Beyond the Waste Converter

These criticisms of the waste converter, however, should not be misunderstood as a total rejection of its validity but serve to show us the two opposing sides of the debate, such that the possibility (or impossibility) and desirability (or undesirability) of usefulness (from the perspective of society with usefulness, the waste converter, and technology as a case study here) are not to the extent that the respective defenders would like us to believe.

Also, the analysis of usefulness, the waste converter, and technology can teach us something important about the ontological principles in existential dialectics, and good examples include the formalness-informalness principle, the absoluteness-relativeness principle, the partiality-totality principle, the predictability-unpredictability principle, the explicability-inexplicability principle, the fiction-reality principle, the finiteness-transfiniteness principle, the preciseness-vagueness principle, the simpleness-complicatedness principle, the openness-hiddenness principle, the denseness-emptiness principle, the change-constancy principle, the order-chaos principle, the slowness-quickness principle, the expansion-contraction principle, the theory-praxis principle, the convention-novelty principle, the evolution-transformation principle, the symmetry-asymmetry principle, the softness-hardness principle, the seriousness-playfulness principle, the activeness-inactivess principle, the selfness-otherness principle, the regression-progression principle, the same-difference principle, the stability-reaction principle, the functionality-nonfunctionality principle, and the intentionality-nonintentionality principle.

For instance, in relation to the formalness-informalness principle, if there is formalness (e.g., the formal logical requirement of soundness and completeness in a system of ideas, as in the use of the waste converter), there is informalness (e.g., the non-formal existence of unsoundness and incompleteness in the use of the waste converter, as shown in the criticisms that its application is limited, that there are competing technologies, that it is not necessarily safer, etc.). And the reverse direction also holds true.

In relation to the absoluteness-relativeness principle, if there is absoluteness (e.g., the absolute view of those enthusiastic advocates of the use of waste converters), there is relativeness (e.g., what is right for the enthusiastic advocates of the use of waste converters is not necessarily so for the critics, because "much of the difficulty inherent in recycling comes from the fact that most products are not designed with recycling in mind"). And the reverse direction also holds true.

In relation to the partiality-totality principle, if there is partiality (e.g., the partial view of of those enthusiastic advocates of the use of waste converters), there is totality (e.g., the more holistic view of the future of waste conversion, such that the whole is not the sum of its parts, that is, it cannot be reduced to the sum of opposing views of those enthusiastic advocates of the use of waste converters and their critics, since there will emerge new views in the future not yet known today, just as there are already views nowadays which do not side with any of them). And the reverse direction also holds true.

In relation to the predictability-unpredictability principle, if there is predictability (e.g., the predictable tendency of those enthusiastic advocates of waste converters to favor their use), there is unpredictability (e.g., the more difficult task to predict exactly the extent to which the use of waste converters will remain influential in a particular future historical era, since there are already competing technologies to challenge waste converters aforecited). And the reverse direction also holds true.

In relation to the explicability-inexplicability principle, if there is explicability (e.g., the explanation by the advocates of the waste converter that it is desirable because of its positive environmental impact), there is inexplicability (e.g., the lack of sufficient explanation by the advocates of the waste converter of why it is necessarily desirable, because some contemporary autoclaves have been so improved that they "employ the same method of sterilization as converters, but do so using larger and more complex equipment that also has higher safety… requirement…"). And the reverse direction also holds true.

In relation to the fiction-reality principle, if there is fiction (e.g., the fictional aspect of the waste converter, insofar as "most products are not designed with recycling in mind," nor is it necessarily safer than a newly improved autoclave), there is reality (e.g., the realistic aspect of the waste converter, insofar as it has certain positive environmental impact). And the reverse direction also holds true.

In relation to the finiteness-transfiniteness principle, if there is finiteness (e.g., the finite number of the categories of waste to be treatable by waste converters, like "solid waste," "hazardous waste," etc.), there is transfiniteness (e.g., the transfinite number of all the indivdiual pieces of "solid waste," "hazardous waste," etc. in history that have ever existed). And the reverse direction also holds true.

In relation to the preciseness-vagueness principle, if there is preciseness (e.g., the precise identification of 5 applications of the waste converter in the article aforecited), there is vagueness (e.g., the vagueness in the identification of the 5 applications of the waste converter in the article aforecited, since it is not clear why there must be only 5, not 8, 9, 10, and so on). And the reverse direction also holds true.

In relation to the simpleness-complicatedness principle, if there is simpleness (e.g., "applications outside of waste treatment centers are increasingly common due to the...simplicity of modern converters"), there is complicatedness (e.g., nowadays "autoclaves employ the same method of sterilization as converters, but do so using larger and more complex equipment"). And the reverse direction also holds true.

In relation to the openness-hiddenness principle, if there is openness (e.g., the open exploration of using the waste converter for waste recycle), there is hiddenness (e.g., the hidden bias in the use of the waste converter, because "most products are not designed with recycling in mind"). And the reverse direction also holds true.

In relation to the denseness-emptiness principle, if there is denseness (e.g., the relatively denser concentration of concerns with using "external water input" in autoclaves), there is emptiness (e.g., the relatively less dense, or more empty, concentration of concerns with using "external water input" in modern converters, which instead "utilize the moisture content already present in the conversion cell to generate steam sterilization conditions"). And the reverse direction also holds true.

In relation to the change-constancy principle, if there is change (e.g., the ever new ways to recycle waste, as shown in the use of the waste converter), there is constancy (e.g., the ever constant existence of different problems in the ever new ways to recycle waste, as shown in the problems with the use of the waste converter pointed out by the critics). And the reverse direction also holds true.

In relation to the order-chaos principle, if there is order (e.g., the more or less orderly appearance of waste treatment, if looked from the vantage point of the neatly organized conversion cycle of the waste converter), there is chaos (e.g,. the more or less chaotic appearance of waste treatment, if looked from the vantage point of the disorderly, unhealthy piles of waste in overused landfills in underdeveloped countries). And the reverse direction also holds true.

In relation to the slowness-quickness principle, if there is slowness (e.g., the relatively slower speed of the "older models" of the waste converter, which "required up to several hours to saturate and sterilize the same load" of waste), there is quickness (e.g., the relatively quicker speed of "the modern converter," which "achieves saturation within 10–15 minutes due to the high degree of pulverization preceding the sterilization phase"). (WK 2012p) And the reverse direction also holds true.

In relation to the expansion-contraction principle, if there is expansion (e.g., the relative expansion of the use of waste converters nowadays, like "upcycling," which is "the process of converting waste materials or useless products into new materials or products of better quality" and "has shown significant growth across the United States" in our time), there is contraction (e.g., the relative decline of the demand on landfills nowadays, since "waste conversion…now allows nearly 99% of all MSW to be reused in some way, thus sharply reducing the demand on landfills" in the U.S.). (WK 2012hh; J. Wang 2011) And the reverse direction also holds true.

In relation to the theory-praxis principle, if there is theory (e.g., the theoretical understanding of "pasteurization" in microbiology), there is praxis (e.g., the practical application of the understanding of "pasteurization" in microbiology to the field of waste treatment, as shown in the use of the waster converter). And the reverse direction also holds true.

In relation to the convention-novelty principle, if there is convention (e.g., the conventional wisdom of using autoclaves for waste treatment), there is novelty (e.g., the alternative novel challenge to the conventional wisdom of using autoclaves for waste treatment, by the new idea of the use of the waste converter instead, which differs from the use of autoclaves). And the reverse direction also holds true.

In relation to the evolution-transformation principle, if there is evolution (e.g., the natural evolution of humans to throw away wastes in the state of nature in a primitive way), there is transformation (e.g., the technical transformation of human ability to deal with wastes by the invention of the modern waste converter). And the reverse direction also holds true.

In relation to the symmetry-asymmetry principle, if there is symmetry (e.g., the existence of different methods of waste treatment, like "waste converters," "plasma-arc gasification," "micro-wave and other irradiation sterilization," and "autoclaves"), there is asymmetry (e.g., waste converters gain more popularity nowadays, when contrasted with"plasma-arc gasification," "micro-wave and other irradiation sterilization," and "autoclaves"). And the reverse direction also holds true.

In relation to the softness-hardness principle, if there is softness (e.g., the application of the waster converter in hospitals), there is hardness (e.g., the critique of the use of the waste converter, as shown in the problems pointed out by the critics as cited). And the reverse direction also holds true.

In relation to the seriousness-playfulness principle, if there is seriousness (e.g., the serious business of those who make good use of the waste converter), there is playfulness (e.g., the playful part of of those who make good use of the waste converter, when they play around with different areas in which it can be used over the years, like hospitals, ships, farms, slaughterhouses, etc.). And the reverse direction also holds true.

In relation to the activeness-inactivess principle, if there is activeness (e.g., the relative activeness of those new autoclaves nowadays which can compete with waste converters, because they "employ the same method of sterilization as converters, but do so using larger and more complex equipment that also has higher safety… requirement…," so these new versions are increasingly used), there is inactiveness (e.g., the relative inactiveness of those older autoclaves nowadays which cannot compete with waste converters, because of their technological obsolescence).

In relation to the selfness-otherness principle, if there is selfness (e.g., the tendency to be mostly centered on one's self or group, as shown by different proponents of competing technologies for waste treatment, since each of them advocates his own preferred type), there is otherness (e.g., the tendency to be mostly centered on others, as shown in the academic works by researchers who try to show the pros and cons of each competing technology for waste treatment, like the use of the waster converter, autoclave, etc., so as to make an advancement in the knowledge of waste treatment). And the reverse direction also holds true.

In relation to the regression-progression principle, if there is regression (e.g., the regression made by waste converters, as shown in the problems pointed out by the critics), there is progression (e.g., the progress made by waste converters, as shown in their contribution to waste disposal aforecited). And the reverse direction also holds true.

In relation to the same-difference principle, if there is similarity in outcome (e.g., the contribution to the molding and control of beliefs, values, and behaviors, regardless of whether this be done by way of waste dumping in overfilled landfills, or by way of the use of waste converters), there is difference in outcome (e.g., the contribution to the molding and control of beliefs, values, and behaviors by way of waste dumping in overfilled landfills for a relatively more individual-materialistic lifeworld—but the contribution to the molding and control of beliefs, values, and behaviors by way of the use of waste converters for a relatively more environment-friendly lifeworld). And the reverse direction also holds true.

In relation to the stability-reaction principle, if there is stability (e.g., the relatively more stable condition of the use of waste converters in modern times), there is reaction (e.g., the use of waste converters in modern times then led to further changes later on, when nowadays, some contemporary autoclaves have been so improved that they "employ the same method of sterilization as converters, but do so using larger and more complex equipment that also has higher safety...requirement...," so these new versions are to compete with waste converters). And the reverse direction also holds true.

In relation to the functionality-nonfunctionality principle, if there is functionality (e.g., the presence of action for which something is fitted, as shown in the dumping of wastes in overused landfills in the older days, as it served the function of waste disposal at that time), there is nonfunctionality (e.g., the relative lack of action for which something is fitted, as shown in the use of waste converters nowadays, so that dumping of wastes in overused landfills can no longer serve the function of waste disposal in our time, due to higher standards for positive environmental impact). And the reverse direction also holds true.

And in relation to the intentionality-nonintentionality principle, if there is intentionality (e.g., the planning or design of something for a certain outcome, as shown in the design of the waste converter for waste recycle in modern time), there is nonintentionality (e.g., the relative absence of planning or design of something for a certain outcome, such that the outcome can even be counter-intentional in being the opposite of what was originally planned, as shown by some contemporary autoclaves, which have been so improved that they "employ the same method of sterilization as converters, but do so using larger and more complex equipment that also has higher safety... requirement," as an alternative technology to compete with the waste converter). And the reverse direction also holds true.

Usefulness and Culture

Usefulness, when examined from the perspective of culture, can cast an illuminating light on its possibility (or impossibility) and desirability (or undesirability), and this can be shown by way of two case studies, namely, (3.5.1) usefulness, zero waste, and resource redesign philosophy, and (3.5.2) usefulness, freeganism, and the ideology of reclaiming waste—to be addressed hereafter, in that order.

Usefulness, Zero Waste, and Resource Redesign Philosophy

An excellent case study here to understand usefulness from the perspective of culture has to do with usefulness, zero waste, and resource redesign philosophy.

The term "zero waste" is "a philosophy that encourages the redesign of resource life cycles so that all products are reused. Any trash sent to landfills and incinerators is minimal. The process recommended is one similar to the way that resources are reused in nature. A working definition of zero waste, often cited by experts in the field originated from a working group of the Zero Waste International Alliance in 2004. The definition is as follows: 'Zero Waste is a goal that is ethical, economical, efficient and visionary, to guide people in changing their lifestyles and practices to emulate sustainable natural cycles, where all discarded materials are designed to become resources for others to use. Zero Waste means designing and managing products and processes to systematically avoid and eliminate the volume and toxicity of waste and materials, conserve and recover all resources, and not burn or bury them. Implementing Zero Waste will eliminate all discharges to land, water or air that are a threat to planetary, human, animal or plant health.'" (WK 2012ii)

A good example is "the cycle of a glass milk bottle. The primary input (or resource) is silica-sand, which is formed into glass and then into a bottle. The bottle is filled with milk and distributed to the consumer. At this point, normal waste methods would see the bottle disposed in a landfill or similar. But with a zero-waste method, the bottle can be saddled at the time of sale with a deposit, which is returned to the bearer upon redemption. The bottle is then washed, refilled, and resold. The only material waste is the wash water, and energy loss has been minimized." (WK 2012ii)

However, as one should expect from any philosophical discussion, there is a heated debate on the very nature of the philosophy of "zero waste"; for illustration, consider the following examples (as summarized in *Sec. 3.5*).

1st Dispute: Zero Waste vs. Recycling

The first dispute to be introduced here has to do with zero waste vs. recycling—as "some claim that the key component to zero waste is recycling, while others reject that notion in favor of reusing high function." (WK 2012ii)

To make the dispute more confusing, the idea of "recycling" is no longer about simply "placing bottles and cans in a recycle bin. The modern version of recycling is more complicated and involves many more elements of financing and government support. For example, a 2007 report by the U.S. Environmental Protection Agency states that the US recycles at a national rate of 33.4% and includes in this figure composted materials. In addition many worldwide commodity industries have been created to handle the materials that are recycled. At the same time, claims of recycling rates have sometimes been exaggerated, for example by the inclusion of soil and organic matter used to cover garbage dumps daily, in the 'recycled' column. In states with recycling incentives, there is constant local pressure to pump up the recycling rate figures." (WK 2012ii)

But "there is one seminal example that brings out the difference between Zero Waste and recycling in stark relief. That example, quoted in Getting To Zero Waste, is the software business. Zero Waste is sensitive to the waste of intellectual effort that would be caused by the need to recreate certain basic inventions of software (called objects in software design) as opposed to copying them over and over whenever needed. The waste would occur as the software developers consume resources while solving problems already solved earlier. The application of Zero Waste analysis is straightforward as it recommends conserving human effort. On the other hand, the usual approach of recycling would be to look for some materials that could be found to reuse. The materials on which software is saved (such as paper or diskettes) is of little significance compared to the saving of human effort and if software is saved electronically, there is no media at all. Thus Zero Waste correctly identifies a wasteful behavior to avoid while recycling has no application." (WK 2012ii)

However, for the critics, "zero waste is poorly supported by the enactment of government laws to enforce the waste hierarchy of refuse, reduce, reuse, recycle and rot (compost). In practice, these laws invariably emphasize destruction and recycling, while the reuse component is marginalized." (WK 2012ii)

2nd Dispute: Zero Waste vs. Dump Capacity

The second dispute to be introduced here has to do with zero waste vs. dump capacity—as some justify Zero Waste by pointing out that "many dumps are currently exceeding carrying capacity," whereas "others counter by pointing out that there are huge tracts of land available throughout the USA and other countries which could be used for dumps." (WK 2012ii)

On the one hand, "the underlying need to move to a society designed along Zero Waste principles arises from the huge waste of resources that is inherent in poorly made, short-lived articles and production processes. The locus of the most egregious wasting takes place as articles are built and processes are run wastefully. The actual placing of a now useless item in a dump is barely the icing on the cake, in terms of the waste it represents. Poorly conceived proposals, that appear with a dismaying regularity on the Internet, to blithely destroy all garbage as a way to solve the garbage problem, make use of the common delusion that it is the garbage itself which is the problem." (WK 2012ii)

On the other hand, the critics call for the need "to convert all or a large portion of existing garbage into oil" or something else, and "one such plan, called Anything Into Oil was promoted by Discover Magazine and Fortune Magazine in 2004," and "Zero Waste analysis...receives no such promotion by the media." (WK 2012ii)

3rd Dispute: Zero Waste vs. Deconstruction

The third dispute to be introduced here has to do with zero waste vs. deconstruction—as some argued that zero waste is "about restructuring production and distribution systems to prevent waste from being manufactured in the first place," whereas others favor instead "deconstruction" as "construction in reverse." (WK 2012ii)

On the one hand, those who prefer construction for Zero Waste argue that "the materials that are still required in the...re-designed, resource-efficient systems will be reused many times as the products that incorporate them are reused....Zero Waste favors the design of buildings as assemblages of high level components, not their creation from rough materials such as lumber, cement or plaster. The details are not worked out yet but to the extent that entire rooms, entire walls, roofs or floors or entire utility systems can be pre-built and installed as completed components, that will be the goal of Zero Waste design....For now, the largest parts that we are able to save tend to be architectural elements, windows, doors, and metals, many of which are being saved and resold by reuse yards such as Urban Ore in Berkeley California. The main parts that still need to be crushed are wood flooring, brick walls, and structural timbers." (WK 2012ii)

On the other hand, the proponents of deconstruction regard it "as construction in reverse," because "it involves carefully taking apart a building to maximize the reuse of materials, thereby reducing waste and conserving resources. Deconstruction can capture materials and some components from the millions of buildings that are existing and that were poorly designed for high level reuse"—and moreover, "until buildings are built as components capable of later dismantling, deconstruction is a stop-gap process that the United States can use to minimize the waste of building materials." (WK 2012ii)

4th Dispute: Zero Waste vs. Market-Based Campaigns

The fourth dispute to be introduced here has to do with zero waste vs. market-based campaigns—as people disagree on the extent to which market-based campaigns can help the Zero Waste movement.

On the one hand, "market-based, legislation-mediated campaigns like Extended Producer Responsibility (EPR) and the Precautionary Principle are among numerous campaigns that have a Zero Waste slogan hung on them by means of claims they will ineluctably lead to policies of Zero Waste." (WK 2012ii)

On the other hand, there are those who question these market-based campaigns, since, for the critics, "at the moment, there is no evidence that EPR will increase reuse, rather than merely moving discard and disposal into private-sector dumping contracts. The Precautionary Principle is put forward to shift liability for proving new chemicals are safe from the public (acting as guinea pig) to the company introducing them. As such, its relation to Zero Waste is dubious. Likewise, many organizations, cities and counties have embraced a Zero Waste slogan while pressing for none of the key Zero Waste changes. In fact, it is common for many such to simply state that recycling is their entire goal. Many commercial or industrial companies claim to embrace Zero Waste but usually mean no more than a major materials recycling effort, having no bearing on product redesign. Examples include Staples, Home Depot, Toyota, General Motors and computer take-back campaigns." (WK 2012ii)

5th *Dispute: Zero Waste vs. Government*

The fifth dispute to be introduced here has to do with zero waste vs. government—as people disagree on the extent to which the state (or government) can help to promote Zero Waste.

On the one hand, there are those who seeks the help of the state (or government) via the enactment of different laws, especially at the regional level, because "policy incrementalism reflects policymakers tendency to build on what already exists and is characterised by minimal disturbance of the current state." (WK 2012ii; M. Landy 2001; P. Lowe 1998)

On the other hand, there are those who argued that "what is needed is a shift from government (formal organisations and procedures of the public sector) to governance (array of governmental and non-governmental institutions). The government alone does not have the cognitive breadth to determine how to reach the ultimate target of Zero Waste which requires the involvement of businesses, NGOs, the public and the state in governing. Governance is 'ultimately concerned with creating the conditions for ordered rule and collective action' and to achieve Zero Waste, a departure from waste management based simply on waste disposal ideology is necessary. The role of the state is to act as central steering mechanism and governance networks that bring together government; public and market actors are viewed as important to achieve Zero Waste similar to many other environmental goals. The Johannesburg World Summit on Sustainable Development emphasized these partnerships between different stakeholders in the environment and the establishment of Public-Private partnerships is seen as the best way to achieve sustainability." (WK 2012ii; E. Davoudi 2005; M. Gandy 1994; R. Rhodes 1997; P. Glasbergen 2007)

6th *Dispute: Zero Waste vs. Nature*

And the sixth dispute to be introduced here has to do with zero waste vs. nature—as people disagree on the extent to which nature is the model for zero waste.

On the one hand, there are those who advocate "the literal application of natural processes" for zero waste, in that they "look to nature as an ultimate model for production and innovative materials." (WK 2012ii)

On the other hand, there are those who favor "the creation of industry-specific more efficient reuse modalities" instead, in that "industrial products are inherently non-natural (such as chemicals and plastics that are mono-molecular) and benefit greatly from industrial methods of reuse, while natural methods requiring degradation and reconstitution are wasteful in that context." (WK 2012ii)

Beyond Zero Waste

This introduction to the debate on Zero Waste does not reject its validity but serves to show us the two opposing sides of the debate, such that the possibility (or impossibility) and desirability (or undesirability) of usefulness (from the perspective of culture with usefulness, zero waste, and resource redesign philosophy as a case study here) are not to the extent that the respective defenders would like us to believe.

Furthermore, the analysis of usefulness, zero waste, and resource redesign philosophy serves as a good lesson about the ontological principles in existential dialectics, and good examples include the formalness-informalness principle, the absoluteness-relativeness principle, the partiality-totality principle, the predictability-unpredictability principle, the explicability-inexplicability principle, the fiction-reality principle, the finiteness-transfiniteness principle, the preciseness-vagueness principle, the simpleness-complicatedness principle, the openness-hiddenness principle, the denseness-emptiness principle, the change-constancy principle, the order-chaos principle, the slowness-quickness principle, the expansion-contraction principle, the theory-praxis principle, the convention-novelty principle, the evolution-transformation principle, the symmetry-asymmetry principle, the softness-hardness principle, the seriousness-playfulness principle, the activeness-inactivess principle, the selfness-otherness principle, the regression-progression principle, the same-difference principle, the stability-reaction principle, the functionality-nonfunctionality principle, the functionality-nonfunctionality principle, and the intentionality-nonintentionality principle.

For instance, in relation to the formalness-informalness principle, if there is formalness (e.g., the formal logical requirement of soundness, consistency, and completeness in a system of ideas, as in the call for Zero Waste), there is informalness (e.g., the non-formal existence of unsoundness, inconsistency, and incompleteness in the call for Zero Waste, as shown in the 6 disputes about Zero Waste aforecited). And the reverse direction also holds true.

In relation to the absoluteness-relativeness principle, if there is absoluteness (e.g., the absolute view of those advocates of Zero Waste "in favor of reusing high function"), there is relativeness (e.g., what is right for those advocates of Zero Waste "in favor of reusing high function" is not necessarily so for others who favor recycling instead). And the reverse direction also holds true.

In relation to the partiality-totality principle, if there is partiality (e.g., the partial view of those advocates of Zero Waste in favor of governance), there is totality (e.g., the more holistic view of the future of Zero Waste, such that the whole is not the sum of its parts, that is, it cannot be reduced to the sum of opposing views of those advocates of Zero Waste in favor of governance and those in favor of government, since there will emerge new views in the future not yet known today, just as there are already others who are critical to both sides nowadays). And the reverse direction also holds true.

In relation to the predictability-unpredictability principle, if there is predictability (e.g., the predictable tendency of those who prefer construction for Zero Waste to argue that "the materials that are still required in the...re-designed, resource-efficient systems will be reused many times as the products that incorporate them are reused"), there is unpredictability (e.g., the more difficult task to predict exactly the extent to which the construction argument for Zero Waste will remain influential in a particular future historical era, because, for the critics, "deconstruction can capture materials and some components from the millions of buildings that are existing and that were poorly designed for high level reuse" and moreover, "until buildings are built as components capable of later dismantling, deconstruction is a stop-gap process that the United States can use to minimize the waste of building materials"). And the reverse direction also holds true.

In relation to the explicability-inexplicability principle, if there is explicability (e.g., the explanation by the advocates for Zero Waste that it is desirable because we can use nature as a model to back it up), there is inexplicability (e.g., the lack of sufficient explanation by the advocates for Zero Waste of why it is necessarily desirable, because there are those who favor "the creation of industry-specific more efficient reuse modalities" instead, in that "industrial products are inherently non-natural [such as chemicals and plastics that are mono-molecular] and benefit greatly from industrial methods of reuse, while natural methods requiring degradation and reconstitution are wasteful in that context"). And the reverse direction also holds true.

In relation to the fiction-reality principle, if there is fiction (e.g., the fictional aspect of market-based campaigns for Zero Waste, insofar as "many organizations, cities and counties have embraced a Zero Waste slogan while pressing for none of the key Zero Waste changes"), there is reality (e.g., the realistic aspect of market-based campaigns for Zero Waste, insofar as "market-based, legislation-mediated campaigns like Extended Producer Responsibility [EPR] and the Precautionary Principle are among numerous campaigns that have a Zero Waste slogan hung on them"). (WK 2012ii) And the reverse direction also holds true.

In relation to the finiteness-transfiniteness principle, if there is finiteness (e.g., the finite number of the categories of materials used in products which can be targeted by Zero Waste, like "steel," "copper," "glass," and "plastics" in bottles, computers, etc.), there is transfiniteness (e.g., the transfinite number of all the individual pieces of "steel," "copper," "glass," and "plastics" in bottles, computers, etc. in history that can be targeted by Zero Waste). And the reverse direction also holds true.

In relation to the preciseness-vagueness principle, if there is preciseness (e.g., the precise identification of 6 disputes in the debate on zero waste in the article aforecited), there is vagueness (e.g., the vagueness in the identification of the 6 disputes in the debate on zero waste in the article aforecited, since it is not clear why there must be only 6, not 8, 9,10, 11, and so on). And the reverse direction also holds true.

In relation to the simpleness-complicatedness principle, if there is simpleness (e.g., the relatively simple analysis of Zero Waste by the advocates), there is complicatedness (e.g., the relatively more complicated analysis of Zero Waste, by challenging the claims and assumptions, as shown in the problems pointed out by the critics as cited). And the reverse direction also holds true.

In relation to the openness-hiddenness principle, if there is openness (e.g., the open exploration by "many commercial or industrial companies" like "Staples, Home Depot, Toyota, General Motors and computer take-back campaigns" to "embrace Zero Waste" in their market-based campaigns), there is hiddenness (e.g., the hidden bias in the market-based campaigns by "many commercial or industrial companies" like "Staples, Home Depot, Toyota, General Motors and computer take-back campaigns" to "embrace Zero Waste," because they "usually mean no more than a major materials recycling effort, having no bearing on product redesign"). And the reverse direction also holds true.

In relation to the denseness-emptiness principle, if there is denseness (e.g., the relatively denser concentration of concerns with the use of "governmental and non-governmental institutions" by those who favor governance for Zero Waste), there is emptiness (e.g., the relatively less dense, or more empty, concentration of concerns with "formal organisations and procedures of the public sector" by those who favor governance for Zero Waste). And the reverse direction also holds true.

In relation to the change-constancy principle, if there is change (e.g., the ever new ways to treat waste, as shown in the idea of Zero Waste), there is constancy (e.g., the ever constant existence of different problems in the ever new ways to treat waste, as shown in the problems with Zero Waste pointed out by the critics). And the reverse direction also holds true.

In relation to the order-chaos principle, if there is order (e.g., the more or less orderly view of waste treatment, if looked from the vantage point of "construction," which favors "the design of buildings as assemblages of high level components, not their creation from rough materials such as lumber, cement or plaster," so that "these re-designed, resource-efficient systems will be reused many times as the products that incorporate them are reused"), there is chaos (e.g,. the more or less chaotic view of waste treatment, if looked from the vantage point of "deconstruction," as shown in the "demolition of traditional buildings" which "has been long done by wrecking ball or bulldozer" into myriad pieces). (WK 2012ii) And the reverse direction also holds true.

In relation to the slowness-quickness principle, if there is slowness (e.g., the relatively slower speed of Zero Waste to be achieved if done by the government alone, because "the government alone does not have the cognitive breadth to determine how to reach the ultimate target of Zero Waste which requires the involvement of businesses, NGOs, the public and the state in governing"), there is quickness (e.g., the relatively quicker speed of Zero Waste to be achieved with "the involvement of businesses, NGOs, the public and the state in governing," etc.). And the reverse direction also holds true.

In relation to the expansion-contraction principle, if there is expansion (e.g., the relative spread of the Zero Waste movement, especially in 1998-2002), there is contraction (e.g., the relative decline of the idea of "recycling" as simply "placing bottles and cans in a recycle bin," especially after the rise of the Zero Waste movement). And the reverse direction also holds true.

In relation to the theory-praxis principle, if there is theory (e.g., the theoretical construction of "Design Efficiency" and "Design Effectiveness" in the Zero Waste Institute), there is praxis (e.g., the practical application of "Design Efficiency" and "Design Effectiveness" in the Zero Waste Institute to the realm of practice, as shown by the Zero Waste International Alliance, which helps a lot of "communities across the globe that have created public policy to promote zero-waste practices"). And the reverse direction also holds true.

In relation to the convention-novelty principle, if there is convention (e.g., the conventional wisdom about waste recycle), there is novelty (e.g., the alternative novel challenge to the conventional wisdom about waste recycle, by the new idea of Zero Waste, which is more than just waste recycle but also waste reuse). And the reverse direction also holds true.

In relation to the evolution-transformation principle, if there is evolution (e.g., the natural evolution of humans to live in caves in the state of nature in a primitive way), there is transformation (e.g., the technical transformation of human ability to cope with shelters by the invention of Zero Waste, which invents "the design of buildings as assemblages of high level components," so that "these re-designed, resource-efficient systems will be reused many times as the products that incorporate them are reused"). And the reverse direction also holds true.

In relation to the symmetry-asymmetry principle, if there is symmetry (e.g., the existence of different entities in nature, like animals, humans, materials, etc.), there is asymmetry (e.g., humans can recycle and reuse materials found in nature in a way that materials cannot recycle and reuse humans, as shown in "the creation of industry-specific more efficient reuse modalities," such that "industrial products are inherently non-natural"). And the reverse direction also holds true.

In relation to the softness-hardness principle, if there is softness (e.g., the appreciation of Zero Waste by its advocates), there is hardness (e.g., the critique of Zero Waste by its critics). And the reverse direction also holds true.

In relation to the seriousness-playfulness principle, if there is seriousness (e.g., the serious business of those participating in the Zero Waste movement), there is playfulness (e.g., the playful part of those participating in the Zero Waste movement, when they play around with different modalities of waste reuse over the years, like "the literal application of natural processes," "the creation of industry-specific more efficient reuse modalities," etc.). And the reverse direction also holds true.

In relation to the activeness-inactivess principle, if there is activeness (e.g., the activeness, in our time, of those wastes in construction like "windows, doors, and metals, many of which are being saved and resold by reuse" in accordance to Zero Waste), there is inactiveness (e.g., the relative inactiveness, in our time, of those wastes in construction like "wood flooring, brick walls, and structural timbers" which cannot be reused in accordance to Zero Waste).

In relation to the selfness-otherness principle, if there is selfness (e.g., the tendency to be mostly centered on one's self or group, as shown by "many commercial or industrial companies" like "Staples, Home Depot, Toyota, General Motors and computer take-back campaigns," which "embrace Zero Waste" but "usually mean no more than a major materials recycling effort, having no bearing on product redesign"), there is otherness (e.g., the tendency to be mostly centered on others, as shown in those academic works which point out the pros and cons of the debate on

Zero Waste for an advancement in the knowledge of reuse). And the reverse direction also holds true.

In relation to the regression-progression principle, if there is regression (e.g., the regression made by Zero Waste, as shown in the disputes pointed out by the critics), there is progression (e.g., the progress made by Zero Waste, as shown in its contribution to a more "zero waste" community aforecited). And the reverse direction also holds true.

In relation to the same-difference principle, if there is similarity in outcome (e.g., the contribution to the molding and control of beliefs, values, and behaviors, regardless of whether this be done by way of "the literal application of natural processes," or by way of "the creation of industry-specific more efficient reuse modalities"), there is difference in outcome (e.g., the contribution to the molding and control of beliefs, values, and behaviors by way of "the literal application of natural processes" for a relatively more naturalistic lifeworld—but the contribution to the molding and control of beliefs, values, and behaviors by way of "the creation of industry-specific more efficient reuse modalities" for a relatively more technophilic lifeworld). And the reverse direction also holds true.

In relation to the stability-reaction principle, if there is stability (e.g., the relatively more stable condition of waste dumping in the older days, when many dumps were not exceeding carrying capacity), there is reaction (e.g., waste dumping in the older days then led to further changes later on, when nowadays, there is an increasing call for reuse like Zero Waste, because "many dumps are currently exceeding carrying capacity"). And the reverse direction also holds true.

In relation to the functionality-nonfunctionality principle, if there is functionality (e.g., the presence of action for which something is fitted, as shown in the practice of waste dumping in the older days, when it served the function of waste treatment at the time), there is nonfunctionality (e.g., the relative lack of action for which something is fitted, as shown in the call for Zero Waste instead nowadays, because "many dumps are currently exceeding carrying capacity," so waste dumping is no longer dominant in serving the essential function of waste treatment in our time). And the reverse direction also holds true.

And in relation to the intentionality-nonintentionality principle, if there is intentionality (e.g., the planning or design of something for a certain outcome, as shown in the design of Zero Waste for more reuse, especially during 1998-2002), there is nonintentionity (e.g., the relative absence of planning or design of something for a certain outcome, such that the outcome can even be counter-intentional in being the opposite of what was originally planned, as shown by "many commercial or industrial companies" like "Staples, Home Depot, Toyota, General Motors and computer take-back campaigns," which "embrace Zero Waste" but "usually mean no more than a major materials recycling effort, having no bearing on product redesign"). (WK 2012ii) And the reverse direction also holds true.

Usefulness, Freeganism, and the Ideology of Reclaiming Waste

Another way to evaluate the possibility (or impossibility) and desirability (or undesirability) of usefulness from the perspective of culture concerns usefulness, freeganism, and the ideology of reclaiming waste.

A contention here is that waste can be reclaimed by using it in an unconventional way, and this ideology is known as "freeganism," which is "the practice of reclaiming and eating food that has been discarded." (WK 2012r)

As an ideology, "freeganism started in the mid 1990s, out of the antiglobalization and environmentalist movements. The movement also has elements of Diggers, an anarchist street theater group based in Haight-Ashbury in San Francisco in the 1960s, that gave away rescued food." (WK 2012r; S. Kurutz 2007)

The ideology of freeganism was described in the manifesto pamphlet "Why Freegan" by former Against Me! drummer Warren Oakes in 1999. The ideology of Freeganism can be analyzed in terms of its major characteristics, as shown below (and summarized in *Sec. 3.6*).

Foraging in Freeganism

The first characteristic of freeganism as an ideology to be addressed here concerns "foraging."

Because of this belief in foraging, "many freegans get free food by pulling it out of the trash (bins), a practice commonly nicknamed 'dumpster diving' in North America and 'skipping' or 'bin diving' in the UK, as well as 'bin raiding' or 'skipitarianism' (so called because the person's diet mostly involves eating out of a skip). Retail suppliers of food such as supermarkets, grocery stores, and restaurants routinely throw away food in perfectly good condition, often because it is approaching its sell-by date (without thereby becoming dangerous), or has damaged packaging. Freegans find food in the garbage of such establishments, which they say allows them to avoid spending money on products that exploit the world's resources, contribute to urban sprawl, treat workers unfairly, or disregard animal rights. By foraging, they believe they are keeping perfectly edible food from adding to landfill clutter and that can feed people and animals who might otherwise go hungry." (WK 2012r; T. Carlson 2006)

Besides foraging for food, "many 'dumpster divers' search for anything that can be recycled or reused, from accessories to power tools in need of small repairs. Some divers collect aluminum cans, which they can then sell for a small profit. Often, these people have all sorts of equipment such as a long pole that they use to move items in the dumpster around." (WK 2012r; A. Brace 2007; N. Willhite 2009)

Also, "instead of buying industrially grown foods, wild foragers find and harvest food and medicinal plants growing in their own communities. Some freegans participate in 'guerrilla' or 'community' gardens, with the stated aim of rebuilding community and reclaiming the capacity to grow one's own food. In order to fertilize those guerrilla gardens, food obtained from dumpster diving is sometimes also reused. In many urban guerrilla gardens, vermiculture is used instead of ordinary composting techniques in order to keep the required infrastructure/room small. Guerrilla gardeners claim to seek an alternative to dependence and participation in what they perceive as an exploitative and ecologically destructive system of global, industrialized corporate food production. Many rural freegans choose to learn about native wild plants which are easily sustainable and either bring favored species home to cultivate or identify wild populations from which to forage. Often rural freegans are also 'homesteaders' who also raise their own dairy livestock and employ alternative energy sources to provide energy for their homesteads, occasionally living 'off the grid' entirely." (WK 2012r)

Sharing in Freeganism

The second characteristic of freeganism as an ideology to be addressed here concerns "sharing."

For instance, "Food Not Bombs recovers food that would otherwise go to waste to serve warm meals on the street to anyone who wants them. The group promotes an ethic of sharing and community, while working to show what they consider to be the injustice of a society in which they claim fighting wars is considered a higher priority than feeding the hungry." (WK 2012r)

Other activities like "Really, Really Free Markets" are "free social events in which freegans can share goods instead of discarding them, share skills, give presents and eat food. A free store is a temporary market where people exchange goods and services outside of a money-based economy. Freegans also advocate sharing travel resources. Internet-based ridesharing reduces but does not eliminate use of cars and all the related resources needed to maintain and operate them." (WK 2012r)

In addition, "community bicycle programs and collectives facilitate community sharing of bicycles, restore found and broken bikes, and teach people how to do their own bike repairs. In the process they build a culture of skill and resource sharing, reuse wasted bikes and bike parts, and create greater access to green transport. In general, co-ops function to provide their local community with additional resources; they are also typically vegan-friendly and local-produce-friendly." (WK 2012r)

Squatting in Freeganism

The third characteristic of freeganism as an ideology to be addressed here concerns "squatting."

In other words, the freegans also believe that "people should not be homeless when unused buildings are available. Freegans consider housing to be a right as opposed to an economic good. As a result of this philosophy, many freegans are involved in squatting. Squatting is the act of someone occupying a building that they do not have any legal claim or ownership over. 'Squatters take a stand against councils and landlords, who would rather keep properties boarded up if they cannot make a sufficient profit from them.' Freegans see this practice as senseless and a counter-productive use of resources." (WK 2012r; A. Corr 1999)

Although no one knows for sure, "there are estimated to be around one million squatters worldwide. Striving for equality and reform, squatting is a political action that has been incorporated in numerous movements. Squatters view the act as a necessity because of the lack of housing available. They believe that there should not be buildings remaining empty when there are people who are in crucial need of a home but lack the resources to legally obtain one. Ultimately, squatting is a way of housing the homeless. However, the buildings that squatters reside in are sometimes used for other purposes as well, such as being changed into community centers that house programs for children, community organizations, and environmental education." (WK 2012r; A. Corr 1999)

Working Less in Freeganism

The fourth characteristic of freeganism as an ideology to be addressed here concerns "working less."

The freegans are known for opposing "the notion of working for the sole purpose of accumulating material items. The need to work is reduced by only purchasing the basic necessities for things such as housing, clothing, and food. Not working resists the idea that joy can only be found through the purchase of material items. Working is seen as sacrificing valuable time to 'take orders from someone else, stress, boredom, monotony, and in many cases risks to physical and psychological well-being.'" (WK 2012r)

Instead, for the freegans, "time could be spent volunteering in service activities, bonding with family, or participating in a number of other endeavors. The concept of voluntary joblessness has been described as means of completing tasks out of love for others while not expecting anything in return for one's services. Working is viewed as a component of a system that has abused our world both socially and ecologically. It is realized that not working at all is not an option for everyone, but that there are ways to limit the need to work as much. Employment does not need to take over or define one's life, and such complete control does not need to be given to supervisors and managers." (WK 2012r; NOUSE 2008)

Veganism in Freeganism

The fifth characteristic of freeganism as an ideology to be addressed here concerns "veganism."

The freegans also address the issue of "veganism, which calls for the avoidance of flesh foods, dairy and eggs, and further extends to avoid the consumption or use of furs, leather, wool, down, and cosmetics and chemical products tested on animals. Vegans choose their diet and lifestyle practices in an attempt to maintain a 'cruelty-free' lifestyle. This lifestyle is based on the ideals of providing 'numerous benefits to animals'lives, to the environment, and to our own health.' A study from the United Nations' report on Livestock and the Environment found that 'a vegan diet can feed many more people than an animal-based diet.' In fact, in 1992 it was estimated that this particular year's food supply would have sufficed to feed approximately 6.3 billion people on a 'purely vegan diet.'" (WK 2012r; E. Josyn 2006)

Lest a misunderstanding occurs, "not all people who identify as freegan are vegan. There are some, because of a confusion with dumpster diver, that consume animal products only if those animal products would otherwise be wasted. 'Vegans are people who avoid products from animal sources or products tested on animals in an effort to avoid harming animals. Freegans take this a step further.'" (WK 2012r)

Humanure in Freeganism

And the sixth characteristic of freeganism as an ideology to be addressed here concerns "humanure."

Interestingly, "as part of the anti-consumerist lifestyle, many freegans practice what is known as collecting and composting human faeces to be used as manure, referred to as 'humanure.' The underlying purpose in this practice is to conserve resources such as electricity and water while furthering an environmentally conscious lifestyle. Rather than setting up plumbing, sewage, and water systems in bathrooms, humanure toilet proponents use simple materials (a few plywood boards, a couple of hinges and screws, and organic material such as sawdust) to create toilet look-alikes that function with much less energy consumption. Humanure composting requires three basic components—a toilet receptacle, organic cover material, and an outdoor composting bin." (WK 2012r; J. Jenkins 2005 & 2005a)

A humanure toilet is not expensive to build, and "an estimated start-up cost for a humanure toilet is about twenty-five dollars. These toilets can also be referred to as sawdust toilets because of the organic material used in the composting process. Maintenance costs are also minimized, as rotted sawdust material can be acquired—if not for free from local lumber sources, then at least for a reasonably low price, while other organic scrap material consists of peat moss, leaf moulds, rice hulls, or grass clippings, all of which are easily accessible." (WK 2012; J. Jenkins 2005)

The use of sawdust is important here, because "sawdust and other organic scraps function as a natural biofilter for the human wastes accrued in the receptacle. Organic scraps can be added along with the sawdust, although it is recommended to always cover excrement and scraps with sawdust regardless of whether other scraps are used. The sawdust also reduces odors, and together these components aid in the composting and detoxifying process." (WK 2012; J. Jenkins 2005)

Problems with the Ideology of Freeganism

However, for the critics, the ideology of freeganism has some major problems. Consider, for illustration, a few examples below (as summarized in *Sec. 3.6*).

Firstly, one main criticism of the ideology of freeganism is that foraging on the street is often looked down upon in mainstream society.

For instance, for the critics, street foragers often do not receive respect from others and are often viewed as trashy, lazy, and less intelligent people who cannot make it in society.

Secondly, another main criticism of the ideology of freeganism is that reclaimed items in dumpsters can be unhealthy and dangerous.

For instance, for the critics, reclaimed items in dumpsters are often used, defective, hazardous, and inferior in quality, because, "when searching for food, a forager may come across food waste that is not entirely sealed from the unwanted waste in the same rubbish sack. This lower quality food is commonly referred to as 'Scree.' As bugs, rodents and other disease carriers also forage in such places, there are risks associated with sourcing and eating such food." (WK 2012r)

Thirdly, still another main criticism of the ideology of freeganism is that it has some extremist roots in religion.

For instance, like many environmentalists, the freegans believe in the idea of "religious obligation to nature in the form of political action," because they uphold the belief that "consumerism leads to exploitation of people and the environment. This conviction that organizing around the environment and human issues is inevitably linked to our moral obligation to nature is what qualifies the Green movement and the freeganism movement as religious. Environmentalism has been described as a 'binding philosophy, the start of environmentalism was a religious reformation; an epiphany; an awakening to desecreated surroundings.' Freeganism is another form of this awakening, and also entails concepts of this religious reformation. The religious gratitude for nature that environmentalists and freegans possess may have stemmed from nature idealism that is the foundation of Transcendentalism." (WK 2012r; V. Scheffer 1991; A. Weiss 2009)

Fourthly, an additional main criticism of the ideology of freeganism is that it is so not practical as romantically assumed by the advocates.

For instance, for the critics, in response to the call for "increasing self-sufficiency," it was found that "one argument against gardening at home is how often people begin a garden and later stop it." (WK 2012q)

And fifthly, another additional main criticism of the ideology of freeganism is that its practice can be criminal.

For instance, for the critics, squatting in many localities are illegal (and criminal); in fact, for this reason, "the criminality of squatting makes it hard to accurately track the number of people involved in this activity." (WK 2012r)

Beyond the Ideology of Freeganism

These criticisms of the ideology of freeganism, however, do not imply a total rejection of its validity but serve to show us the two opposing sides of the debate, such that the possibility (or impossibility) and desirability (or undesirability) of usefulness (from the perspective of culture with usefulness, freeganism, and the ideology of reclaiming waste as a case study here) are not to the extent that the respective defenders would like us to believe.

Besides, the analysis of usefulness, freeganism, and the ideology of reclaiming waste reveals the logic of the ontological principles in existential dialectics, and good examples include the formalness-informalness principle, the absoluteness-relativeness principle, the partiality-totality principle, the predictability-unpredictability principle, the explicability-inexplicability principle, the fiction-reality principle, the finiteness-transfiniteness principle, the preciseness-vagueness principle, the simpleness-complicatedness principle, the openness-hiddenness principle, the denseness-emptiness principle, the change-constancy principle, the order-chaos principle, the slowness-quickness principle, the expansion-contraction principle, the theory-praxis principle, the convention-novelty principle, the evolution-transformation principle, the symmetry-asymmetry principle, the softness-hardness principle, the seriousness-playfulness principle, the activeness-inactivess principle, the selfness-otherness principle, the regression-progression principle, the same-difference principle, the stability-reaction principle, the functionality-nonfunctionality principle, and the intentionality-nonintentionality principle.

For instance, in relation to the formalness-informalness principle, if there is formalness (e.g., the formal logical requirement of soundness and completeness in a system of ideas, as in the ideology of freeganism), there is informalness (e.g., the non-formal existence of unsoundness and incompleteness in the ideology of freeganism, as shown in the criticisms that foraging on the street is often looked down upon by those in mainstream society, that reclaimed items in dumpsters can be unhealthy and dangerous, that it has some extremist roots in religion, that it is not so practical as romantically assumed by the advocates, that its practice can be criminal, etc.). And the reverse direction also holds true.

In relation to the absoluteness-relativeness principle, if there is absoluteness (e.g., the absolute religious view of freeganism in regard to the idea of "religious obligation to nature in the form of political action"), there is relativeness (e.g., what is right for the religious view of freeganism in regard to the idea of "religious obligation to nature in the form of political action" is not necessarily so for the critics, who question their religious fanaticism). And the reverse direction also holds true.

In relation to the partiality-totality principle, if there is partiality (e.g., the partial view of freeganism on reclaiming wastes), there is totality (e.g., the more holistic view of the future of reclaiming wastes, such that the whole is not the sum of its parts, that is, it cannot be reduced to the sum of opposing views of the freegans and the critics, since there will emerge new views in the future not yet known today, just as there are already other views nowadays which do not side with either). And the reverse direction also holds true.

In relation to the predictability-unpredictability principle, if there is predictability (e.g., the predictable tendency of the freegans to practice squatting), there is unpredictability (e.g., the more difficult task to predict exactly the extent to which squatting is pervasive on a particular future occasion, since, for the critics, squatting in many localities are illegal and criminal, so for this reason, "the criminality of squatting makes it hard to accurately track the number of people involved in this activity"). And the reverse direction also holds true.

In relation to the explicability-inexplicability principle, if there is explicability (e.g., the explanation by the advocates of freeganism that it is desirable because of the benefits of reclaiming wastes), there is inexplicability (e.g., the lack of sufficient explanation by the advocates of freeganism of why it is necessarily desirable because reclaimed items in dumpsters can be unhealthy and dangerous). And the reverse direction also holds true.

In relation to the fiction-reality principle, if there is fiction (e.g., the fictional aspect of freeganism, insofar as reclaimed items in dumpsters can be unhealthy and dangerous, it has an extremist roots in religion, and it is not so practical as romantically assumed by the advocates), there is reality (e.g., the realistic aspect of freeganism, insofar as it has contributed to the efforts of reclaiming wastes). And the reverse direction also holds true.

In relation to the finiteness-transfiniteness principle, if there is finiteness (e.g., the finite number of the categories of food suppliers like "supermarkets, grocery stores, and restaurants" targeted by dumpster divers for food wastes), there is transfiniteness (e.g., the transfinite number of all the individual items of food wastes in history that have ever been discarded by "supermarkets, grocery stores, and restaurants"). And the reverse direction also holds true.

In relation to the preciseness-vagueness principle, if there is preciseness (e.g., the precise identification of 6 characteristics of freeganism in the article aforecited), there is vagueness (e.g., the vagueness in the identification of the 6 characteristics of freeganism in the article aforecited, since it is not clear why there must be only 6, not 8, 9, 10, and so on). And the reverse direction also holds true.

In relation to the simpleness-complicatedness principle, if there is simpleness (e.g., the relatively simple analysis of the ideology of freeganism by its advocates), there is complicatedness (e.g., the relatively more complicated analysis of the ideology of freeganism, by challenging the claims and assumptions, as shown in the problems pointed out by the critics as cited). And the reverse direction also holds true.

In relation to the openness-hiddenness principle, if there is openness (e.g., the open exploration of the possibility of "reclaiming and eating food that has been discarded" by the freegans), there is hiddenness (e.g., the hidden bias in freeganism, because of its hidden extremist roots in religion). And the reverse direction also holds true.

In relation to the denseness-emptiness principle, if there is denseness (e.g., the relatively denser concentration of concerns with working and consuming less among the freegans), there is emptiness (e.g., the relatively less dense, or more empty, concentration of concerns with working and consuming less among the capitalists). And the reverse direction also holds true.

In relation to the change-constancy principle, if there is change (e.g., the ever new ways to reclaim wastes, as shown in freeganism), there is constancy (e.g., the ever constant existence of different problems in the ever new ways to reclaim wastes, as shown in the problems with freeganism pointed out by the critics). And the reverse direction also holds true.

In relation to the order-chaos principle, if there is order (e.g., the more or less orderly appearance of food wastes, if looked from the vantage point of them being contained inside dumpsters of food suppliers like "supermarkets, grocery stores, and restaurants"), there is chaos (e.g,. the more or less chaotic appearance of food wastes, if looked from the vantage point of them being thrown away on the streets and in different dump sites all over the world). And the reverse direction also holds true.

In relation to the slowness-quickness principle, if there is slowness (e.g., the relatively slower speed of unused buildings to be occupied if based on a free market economy), there is quickness (e.g., the relatively quicker speed of unused buildings to be occupied if based on an alternative practice known as squatting). And the reverse direction also holds true.

In relation to the expansion-contraction principle, if there is expansion (e.g., the relative spread of the practice of freeganism in the developed world nowadays), there is contraction (e.g., the relative decline of environment-unfriendly practices in the developed world nowadays). And the reverse direction also holds true.

In relation to the theory-praxis principle, if there is theory (e.g., the theoretical construction of "anti-globalization" worldviews against corporate capitalism in contemporary times), there is praxis (e.g., the practical application of "anti-globalization" worldviews against corporate capitalism in contemporary times to the field of waste treatment, like freeganism). And the reverse direction also holds true.

In relation to the convention-novelty principle, if there is convention (e.g., the conventional wisdom about work ethics for working more to earn a good living), there is novelty (e.g., the alternative novel challenge to the conventional wisdom about work ethics for working more to earn a good living, by the new idea of "working less" instead in freeganism for the anti-capitalist movement). And the reverse direction also holds true.

In relation to the evolution-transformation principle, if there is evolution (e.g., the natural evolution of humans to live in caves in the state of nature in a primitive way), there is transformation (e.g., the technical transformation of human ability to live by the invention of squatting in freeganism). And the reverse direction also holds true.

In relation to the symmetry-asymmetry principle, if there is symmetry (e.g., the existence of different social classes, like the poor and the rich), there is asymmetry (e.g., the freegans live by reclaiming wastes but the rich live by wastefully consuming resources). And the reverse direction also holds true.

In relation to the softness-hardness principle, if there is softness (e.g., the appreciation of environmentalism by the freegans), there is hardness (e.g., the critique of environmentalism by the market-oriented capitalists). And the reverse direction also holds true.

In relation to the seriousness-playfulness principle, if there is seriousness (e.g., the serious business of those who practice freeganism), there is playfulness (e.g., the playful part of those who practice freeganism, when they play around with different practices over time, like the 6 characteristics aforediscussed). And the reverse direction also holds true.

In relation to the activeness-inactivess principle, if there is activeness (e.g., the activeness of those in the corporate capitalist world who are driven to work, work, and work for a good living), there is inactiveness (e.g., the relative inactiveness of those in the freeganist world who are willing to consume and work less instead, although they spend time on something else).

In relation to the selfness-otherness principle, if there is selfness (e.g., the tendency to be mostly centered on one's self or group, as shown in the possession of private property by the capitalists), there is otherness (e.g., the tendency to be mostly centered on others, as shown in the practice of sharing by the freegans). And the reverse direction also holds true.

In relation to the regression-progression principle, if there is regression (e.g., the regression made by freeganism, as shown in the problems pointed out by the critics), there is progression (e.g., the progress made by freeganism, as shown in its contribution to the reclaiming of wastes aforecited). And the reverse direction also holds true.

In relation to the same-difference principle, if there is similarity in outcome (e.g., the contribution to the molding and control of beliefs, values, and behaviors, regardless of whether this be done by way of freeganism, or by way of corporate capitalism), there is difference in outcome (e.g., the contribution to the molding and control of beliefs, values, and behaviors by way of freeganism for a relatively more egalitarian lifeworld—but the contribution to the molding and control of beliefs, values, and behaviors by way of corporate capitalism for a relatively more elitist lifeworld). And the reverse direction also holds true.

In relation to the stability-reaction principle, if there is stability (e.g., the relatively more stable condition of the practice of corporate capitalism in the older days, when "councils and landlords…would rather keep properties boarded up if they cannot make a sufficient profit from them," and good buildings are left unused, while there were homeless on the streets), there is reaction (e.g., the practice of corporate capitalism in the older days then led to further changes later on, as shown in the rise of freeganism nowadays, since "freegans see this practice [of leaving good buildings unused when there are homeless on the streets] as senseless and a counter-productive use of resources"). (WK 2012r; NOUSE 2008) And the reverse direction also holds true.

In relation to the functionality-nonfunctionality principle, if there is functionality (e.g., the presence of action for which something is fitted, as shown in the practice of "councils and landlords" in the older days who "would rather keep properties boarded up if they cannot make a sufficient profit from them," because this serves the function of perpetuating corporate capitalism at the time), there is nonfunctionality (e.g., the relative lack of action for which something is fitted, as shown in the practice of squatting by the freegans nowadays, such that the practice of "councils and landlords" in the older days for leaving good buildings unused is no longer so functional in our time, because of different standards and expectations, due to "the anti-globalization movement"). And the reverse direction also holds true.

And in relation to the intentionality-nonintentionality principle, if there is intention (e.g., the planning or design of something for a certain outcome, as shown in the design of freeganism for a better world based on reclaiming wastes), there is nonintention (e.g., the relative absence of planning or design of something for a certain outcome, such that the outcome can even be counter-intentional in being the opposite of what was originally planned, as shown in the unintended consequences of freeganism when implemented, in that foragers on the street suffer from disrespect by those in mainstream society, that reclaimed items in dumpsters can be unhealthy and dangerous, that it is not so practical as romantically assumed by the advocates, and that its practice can be criminal, etc.). And the reverse direction also holds true.

The Undesirability of Usefulness

This comprehensive analysis of the possibility (or impossibility) and desirability (or undesirability) of usefulness in the context of waste, from the perspectives of the mind, nature, society, and culture, is to show us the different ways in which it is both possible and desirable, but not to the extent that the spokespersons from each side would like us to believe.

Thus, usefulness has its desirability, just as it has its undesirability too, as the other side of the same coin.

As a reminder, the other side of the story concerns uselessness, the opposite of usefulness, which was already addressed in the previous chapter, Chapter Two.

These dual analyses should not be underestimated, because those on the side of usefulness often downgrade uselessness merely as the other end of waste without really appreciating it from the vantage point of uselessness too—and vice versa.

In any event, what then can now be concluded about the future of waste, at the end of this long intellectual journey about uselessness (in Chapter Two) and usefulness (in this chapter)?

This is the question which the next chapter, Chapter Four on the future of waste, is to answer—so we now turn to chapter four to address this question.

Table 3.1. Usefulness and Its Double Sides

- **Usefulness and the Mind**
  —Ex: usefulness, EPR, and the waste-recycling attitude
  —Ex: usefulness, dumpster-diving, and the waste-collecting habit

- **Usefulness and Nature**
  —Ex: usefulness, waste-to-energy, and sustainability
  —Ex: usefulness, waste, and animal products

- **Usefulness and Society**
  —Ex: usefulness, waste recycle, and the cost-benefit analysis
  —Ex: usefulness, the waste converter, and technology

- **Usefulness and Culture**
  —Ex: usefulness, zero waste, and resource redesign philosophy
  —Ex: usefulness, freeganism, and the ideology of reclaiming waste

Notes: The examples in the categories are solely illustrative (not exhaustive), and the comparison is relative (not absolute), nor are they necessarily mutually exclusive. And some can be easily re-classified elsewhere. As generalities, they allow exceptions.
Source: A summary of Ch.3 of *FPHW*

Table 3.2. Usefulness, EPR, and the Waste-Recycling Attitude

- **Goals of EPR**
  - —Ex: to "encourage companies to design products for reuse, recyclability, and materials reduction"
  - —Ex: to "correct…market signals to the consumer by incorporating waste management costs into the product's price"
  - —Ex: to "promot[e]…innovation in recycling technology"

- **Approaches to EPR**
  - —Ex: "mandatory"
  - —Ex: "negotiated"
  - —Ex: "voluntary"

- **Problems with EPR**
  - —Firstly, one main criticism of EPR is that its implementation can be limited for complex products.
  - —Secondly, another main criticism of EPR is that its application is not effective, even in relevant cases.
  - —Thirdly, still another main criticism of EPR is that it focuses too much on producer responsibility, not consumer responsibility.
  - —Fourthly, an additional main criticism of EPR is that its measurements are not consistent over time.
  - —Fifthly, another additional main criticism of EPR is that it can be counter-productive.
  - —And sixthly, still another additional main criticism of EPR is that it can be misused for something else.

Notes: The examples in the categories are solely illustrative (not exhaustive), and the comparison is relative (not absolute), nor are they necessarily mutually exclusive. And some can be easily re-classified elsewhere. As generalities, they allow exceptions.
Source: From *Sec. 3.2.1* of *FPHW*. See book for citations.

Table 3.3. Usefulness, Waste-to-Energy, and Sustainability

- **Resource Recovery Methods**
  —Ex: "biological processing"
  —Ex: "source-separated curbside collection"

- **The Case of Waste-to-Energy**
  —Ex: "WtE methods"
  —Ex: "global WtE developments"

- **Problems with Waste-to Energy**
  —Firstly, one main criticism of waste-to-energy is that it can be polluting.
  —Secondly, another main criticism of waste-to-energy is that it can use up valuable resources.
  —Thirdly, still another main criticism of waste-to-energy is that incineration competes with other waste-treatment activities.
  —Fourthly, an additional main criticism of waste-to-energy is that each WtE method has its own costs and benefits (or advantages and disadvantages).
  —And fifthly, another additional main criticism of waste-to-energy is that energy efficiency has its limits.

Notes: The examples in the categories are solely illustrative (not exhaustive), and the comparison is relative (not absolute), nor are they necessarily mutually exclusive. And some can be easily re-classified elsewhere. As generalities, they allow exceptions.
Source: From *Sec. 3.3.1* of *FPHW*. See book for more details and citations.

Table 3.4. Usefulness, Waste Recycle, and the Cost-Benefit Analysis

---

• The first disagreement between the proponents and opponents of waste recycle has to do with recycling vs. other methods.

• The second disagreement between the proponents and opponents of waste recycle has to do with economic vs. fiscal measurement.

• The third disagreement between the proponents and opponents of waste recycle has to do with whether the amount of energy saved in recycling is more than the amount of energy spent in the process.

• The fourth disagreement between the proponents and opponents of waste recycle has to do with the amount of money saved.

• The fifth disagreement between the proponents and opponents of waste recycle has to do with working conditions.

• The sixth disagreement between the proponents and opponents of waste recycle has to do with environmental impact.

• The seventh disagreement between the proponents and opponents of waste recycle has to do with income loss and social costs.

• And the eighth disagreement between the proponents and opponents of waste recycle has to do with "public participation in recycling programmes."

Notes: The examples in the categories are solely illustrative (not exhaustive), and the comparison is relative (not absolute), nor are they necessarily mutually exclusive. And some can be easily re-classified elsewhere. As generalities, they allow exceptions.
Source: From *Sec. 3.4.1* of *FPHW.* See book for more details and citations.

Table 3.5. Usefulness, Zero Waste, and Resource Redesign Philosophy

• The first dispute to be introduced here has to do with zero waste vs. recycling—as "some claim that the key component to zero waste is recycling, while others reject that notion in favor of reusing high function." (WK 2012ii)

• The second dispute to be introduced here has to do with zero waste vs. dump capacity—as some justify Zero Waste by pointing out that "many dumps are currently exceeding carrying capacity," whereas "others counter by pointing out that there are huge tracts of land available throughout the USA and other countries which could be used for dumps." (WK 2012ii)

• The third dispute to be introduced here has to do with zero waste vs. deconstruction—as some argued that zero waste is "about restructuring production and distribution systems to prevent waste from being manufactured in the first place," whereas others favor instead "deconstruction" as "construction in reverse." (WK 2012ii)

• The fourth dispute to be introduced here has to do with zero waste vs. market-based campaigns—as people disagree on the extent to which market-based campaigns can help the Zero Waste movement.

• The fifth dispute to be introduced here has to do with zero waste vs. government—as people disagree on the extent to which the state (or government) can help to promote Zero Waste.

• And the sixth dispute to be introduced here has to do with zero waste vs. nature—as people disagree on the extent to which nature is the model for zero waste.

Notes: The examples in the categories are solely illustrative (not exhaustive), and the comparison is relative (not absolute), nor are they necessarily mutually exclusive. And some can be easily re-classified elsewhere. As generalities, they allow exceptions.
Source: From *Sec. 3.5.1* of *FPHW*. See book for more details and citations.

Table 3.6. Usefulness, Freeganism, and the Ideology of Reclaiming Waste

- **Main Characteristics of Freeganism**
  —Ex: "foraging"
  —Ex: "sharing"
  —Ex: "squatting"
  —Ex: "working less"
  —Ex: "veganism"
  —Ex: "humanure"

- **Problems with the Ideology of Freeganism**
  —Firstly, one main criticism of the ideology of freeganism is that foraging on the street is often looked down upon in mainstream society.
  —Secondly, another main criticism of the ideology of freeganism is that reclaimed items in dumpsters can be unhealthy and dangerous.
  —Thirdly, still another main criticism of the ideology of freeganism is that it has some extremist roots in religion.
  —Fourthly, an additional main criticism of the ideology of freeganism is that it is so not practical as romantically assumed by the advocates.
  —And fifthly, another additional main criticism of the ideology of freeganism is that its practice can be criminal.

Notes: The examples in the categories are solely illustrative (not exhaustive), and the comparison is relative (not absolute), nor are they necessarily mutually exclusive. And some can be easily re-classified elsewhere. As generalities, they allow exceptions.
Source: From *Sec. 3.5.2* of *FPHW.* See book for more details and citations.

• Part Four •

Conclusion

CHAPTER 4
CONCLUSION—THE FUTURE OF WASTE

Pessimism is a waste of time.
__Norman Cousins (BRA 2012)

Beyond Uselessness and Usefulness

The analyses of uselessness (in Chapter Two) and usefulness (in Chapter Three) are to reveal the extent of their possibility (or impossibility) and desirability (or undesirability) in the study of waste.

These two analyses are essential for the evaluation of the opposing practices of waste treatment (and other views as already introduced in *Sec. 1.5* and the rest of the book).

As already indicated in *Sec.1.1*, this challenge to the opposing views of waste does not imply that waste has no practical value, or that those interdisciplinary fields (related to waste) like epidemiology, global warming, waste management, low-carbon economics, ethical consumerism, resource recovery, freeganism, environmental justice, space debris, and so on are unimportant. Of course, neither of these extreme views is reasonable.

Rather, this book offers an alternative (better) way to understand the future of waste, especially in the dialectic context of uselessness and usefulness—while learning from different approaches in the literature but without favoring any one of them (nor integrating them, since they are not necessarily compatible with each other).

More specifically, this book offers a new theory (that is, *the transfigurative theory of waste*) to go beyond the existing approaches in a novel way.

As a recap from Chapter One, my *transfigurative theory of waste* has four major distinct features.

Firstly, my theory makes good use of all theoretical approaches to the study of waste in relation to uselessness and usefulness, be they about the mental argument, the natural argument, the societal argument, or the cultural argument—especially from the most comprehensive combined perspectives of the mind, nature, society, and culture (as already dealt with in Chapter Two and Chapter Three).

Secondly, just like many other theories of mine in my previous books, my theory here does not heavily favor any specific theory over others in the literature, nor trying to integrate them (as they are not necessarily compatible with each other).

Thirdly, my theory treats the issue of waste in the distinctive dialectic context of no uselessness without usefulness, and vice versa.

And fourthly, it contains twenty-nine major theses, namely, (a) the first thesis: the formalness-informalness principle, (b) the second thesis: the absoluteness-relativeness principle, (c) the third thesis: the partiality-totality principle, (d) the fourth thesis: the predictability-unpredictability principle, (e) the fifth thesis: the explicability-inexplicability principle, (f) the sixth thesis: the fiction-reality principle, (g) the seventh thesis: the finiteness-transfiniteness principle, (h) the eighth thesis: the preciseness-vagueness principle, (i) the ninth thesis: the simpleness-complicatedness principle, (j) the tenth thesis: the openness-hiddenness principle, (k) the eleventh thesis: the denseness-emptiness principle, (l) the twelfth thesis: the change-constancy principle, (m) the thirteenth thesis: the order-chaos principle, (n) the fourteenth thesis: the slowness-quickness principle, (o) the fifteenth thesis: the expansion-contraction principle, (p) the sixteenth thesis: the theory-praxis principle, (q) the seventeenth thesis: the convention-novelty principle, (r) the eighteenth thesis: the evolution-transformation principle, (s) the nineteenth thesis: the symmetry-asymmetry principle, (t) the twentieth thesis: the softness-hardness principle, (u) the twenty-first thesis: the seriousness-playfulness principle, (v) the twenty-second thesis: the activeness-inactiveness principle, (w) the twenty-third thesis: the selfness-otherness principle, (x) the twenty-fourth thesis: the regression-progression principle, (y) the twenty-fifth thesis: the sameness-difference principle, (z) the twenty-sixth thesis: the stability-reaction principle, (aa) the twenty-seventh thesis: the functionality-nonfunctionality principle, (bb) the twenty-eighth thesis: the intentionality-nonintentionality principle, and (cc) the twenty-ninth thesis: the post-human transfiguration—to be elaborated in the rest of the book, with a summary in the concluding chapter.

Because of the last thesis on "the post-human transfiguration," the word "transfigurative" in the title of my theory are suggestive, because it refers to the striking changes of waste over time—which then require a multifaceted exploration of the dialectic relations between uselessness and usefulness, as well as an understanding of the different views regarding these dialectic relations in the study of waste (together with other debates as will be discussed in the rest of the book)—and in the end it calls for a need to transcend them by future advanced humans and later post-humans in the context of the great transformations of waste in the future world (as will be discussed in the rest of this chapter).

With this clarification in mind—the seminal project here, if successful, will fundamentally change the way that we think about uselessness and usefulness in the study of waste together with other debates as already discussed in the rest of the book), from the combined perspectives of the mind, nature, society, and culture, with enormous implications for the human future and what I originally called its "post-human" fate.

In Relation to Method

The first six theses are in relation to method in existential dialectics, as shown below.

1st Thesis: The Formalness-Informalness Principle

The first thesis refers to the formalness-informalness principle (on the formal requirements of logical systems) in the category of "method" in existential dialectics, which was first proposed in *FPHML* (and later, *FPHH*, *FPHGAM*, *FPHSEM*, and other books of mine), in that the formal requirements of a logical system (e.g., consistency, soundness, and completeness) have both usefulness and non-usefulness, to the extent that, if there are formal systems requiring them, there are alternative ones which do not.

The formalness-informalness principle reveals something interesting in existential dialectics, in that it goes beyond both classical and non-classical logics (while learning something from both).

Thus, it does *not* exclude classical logics but simply goes beyond both classical and non-classical logics, while learning something from each. There is no formalness without informalness—and vice versa.

For illustration, existential dialectics can make use of both classical logics under certain conditions (e.g., especially, though not exclusively, when they are clear-cut, etc.) and non-classical logics under alternative conditions (especially, though not exclusively, when they are "unknown," "irrelevant," "ambiguous," "possible," with "different degrees of truth," empirically inconsistent in a desirable way, etc.). (WK 2008u)

For instance, on the one hand, there is formalness (e.g., the formal logical requirement of soundness and completeness in a system of ideas, as in the study of biodegradable waste and epidemics, in *Sec. 2.2.1*; the formal logical requirement of soundness, consistency, and completeness in a system of ideas, as in the study on clinical waste and diseases, in *Sec. 2.2.2*; the formal logical requirement of soundness and completeness in a system of ideas, as in the study on litter and the environment, in *Sec. 2.3.1*; the formal logical requirement of soundness and completeness in a system of ideas, as in the study on space debris, in *Sec. 2.3.2*; the formal logical requirement of soundness and completeness in a system of ideas, as in the creation of the Basel Convention, in *Sec. 2.4.1*; the formal logical requirement of soundness and completeness in a system of ideas, as in the launch of the environmental justice movement, in *Sec. 2.4.2*; the formal logical requirement of soundness, consistency, and completeness in a system of ideas, as in ethcal consumerism, in *Sec. 2.5.1*; the formal logical requirement of soundness, completeness, and consistency in a system of ideas, as in the simple living movement, in *Sec. 2.5.2*; the formal logical requirement of soundness, completeness, and consistency in a system of ideas, as in EPR, in *Sec. 3.2.1*; the formal logical requirement of soundness and completeness in a system of ideas, as in the practice of dumpster diving, in *Sec. 3.2.2*; the formal logical requirement of soundness and completeness in a system of ideas, as in the idea of WtE, in *Sec. 3.3.1*; the formal logical requirement of soundness and completeness in a system of ideas, as in the use of animal products and animal by-products, in *Sec. 3.3.2*; the formal logical requirement of soundness, consistency, and completeness in a system of ideas, as in the debate on the cost-benefit analysis of waste recycle, in *Sec. 3.4.1*; the formal logical requirement of soundness and completeness in a system of ideas, as in the use of the waste converter, in *Sec. 3.4.2*; the formal logical requirement of soundness, consistency, and completeness in a system of ideas, as in the call for Zero Waste, in *Sec. 3.5.1*; and the formal logical requirement of soundness and completeness in a system of ideas, as in the ideology of freeganism, in *Sec. 3.5.2*).

On the other hand, there is also informalness (e.g., the non-formal existence of unsoundness and incompleteness in the study of biodegradable waste and epidemics, as shown in the criticisms that epidemics can produce desirable effects instead, that epidemics can be stopped or prevented, that it is not clear to what extent exactly biodegradable waste and biowaste have really caused epidemics in history, etc., in *Sec. 2.2.1*; the non-formal existence of unsoundness, inconsistency, and incompleteness in the study on clinical waste and diseases, as shown in the criticisms that the governmental regulation of clinical waste disposal can be costly and thus produce incentives to violate it, that there are no universal agreement about its enforcement, that the methods of clinical waste disposal have their own limits, that what constitutes "clinical waste" [or "medical waste"] depends on different definitions and has crucial implications in practice, etc., in *Sec. 2.2.2*; the non-formal existence of unsoundness and incompleteness in the study on litter and the environment, as shown in the criticisms that the prevention of littering is too complicated in practice, that many cases of littering are unintentional [which is subject to different uncertain circumstances], that each proposed solution has its own disadvantages, that there is no universal regulation of littering behavior, etc., in *Sec. 2.3.1*; the non-formal existence of unsoundness and incompleteness in the study on space debris, as shown in the criticisms that the measurement of space debris is not an accurate business in the first place, that to minimize space debris by preventing space collisions is easier said than done technologically, that many of the proposed solutions are too expensive to be cost-effective, that there is no international treaty banning space debris, that there are simply too many pieces of space debris to be able to cope with them all or even most of them, etc., in *Sec. 2.3.2*; the non-formal existence of unsoundness and incompleteness in the creation of the Basel Convention, as shown in the criticisms that what constitutes hazardous waste is subject to contention, that there are questionable rules about "obligations" in the Basel Convention, that there is no enforcement provisions in the Basel Convention, that it does not go far enough to protect less developed countries from the spread of hazardous waste, etc., in *Sec. 2.4.1*; the non-formal existence of unsoundness and incompleteness in the launch of the environmental justice movement, as shown in the criticisms that what constitutes fairness or its opposite is in the eyes of the beholder, that the environmental movement has been elitist in producing the opposite effects, that the legal process is too costly, that there is no universal agreement on the definition of "environmental justice" and "environmental injustice," etc., in *Sec. 2.4.2*; the non-formal existence of unsoundness, inconsistency, and incompleteness in ethical consumerism, as shown in the criticisms that the actual effect of ethical consumerism is limited, that moral boycott can be based on incorrect or limited information, that it does not give sufficient attention to the power structure of society, that it can be counter-productive in encouraging immoral behavior instead, that ethical consumerism is a mistake in thinking, that there are inconsistent definitions of what constitutes something "moral" in the first place, etc., in *Sec. 2.5.1*; the non-formal existence of unsoundness, incompleteness, and inconsistency in the simple living movement, as shown in the criticisms that simple living can be decadent, that simple living is not possible for everyone, that it is ideological, in that simple lifestyle has been used by different groups for their ulterior purposes, that it is not so practical as romantically assumed by the advocates, that different arguments by different advocates in the movement can be mutually inconsistent or contradictory, etc., in *Sec. 2.5.2*; the non-formal existence of unsoundness, incompleteness, and inconsistency in EPR, as shown in the criticisms that its implementation can be limited for complex electronic products, that its application is not effective, that it focuses too much on producer responsibility, that its measurements are not consistent over time, that it can be counter-productive, that it can be misused for something else, etc., in *Sec. 3.2.1*; the non-formal existence of unsoundness and incompleteness in the practice of dumpster diving, as shown in the criticisms that it can be illegal, that the practice of dumpster diving is not well respected by those in mainstream society, that it is unhealthy, that it is dangerous, that the items in dumpsters are of inferior quality, etc., in *Sec. 3.2.2*; the non-formal existence of unsoundness and incompleteness in the idea of WtE, as shown in the criticisms that it can be polluting, that it can use up valuable resources, that incineration competes with other waste-treating activities, that each WtE method has its own costs and benefits, that energy efficiency has its limits, etc., in *Sec. 3.3.1*; the non-formal existence of unsoundness and incompleteness in the use of animal products and animal by-products, as shown in the criticisms that the use of animal products face competition from alternative sources, that their use is prohibited by certain groups for different moral and religious reasons, that their use is also not recommended for health reason, etc., in *Sec. 3.3.2*; the non-formal existence of unsoundness, inconsistency, and incompleteness in the debate on the cost-benefit analysis of waste recycle, as shown in the criticisms by the critics in regard to the 8 issues afore-discussed, in *Sec. 3.4.1*; the non-formal existence of unsoundness and incompleteness in the use of the waste converter, as shown in the criticisms that its application is limited, that there are competing technologies, that it is not necessarily safer, etc., in *Sec. 3.4.2*; the non-formal existence of unsoundness, inconsistency, and incompleteness in the call for Zero Waste, as shown in the 6 disputes about Zero Waste aforecited, in *Sec. 3.5.1*; and the non-formal existence of unsoundness and incompleteness in the ideology of freeganism, as shown in the criticisms that foraging on the street is often looked down upon by those in mainstream society, that reclaimed items in dumpsters can be unhealthy and dangerous, that it has some extremist roots in religion, that it is not so practical as romantically assumed by the advocates, that its practice can be criminal, etc., in *Sec. 3.5.2*).

And the reverse direction also holds true.

However, it should be stressed (as this is something that I regularly did in my previous books, *almost verbatim*) that there are different shades of gray (or different degrees of truth) in the two opposites, and the classification is not necessarily mutually exclusive either. In fact, even when some combinations of the two occur, they only end up sharing the same dialectic relationship, but in a different degree.

Also, there are some other relationships (as a kind of family resemblance) which have something in common with the principle, but they are not exactly the same but only more or less comparable, which varies from case to case.

Good examples of family resemblance in relation to the principle are logicalness-nonlogicalness, rationality-nonrationality, etc.

2nd Thesis: The Absoluteness-Relativeness Principle

The second thesis refers to the absoluteness-relativeness principle (on the multiplicity of things) in the category of "method" in existential dialectics, which was first proposed in *FPHK*, in that there is the multiplicity of things in reality, be they about entities, qualities (or properties), and relationships. If there is something absolute, there is likewise something relative. And there is no absoluteness without relativeness—and vice versa.

Both absoluteness and relativeness here are also relevant to different modalities often cited in the literature on ontology, such as possibility (e.g., something "can" happen) and its opposite (e.g., impossibility), probability (e.g., something "will" happen) and its opposite (e.g., improbability), and necessity (e.g., something "should" happen) and its opposite (e.g., contingency).

For instance, on the one hand, there is absoluteness (e.g., the absolute view of the Imperial Japanese Army Air Service when bombing Ningbo with fleas carrying the bubonic plague in 1940, in *Sec. 2.2.1*; the absolute view of those "private waste contractors to dump illegally to avoid high fees" in the 1980s aforecited, in *Sec. 2.2.2*; the absolute view of those "Dutch police and local supervisors [known as *buitengewoon opsporingsambtenaar*, or BOA]," who "fine citizens for throwing away cans, bottles or wrappers onto the street," in *Sec. 2.3.1*; the absolute view of the ITU, which "has placed increasingly strict requirements on the station-keeping ability of new satellites and demands that the owners guarantee their ability to safely move the satellites out of their orbital slots at the end of their lifetimes," in *Sec. 2.3.2*; the absolute view of "Greenpeace and key European countries such as Denmark" for "the Basel Ban Amendment to the Basel Convention," in *Sec. 2.4.1*; the absolute view of the advocates of the Environmental Justice movement and the Civil Rights Movement for "social justice, equal protection, and an end to institutional discrimination," in *Sec. 2.4.2*; the absolute view of the advocates of ethical consumerism in regard to the moral basis, in *Sec. 2.5.1*; the absolute view of those people in the simple living movement who "eschew all technology" and "are often referred to as…neo-Luddites," in *Sec. 2.5.2*; the absolute view of EPR in terms of focusing on producer responsibility, in *Sec. 3.2.1*; the absolute view of "a law issued in 2000" in Italy which "declared dumpster diving to be legal," in *Sec. 3.2.2*; the absolute view of those who favor incineration, in *Sec. 3.3.1*; the absolute view of certain groups for vegetarian, kosher, and halaal diets against the inclusion of some non-human animal products, in *Sec. 3.3.2*; the absolute view of those "businesses" which "ignore them [negative externalities] despite the costs imposed on society," in *Sec. 3.4.1*; the absolute view of those enthusiastic advocates of the use of waste converters, in *Sec. 3.4.2*; the absolute view of those advocates of Zero Waste "in favor of reusing high function," in *Sec. 3.5.1*; and the absolute religious view of freeganism in regard to the idea of "religious obligation to nature in the form of political action," in *Sec. 3.5.2*).

On the other hand, there is relativeness (e.g., what is right for the Imperial Japanese Army Air Service to bomb Ningbo with fleas carrying the bubonic plague in 1940 is not necessarily so for the Chinese victims and their family members, in *Sec. 2.2.1*; what is right for those "private waste contractors to dump illegally to avoid high fees" in the 1980s aforecited is not necessarily so for the federal government to enforce proper clinical waste disposal under the law, in *Sec. 2.2.2*; what is right for those "Dutch police and local supervisors [known as *buitengewoon opsporingsambtenaar*, or BOA]," who "fine citizens for throwing away cans, bottles or wrappers onto the street," is not necessarily so for those citizens who are fined but disagree about the punishment, in *Sec. 2.3.1*; what is right for the ITU in regard to its "strict requirements on the station-keeping ability of new satellites" is not necessarily so for the critics, since "studies have suggested that even the existing ITU requirements are not enough to have a major effect on collision frequency," in *Sec. 2.3.2*; what is right for "Greenpeace and key European countries such as Denmark" for "the Basel Ban Amendment to the Basel Convention" is not necessarily so for "a number of industry groups as well as nations including Australia and Canada" against the amendment, in *Sec. 2.4.1*; what is right for the advocates of the Environmental Justice movement and the Civil Rights Movement in regard to "social justice, equal protection, and an end to institutional discrimination" is not necessarily so for the opponents like some conservative industry groups, white supremacists, etc., in *Sec. 2.4.2*; what is right for the advocates of ethical consumerism in regard to the moral basis is not necessarily so for the critics, who argued that moral boycott can be based on incorrect or limited information, that it can be counter-productive in encouraging immoral behavior instead, that ethical consumerism is a mistake in thinking, etc., in *Sec. 2.5.1*; what is right for those people in the simple living movement who "eschew all technology" and "are often referred to as...neo-Luddites" is not necessarily so for others in the movement, who have the opposite interpretation, in that they "see cutting-edge technologies as a way to make a simple lifestyle within mainstream culture easier and more sustainable," in *Sec. 2.5.2*; what is right for those in EPR to focus on producer responsibility is not necessarily so for the critics, who ask for a focus on consumer responsibility too, in *Sec. 3.2.1*; what is right for "a law issued in 2000" in Italy which "declared dumpster diving to be legal" is not necessarily so for the law in Germany where "a dumpster's contents are regarded as the property of the dumpster's owner," so "taking items from a dumpster is viewed as theft," in *Sec. 3.2.2*; what is right for those who favor incineration is not necessarily so for the critics who are concerned that "incinerators destroy valuable resources" and that "they may reduce the incentives for recycling and waste minimization activities," in *Sec. 3.3.1*; what is right for certain groups like those for vegetarian, kosher, and halaal diets in regard to their prohibition of the inclusion of some non-human animal products is not necessarily so for others who consumer foods based on animal products and animal by-products, in *Sec. 3.3.2*; what is good for those "businesses" which "ignore them [negative externalities] despite the costs imposed on society" is not necessarily so for the critics, who, "to make such non-fiscal benefits economically relevant,...have pushed for legislative action to increase the demand for recycled materials," in *Sec. 3.4.1*; what is right for the enthusiastic advocates of the use of waste converters is not necessarily so for the critics, because "much of the difficulty inherent in recycling comes from the fact that most products are not designed with recycling in mind," in *Sec. 3.4.2*; what is right for those advocates of Zero Waste "in favor of reusing high function" is not necessarily so for others who favor recycling instead, in *Sec. 3.5.1*; and what is right for the religious view of freeganism in regard to the idea of "religious obligation to nature in the form of political action" is not necessarily so for the critics, who question their religious fanaticism, in *Sec. 3.5.2*).

And the reverse direction also holds true.

However, it should be stressed (as this is something that I regularly did in my previous books, *almost verbatim*) that there are different shades of gray (or different degrees of truth) in the two opposites, and the classification is not necessarily mutually exclusive either. In fact, even when some combinations of the two occur, they only end up sharing the same dialectic relationship, but in a different degree.

Also, there are some other relationships (as a kind of family resemblance) which have something in common with the principle, but they are not exactly the same but only more or less comparable, which varies from case to case.

Good examples of family resemblance in relation to the absoluteness-relativeness principle are internalness-externalness, immanence-transcendence, internality-externality, isolatedness-interconnectiveness, autonomy-interdependency, universalness-relativeness, independence-dependence, and so forth.

3rd Thesis: The Partiality-Totality Principle

The third thesis refers to the partiality-totality principle (on the relationships between whole and parts) in the category of "method" in existential dialectics, which was first proposed in *FC* and later *FPHGEOL*, based on earlier works like *FHC* and *FPHC*), in that any analysis of a phenomenon in the world requires both the examination of the individual parts and the whole of them (with the whole as being not the sum of its parts), without privileging the whole as more important than the parts, or vice-versa (that is, the parts as more important than the whole).

In fact, this principle targets two kinds of misleading methodologies as used over the ages, that is, again in the absence of better words, what I called (in *FHC*, *FPHC* and *FC*) "reductionism" and its opposite, "reverse-reductionism," which can take four major forms in relation to concept, theory, methodology, and ontology (as summarized in *Sec. 1.8*).

Thus, reductionism and reverse-reductionism can be (i) conceptual, (ii) theoretical, (iii) methodological, and (iv) ontological. And the whole is not the sum of its parts, and there is no partiality without totality—and vice versa.

Works on emergence (as in *FC*), gestalt psychology (as in *FPHGEOL*), and the hermeneutic circle (as in *FPHDA*), just to name a few of them, have contributed to the understanding of whole and parts.

For instance, on the one hand, there is partiality (e.g., the partial view of the bubonic plague as a biological weapon by the Imperial Japanese Army Air Service when bombing Ningbo with fleas carrying the bubonic plague in 1940, in *Sec. 2.2.1*; the partial view of those "private waste contractors to dump illegally to avoid high fees" in the 1980s aforecited, in *Sec. 2.2.2*; the partial view of litter by the advocates of "Clean Up the World," which is a worldwide anti-litter campaign, in *Sec. 2.3.1*; the partial view of the ITU in regard to its "strict requirements on the station-keeping ability of new satellites" to manage space debris, in *Sec. 2.3.2*; the partial view of "Greenpeace and key European countries such as Denmark" for "the Basel Ban Amendment to the Basel Convention," in *Sec. 2.4.1*; the partial view of the advocates of the Environmental Justice movement and the Civil Rights Movement for "social justice, equal protection, and an end to institutional discrimination," in *Sec. 2.4.2*; the partial view of the advocates of ethical consumerism in regard to the moral basis, in *Sec. 2.5.1*; the partial view of those people in the simple living movement who "eschew all technology" and "are often referred to as…neo-Luddites," in *Sec. 2.5.2*; the partial view of EPR in terms of its focus on the producer side for waste recycling, in *Sec. 3.2.1*; the partial view of "a law issued in 2000" in Italy which "declared dumpster diving to be legal," in *Sec. 3.2.2*; the partial view of the advocates of each WtE method, in *Sec. 3.3.1*; the partial view of certain groups for vegetarian, kosher, and halaal diets against the inclusion of some non-human animal products, in *Sec. 3.3.2*; the partial view of those "businesses" which "ignore them [negative externalities] despite the costs imposed on society," in *Sec. 3.4.1*; the partial view of of those enthusiastic advocates of the use of waste converters, in *Sec. 3.4.2*; the partial view of those advocates of Zero Waste in favor of governance, in *Sec. 3.5.1*; and the partial view of freeganism on reclaiming wastes, in *Sec. 3.5.2*).

On the other hand, there is also totality (e.g., the more holistic view of the bubonic plague, such that the whole is not the sum of its parts, that is, it cannot be reduced to the sum of opposing views of the Imperial Japanese Army Air Service, the Chinese victims, etc., since there will emerge new views in the future not yet known today, just as there are already alternative views nowadays which question the claims by the Chinese Communist Party and their critics about the incident, in *Sec. 2.2.1*; the more holistic view of clinical waste disposal, such that the whole is not the sum of its parts, that is, it cannot be reduced to the sum of opposing views of those "private waste contractors" to dump illegally to avoid high fees" in the 1980s, the federal government, etc., since there will emerge new views in the future not yet known today, just as there are already criticisms against the two sides nowadays, in *Sec. 2.2.2*; the more holistic view of litter, such that the whole is not the sum of its parts, that is, it cannot be reduced to the sum of opposing views held by the advocates of "Clean Up the World" and their critics, since there will emerge new views in the future not yet known today, just as there are already critics about both sides nowadays, in *Sec. 2.3.1*; the more holistic view of the future of space debris, such that the whole is not the sum of its parts, that is, it cannot be reduced to the sum of the opposing views of the ITU, the skeptics, and others, since there will emerge new views in the future not yet known today, just as there are already criticisms of them all nowadays, in *Sec. 2.3.2*; the more holistic view of the future of the Basel Convention, such that the whole is not the sum of its parts, that is, it cannot be reduced to the sum of the opposing views of the pros and cons of "the Basel Ban Amendment to the Basel Convention," since there will emerge new views in the future not yet known today, just as there are already criticisms against both groups nowadays, in *Sec. 2.4.1*; the more holistic view of the future of environmental justice, such that the whole is not the sum of its parts, that is, it cannot be reduced to the sum of opposing views of the advocates of the Environmental Justice movement and their critics, since there will emerge new views in the future not yet known today, just as there are already criticisms against them all nowadays, in *Sec. 2.4.2*; the more holistic view of the future of ethical consumerism, such that the whole is not the sum of its parts, that is, it cannot be reduced to the sum of opposing views of the advocates and the critics, since there will emerge new views in the future not yet known today, just as there already some nowadays who do not side with either of the two opposing sides, in *Sec. 2.5.1*; the more holistic view of the future of technology, such that the whole is not the sum of its parts, that is, it cannot be reduced to the sum of opposing views of the neo-Luddites and their critics, since there will emerge new views in the future not yet known today, just as there are already different views nowdays which question both sides of the debate, in *Sec. 2.5.2*; the more holistic view of waste recycling, such that the whole is not the sum of its parts, that is, it cannot be reduced to the sum of opposing views of those for EPR and their critics, since there will emerge new views in the future not yet known today, just as there are already criticisms against them all nowadays, in *Sec. 3.2.1*; the more holistic view of the permissability of dumpster diving, such that the whole is not the sum of its parts, that is, it cannot be reduced to the sum of opposing views under the law in Italy, Germany, etc., since there will emerge new views in the future not yet known today, just as there are already other views which are critical to them all nowadays, in *Sec. 3.2.2*;

the more holistic view of WtE, such that the whole is not the sum of its parts, that is, it cannot be reduced to the sum of opposing views of the different advocates of different WtE methods and their critics, since there will emerge new views in the future not yet known today, just as there are already criticisms against each of them nowadays, in *Sec. 3.3.1*; the more holistic view of the future of non-human animal products, such that the whole is not the sum of its parts, that is, it cannot be reduced to the sum of opposing views of certain groups for vegetarian, kosher, and halaal diets and those for the inclusion of some non-human animal products, since there will emerge new views in the future not yet known today, just as there are already those nowadays who do not side with either side completely, in *Sec. 3.3.2*; the more holistic view of waste recycle, such that the whole is not the sum of its parts, that is, it cannot be reduced to the sum of opposing views of those businesses who ignore negative externalities and their critics for waste recycle, since there will emerge new views in the future not yet known today, just as there are already other views which do not side with the opposing sides nowadays, in *Sec. 3.4.1*; the more holistic view of the future of waste conversion, such that the whole is not the sum of its parts, that is, it cannot be reduced to the sum of opposing views of those enthusiastic advocates of the use of waste converters and their critics, since there will emerge new views in the future not yet known today, just as there are already views nowadays which do not side with any of them, in *Sec. 3.4.2*; the more holistic view of the future of Zero Waste, such that the whole is not the sum of its parts, that is, it cannot be reduced to the sum of opposing views of those advocates of Zero Waste in favor of governance and those in favor of government, since there will emerge new views in the future not yet known today, just as there are already others who are critical to both sides nowadays, in *Sec. 3.5.1*; and the more holistic view of the future of reclaiming wastes, such that the whole is not the sum of its parts, that is, it cannot be reduced to the sum of opposing views of the freegans and the critics, since there will emerge new views in the future not yet known today, just as there are already other views nowadays which do not side with either, in *Sec. 3.5.2*).

And the reverse direction also holds true.

However, it should be stressed (as this is something that I regularly did in my previous books, *almost verbatim*) that there are different shades of gray (or different degrees of truth) in the two opposites, and the classification is not necessarily mutually exclusive either. In fact, even when some combinations of the two occur, they only end up sharing the same dialectic relationship, but in a different degree.

Also, there are some other relationships (as a kind of family resemblance) which have something in common with the principle, but they are not exactly the same but only more or less comparable, which varies from case to case.

Good examples of family resemblance in relation to the principle are individualisticness-holisticness, partness-wholeness, analysis-synthesis, and the like.

4th Thesis: The Predictability-Unpredictability Principle

The fourth thesis is called the predictability-unpredictability principle (on the occurrence of events) in the category of "method" in existential dialectics (which was first proposed in *FC* and, later, other books of mine), in that both predictability and unpredictability have a major role to play in the occurrence of things, so that neither determinism nor indeterminism wins the centuries-old fight. There is no predictability without unpredictability—and vice versa.

There are events which are predictable, just as there are those which are not. Or what is regarded as unpredictable at one point in time may turn out to be predictable later, and, conversely, what is deemed as predictable may turn out to not be so predictable. Even in predictability, outcomes are subject to uncertainty, the degree of which varies from case to case.

Works on complexity theory (as in *FC*), quantum mechanics (as in *FPHST* and *FPHG*), frequency probability, subjective (or Bayesian) probability, and "special causes" (as in *FPHGAM*), just to name a few of them, have contributed to the understanding of uncertainity.

For instance, on the one hand, there is predictability (e.g., the predictable tendency of scientists to use antibiotics like "streptomycin," "gentamicin," and "tetracyclines" for the treatment of bubonic plague, in *Sec. 2.2.1*; the predictable tendency of the federal government to enforce proper clinical waste disposal under a particular law like the MWTA, in *Sec. 2.2.2*; the predictable tendency of the advocates of "Clean Up the World" to campaign against litter, in *Sec. 2.3.1*; the predictable tendency of "the United Nations Committee on the Peaceful Uses of Outer Space" or COPUOS to provide "voluntary guidelines" for space debris management in the foreseeable future, in *Sec. 2.3.2*; the predictable tendency of the parties in the Basel Convention to advocate "a general prohibition on the exportation or importation of wastes between Parties and non-Parties," in *Sec. 2.4.1*; the predictable tendency of the Environmental Justice movement nowadays to encourage the poor to participate, in *Sec. 2.4.2*; the predictable tendency of the proponents of ethical consumerism to advocate moral or positive buying, in *Sec. 2.5.1*; the predictable tendency of those in the simple living movement to advocate simple lifestyle, in *Sec. 2.5.2*; the predictable tendency of those for EPR to advocate producer responsibility, in *Sec. 3.2.1*; the predictable tendency of dumpster divers to "occasionally get in trouble for trespassing" if "dumpsters are...located on private premises," in *Sec. 3.2.2*; the predictable tendency of "incineration" to be "the most common WtE implementation" in the foreseeable future, in *Sec. 3.3.1*; the predictable tendency of AAFCO to claim that "using the term 'human grade' is 'false and misleading'" in the foreseeable future, in *Sec. 3.3.2*; the predictable tendency of those in "the study conducted by the Technical University of Denmark" to argue that "in 83 percent of cases, recycling is the most efficient method to dispose of household waste," in *Sec. 3.4.1*; the predictable tendency of those enthusiastic advocates of waste converters to favor their use, in *Sec. 3.4.2*; the predictable tendency of those who prefer construction for Zero Waste to argue that "the materials that are still required in the...re-designed, resource-efficient systems will be reused many times as the products that incorporate them are reused," in *Sec. 3.5.1*; and the predictable tendency of the freegans to practice squatting, in *Sec. 3.5.2*).

On the other hand, there is unpredictability (e.g., the more difficult task to predict exactly the extent to which the use of antibiotics like "streptomycin," "gentamicin," and "tetracyclines" will be successful on a particular future occasion of plague occurrence, since they are not totally effective as "mortality associated with treated cases of bubonic plague" still happens, in *Sec. 2.2.1*; the more difficult task to predict exactly the extent to which the federal government will be successful in enforcing proper clinical waste disposal under a particular law in a particular future era, since different laws like the MWTA are passed and then repealed or revised over time in history, in *Sec. 2.2.2*; the more difficult task to predict exactly the extent to which the advocates of "Clean Up the World" will be successful in the campaign against litter on a particular future occasion, because many cases of littering are unintentional [which is subject to different uncertain circumstances], so it is difficult to control littering behavior, in *Sec. 2.3.1*; the more difficult task to predict exactly the extent to which COPUOS will be successful in managing space debris on a particular farther future occasion, since "there is no international treaty mandating behaviour to minimize space debris," in *Sec. 2.3.2*; the more difficult task to predict exactly the extent to which the Basel Convention will be successful in banning "the exportation or importation of wastes between Parties and non-Parties" on a particular future occasion, because it "contains no enforcement provisions," in *Sec. 2.4.1*; the more difficult task to predict exactly the extent to which the Environmental Justice Movement nowadays will be successful in encouraging the poor to participate in a particular future historical era, because of the problems that it has been plagued with "three different forms" of elitism, that it is not cost-effective, etc., in *Sec. 2.4.2*; the more difficult task to predict exactly the extent to which ethical consumerism will be effective on a particular future occasion, since, for the critics, the actual effect of ethical consumerism is limited in niche markets, that moral boycott can be based on incorrect or limited information, etc., in *Sec. 2.5.1*; the more difficult task to predict exactly the extent to which simple lifestyle will remain influential in a particular future historical era, since there are already different criticisms against it nowadays, in *Sec. 2.5.2*; the more difficult task to predict exactly the extent to which EPR will be successful on a particular future occasion, because of the numerous problems pointed out by the critics aforecited, in *Sec. 3.2.1*; the more difficult task to predict exactly the extent to which dumpster divers "get in trouble for trespassing" on a particular future occasion, because "the law is enforced with varying degrees of rigor" or can even be ignored by police, depending on different circumstances, in *Sec. 3.2.2*; the more difficult task to predict exactly the extent to which "incineration" will remain "the most common WtE implementation" in a particular more distant future historical era, since there are already competing technologies and waste-treatment alternatives nowadays, in *Sec. 3.3.1*; the more difficult task to predict exactly the extent to which the term "human grade" will not be used "on websites or other advertising," because "the AAFCO only governs the pet food packaging and has no authority over websites or advertising," although things may change later, in *Sec. 3.3.2*; the more difficult task to predict exactly the extent to which recycling is the most efficient method on a particular future occasion, since, in another study in 2004 by "the Danish Environmental Assessment Institute," which "concluded that incineration was the most effective method for disposing of drink containers, even aluminium ones" instead, in *Sec. 3.4.1*; the more difficult task to predict exactly the extent to which the use of waste converters will remain influential in a particular future historical era, since there are already competing technologies to challenge waste converters aforecited, in *Sec. 3.4.2*; the more difficult task to predict exactly the extent to which the construction argument for Zero Waste will remain influential in a particular future historical era, because, for the critics, "deconstruction can capture materials and some components from the millions of buildings that are existing and that were poorly designed for high level reuse" and moreover, "until buildings are built as components capable of later dismantling, deconstruction is a stop-gap process that the United States can use to minimize the waste of building materials," in *Sec. 3.5.1*; and the more difficult task to predict exactly the extent to which squatting is pervasive on a particular future occasion, since, for the critics, squatting in many localities are illegal and criminal, so for this reason, "the criminality of squatting makes it hard to accurately track the number of people involved in this activity," in *Sec. 3.5.2*).

And the reverse direction also holds true.

Again, it should be stressed, however, that there are different shades of gray (or different degrees of truth) in the two opposites, and the classification is not necessarily mutually exclusive either. In fact, even when some combinations of the two occur, they only end up sharing the same dialectic relationship, but in a different degree.

Also, there are some other relationships (as a kind of family resemblance) which have something in common with the principle, but they are not exactly the same but only more or less comparable, which varies from case to case.

Good examples of family resemblance in relation to the principle are sureness-arbitrariness, deterministicness-randomness, likeliness-riskiness, sureness-doubtfulness, deterministicness-indeterministicness, certainty-uncertainty, and so forth.

5th Thesis: The Explicability-Inexplicability Principle

The fifth thesis is called the explicability-inexplicability principle (on the underlying mechanisms of things) in the category of "method" in existential dialectics (which was proposed in *FPHU* and other books of mine), in that both explicability and inexplicability are part of the understanding of things. There is no explicability without inexplicability—and vice versa.

This principle tells us the dual nature of the research dilemma, in that, if reality can be explained in some ways, it also has its other ways which are not quite explainable, at a given point in time.

For instance, on the one hand, there is explicability (e.g., the explanation by the Imperial Japanese Army Air Service to bomb Ningbo with fleas carrying the bubonic plague in 1940 because it is an effective biological weapon, in *Sec 2.2.1*; the explanation by the advocates of the MWTA that proper clinical waste disposal is important to avoid "the spread of infectious disease," in *Sec. 2.2.2*; the explanation by "Dutch police and local supervisors [known as *buitengewoon opsporingsambtenaar*, or BOA]," who "fine citizens for throwing away cans, bottles or wrappers onto the street" on the basis of anti-litter enforcement, in *Sec. 2.3.1*; the explanation, by the advocates of different ways of "external removal" to manage space debris, on the basis of their technological feasibility, in *Sec. 2.3.2*; the explanation by those in "the 1988 Koko case," in that "5 ships transported 8,000 barrels of hazardous waste from Italy to the small town of Koko in Nigeria in exchange for $100 monthly rent which was paid to a Nigerian for the use of his farmland," in *Sec. 2.4.1*; the explanation, by the advocates of the Environmental Justice Movement, that it is desirable because of its campaign to end environmental discrimination, in *Sec. 2.4.2*; the explanation, by the advocates of ethical consumerism, that it is desirable because of its positive effects on waste reduction, animal rights, etc., in *Sec. 2.5.1*; the explanation by those in the Judeo-Christian tradition that simple living is desirable because of its religious promise, in *Sec. 2.5.2*; the explanation by the proponents of EPR that it is desirable because of its contribution to waste recycling, in *Sec. 3.2.1*; the explanation, by dumpster divers, that dumpster diving is desirable because of the value of waste resuse, in *Sec. 3.2.2*; the explanation by the advocates of incineration that it is desirable because of its energy creation, in *Sec. 3.3.1*; the explanation by the advocates of animal products that they are desirable for food consumption, in *Sec. 3.3.2*; the explanation by the advocates of waste recycle that it is desirable for environmental sustainability, in *Sec. 3.4.1*; the explanation by the advocates of the waste converter that it is desirable because of its positive environmental impact, in *Sec. 3.4.2*; the explanation by the advocates for Zero Waste that it is desirable because we can use nature as a model to back it up, in *Sec. 3.5.1*; and the explanation by the advocates of freeganism that it is desirable because of the benefits of reclaiming wastes, in *Sec. 3.5.2*).

On the other hand, there is inexplicability (e.g., the lack of sufficient explanation by the Imperial Japanese Army Air Service of why bombing Ningbo with fleas carrying the bubonic plague in 1940 is necessarily desirable, especially from the vantage point of the Chinese victims and their family members, in *Sec. 2.2.1*; the lack of sufficient explanation by the advocates of the MWTA of why proper clinical waste disposal is necessarily desirable, since, in the 1980s, "the high cost [~$1500/ton] associated with the legal disposal of the waste" gave incentives to "private waste contractors to dump illegally to avoid high fees," in *Sec. 2.2.2*; the lack of sufficient explanation by "Dutch police and local supervisors [known as *buitengewoon opsporingsambtenaar*, or BOA]" of why fining "citizens for throwing away cans, bottles or wrappers onto the street" is necessarily desirable, because those citizens who are fined can disagree, which also explains why they litter in the first place, in *Sec. 2.3.1*; the lack of sufficient explanation, by the advocates of different ways of "external removal" to manage space debris, of why these solutions are necessarily desirable, since for the critics, "as of 2006, the cost of launching any of these solutions [in terms of external removal, for example] is about the same as launching any spacecraft," and "none of the existing solutions are currently cost-effective," in *Sec. 2.3.2*; the lack of sufficient explanationby by those in "the 1988 Koko case of why the transport of 8,000 barrels of hazardous waste from Italy to the small town of Koko in Nigeria is necessarily desirable, since, for the critics, incidents like this "have been deemed 'Toxic Colonialism' by many developing countries," in *Sec. 2.4.1*; the lack of sufficient explanation, by the advocates of the Environmental Justice Movement, of why its campaign is necessarily desirable, since, for the critics, it has been plagued with "three different forms" of elitism, with opposite effects, in *Sec. 2.4.2*; the lack of sufficient explanation, by the advocates of ethical consumerism, of why it is necessarily desirable because, for the critics, it can be counter-productive in encouraging immoral behavior instead, and so on, in *Sec. 2.5.1*; the lack of sufficient explanation by those in the Judeo-Christian tradition of why simple living is necessarily desirable, because, for the critics like Friedrich Nietzsche, asceticism in the Judeo-Christian tradition is essentially life-denying, in *Sec. 2.5.2*; the lack of sufficient explanation by the proponents of EPR of why it is necessarily desirable, because "other critics are concerned that manufacturers may use takeback programs to take secondhand electronics off the reuse market, by shredding rather than reusing or repairing goods that come in for recycling," together with other problems aforecited, in *Sec. 3.2.1*; the lack of sufficient explanation by dumpster divers of why dumpster diving is necessarily desirable, because, for the critics, it is unhealthy and dangerous, or the items in dumpsters are of inferior quality, etc., in *Sec. 3.2.2*; the lack of sufficient explanation by the advocates of incineration of why it is necessarily desirable, since, for the critics, "incinerators destroy valuable resources" and "they may reduce the incentives for recycling and waste minimization activities," besides other problems, in *Sec. 3.3.1*; the lack of sufficient explanation by the advocates of animal products of why they are necessarily desirable, because, for the critics, their use is prohibited by certain groups for different moral and religious reasons, like those for vegetarian, kosher, and halaal diets only, in *Sec. 3.3.2*; the lack of sufficient explanation by the advocates of waste recycle of why it is necessarily desirable, since, for the critics, "there is a net loss to the poor and possibly the whole of a society," because "traditional job of recycling is performed by the entrepreneurial poor such as the karung guni, Zabaleen, the rag-and-bone man, waste picker, and junk man," so "with the creation of large recycling organizations that may be profitable…, the poor are more likely to be driven out of the recycling and the remanufacturing market," in *Sec. 3.4.1*; the lack of sufficient explanation by the advocates of the waste converter of why it is necessarily desirable, because some contemporary autoclaves have been so improved that they "employ the same method of sterilization as converters, but do so using larger and more complex equipment that also has higher safety… requirement…," in *Sec. 3.4.2*; the lack of sufficient explanation by the advocates for Zero Waste of why it is necessarily desirable, because there are those who favor "the creation of industry-specific more efficient reuse modalities" instead, in that "industrial products are inherently non-natural [such as chemicals and plastics that are mono-molecular] and benefit greatly from industrial methods of reuse, while natural methods requiring degradation and reconstitution are wasteful in that context," in *Sec. 3.5.1*; and the lack of sufficient explanation by the advocates of freeganism of why it is necessarily desirable because reclaimed items in dumpsters can be unhealthy and dangerous, in *Sec. 3.5.2*).

And the reverse direction also holds true.

Once more, it should be stressed, however, that there are different shades of gray (or different degrees of truth) in the two opposites, and the classification is not necessarily mutually exclusive either. In fact, even when some combinations of the two occur, they only end up sharing the same dialectic relationship, but in a different degree.

Also, there are some other relationships (as a kind of family resemblance) which have something in common with the principle, but they are not exactly the same but only more or less comparable, which varies from case to case.

Good examples of family resemblance in relation to the principle are underlyingness-regularness, causation-regularity, causation-correlation, explanation-description, explanation-narration, explanation-interpretation, and so forth.

6th Thesis: The Fiction-Reality Principle

The sixth thesis is called the fiction-reality principle (on the different faces of reality) in the category of "method" in existential dialectics (as first proposed in *FPHLIT*), in that both fiction and reality co-exist. There is no reality without fiction—and vice versa.

This principle tells us the different categories of reality, in that, if there is reality, there is also some fiction in it—just as, if there is fiction, there is also some reality in it.

For instance, on the one hand, there is fiction (e.g., the fictional aspect of the conventional view about the bubonic plague, insofar as it is regarded as undesirable, as questioned by the Imperial Japanese Army Air Service in 1940 in the decision to use it as an effective biological weapon instead, in *Sec. 2.2.1*; the fictional aspect of the MWTA, insofar as "the Act was repealed in 1991," because the federal government was no longer deemed necessary to enforce propper clinical waste disposal, so the enforcement was then passed to the states instead, in *Sec. 2.2.2*; the fictional aspect of anti-litter solutions, insofar as each proposed solution has its own disadvantages and does not often work as claimed, in *Sec. 2.3.1*; the fictional aspect of the measurement of space debris, insofar as "GEO orbit is too distant to make accurate measurements of the existing debris field for objects under 1 metre [3 ft 3 in], so the precise nature of the existing problem is not well known," in *Sec. 2.3.2*; the fictional aspect of the definition of "hazardous waste" in the Basel Convention, insofar as it is subject to contention, because it does not include some important types of hazardous waste like "radioactive waste," "wastes from the normal operation of ships," and certain "'commodities' which are not considered wastes," etc., in *Sec. 2.4.1*; the fictional aspect of the Environmental Justice Movement, insofar as it has been plagued with "three different forms" of elitism, with opposite effects, in *Sec. 2.4.2*; the fictional aspect of ethical consumerism, insofar as it can be "a catastrophic mistake" as pointed out by George Monbiot, in *Sec. 2.5.1*; the fictional aspect of "advertising," insofar as it misleadingly glorifies or encourages "a consumerist mentality," in *Sec. 2.5.2*; the fictional aspect of EPR, insofar as it "has rarely been consistently quantified," in *Sec. 3.2.1*; the fictional aspect of dumpster diving, insofar as it is unhealthy and dangerous, the items in dumpsters are of inferior quality, etc., in *Sec. 3.2.2*; the fictional aspect of incineration as a safe WtE method, insofar as its proponents tried to downplay the problem "associated with incinerating MSW to make electrical energy," because of "the potential for pollutants to enter the atmosphere with the flue gases from the boiler," and "these pollutants can be acidic and in the 1980s were reported to cause environmental damage by turning rain into acid rain," in *Sec. 3.3.1*; the fictional aspect of certain new types of pet food which claim "the use of human-grade meat sources only, with no animal meat by-products," because "AAFCO has declared that using the term 'human grade' is 'false and misleading,'" in *Sec. 3.3.2*; the fictional aspect of the fiscal measurement of waste recycle, insofar as it ignores "negative externalities" like pollution, in *Sec. 3.4.1*; the fictional aspect of the waste converter, insofar as "most products are not designed with recycling in mind," nor is it necessarily safer than a newly improved autoclave, in *Sec. 3.4.2*; the fictional aspect of market-based campaigns for Zero Waste, insofar as "many organizations, cities and counties have embraced a Zero Waste slogan while pressing for none of the key Zero Waste changes," in *Sec. 3.5.1*; and the fictional aspect of freeganism, insofar as reclaimed items in dumpsters can be unhealthy and dangerous, it has an extremist roots in religion, and it is not so practical as romantically assumed by the advocates, in *Sec. 3.5.2*).

On the other hand, there is reality (e.g., the realistic aspect of the bubonic plague, insofar as it can have destructive impact on its victims, in *Sec. 2.2.1*; the realistic aspect of the MWTA, insofar as the Act was in force between 1988 and 1991, in *Sec. 2.2.2*; the realistic aspect of anti-litter solutions, insofar as each proposed solution has some impact on cleaning up litter, in *Sec. 2.3.1*; the realistic aspect of the measurement of space debris, insofar as "the mid-2009 update to the NASA debris FAQ, had estimated that "the number of...debris items smaller than 1 centimetre [0.39 in] exceeds tens of millions," in *Sec. 2.3.2*; the realistic aspect of the definition of "hazardous waste" in the Basel Convention, insofar as it includes many, though not all, types of "hazardous waste," in *Sec. 2.4.1*; the realistic aspect of the Environmental Justice Movement, insofar as it has received help from the Civil Rights Movement, in *Sec. 2.4.2*; the realistic aspect of ethical consumerism, insofar as it can contribute to the buying of ethical products, in *Sec. 2.5.1*; the realistic aspect of "advertising," insofar as it contributes to the awareness of some products on sale, in *Sec. 2.5.2*; the realistic aspect of EPR, insofar as it contributes to the awareness of producer responsibility, in *Sec. 3.2.1*; the realistic aspect of dumpster diving, insofar as it can save by reusing the discarded items, in *Sec. 3.2.2*; the realistic aspect of incineration as a safe WtE method, insofar as it has contributed to the creation of electrical energy, in *Sec. 3.3.1*; the realistic aspect of certain new types of pet food, insofar as they are useful for the advocates of pet food consumption, in *Sec. 3.3.2*; the realistic aspect of the fiscal measurement of waste recycle, insofar as it includes the costs and benefits of waste recycle for an individual business, but not society, in *Sec. 3.4.1*; the realistic aspect of the waste converter, insofar as it has certain positive environmental impact, in *Sec. 3.4.2*; the realistic aspect of market-based campaigns for Zero Waste, insofar as "market-based, legislation-mediated campaigns like Extended Producer Responsibility [EPR] and the Precautionary Principle are among numerous campaigns that have a Zero Waste slogan hung on them," in *Sec. 3.5.1*; and the realistic aspect of freeganism, insofar as it has contributed to the efforts of reclaiming wastes, in *Sec. 3.5.2*.).

And the reverse direction also holds true.

Once more, it should be stressed, however, that there are different shades of gray (or different degrees of truth) in the two opposites, and the classification is not necessarily mutually exclusive either. In fact, even when some combinations of the two occur, they only end up sharing the same dialectic relationship, but in a different degree.

Also, there are some other relationships (as a kind of family resemblance) which have something in common with the principle, but they are not exactly the same but only more or less comparable, which varies from case to case.

Good examples of family resemblance in relation to the principle are fiction-nonfiction, fiction-fact, imagination-fact, imagination-nonfiction, subjectivity-objectivity, and so forth.

In Relation to Structure

The next five theses are in relation to structure in existential dialectics, as shown below.

7th Thesis: The Finiteness-Transfiniteness Principle

The seventh thesis is called the finiteness-transfiniteness principle (on the nature of numbers) in the category of "structure" in existential dialectics (as first proposed in *FPHG*), in that, if there are finite things, there are likewise transfinite ones—and vice versa.

To avoid confusion, my usage of the word "transfinite" here differs radically from the one used by Cantor (and other mathematicians) for "relative" infinity—and is more limited, in light of the problems confronting any attempt to understand the idea of infinity, be it by intuition, imagination, and conception (as detailedly analyzed in *Sec. 2.2.3* of *FPHG*).

Instead, by "transfinity," I allow numbers which can be many times larger—or smaller, for that matter—than the finite things that we encounter in daily life, but they do not have to be related to the idea of infinity at all (which may not exist).

Of course, there may be some *borderline* cases, in which it is not clear whether the number in question is transfinite (in my usage) or simply a mathematical convenience. A good example of a borderline case is the Planck unit of length for "the smallest space possibly measured in nature," which is "less than billionths of trillionths of trillionths of an inch" (or something like 1.6×10^{-35} meters). (P. Baofu 2006a; N. McAleer 1987: 219; WKV 2008; D. Corbett 2008a)

That said—my usage of "transfinity" can also resolve (or better, dissolve) an age-old problem in philosophy known as "Zeno's paradoxes" (as already explained in *Sec. 2.2.3* of *FPHG*).

In any event, in relation to the finiteness-transfiniteness principle, on the other hand, there is finiteness (e.g., the finite number of the categories of epidemics which were perpuetuated by zoonotic diseases, "circulating mainly among small rodents and their fleas," like "the Bubonic Plague, cholera and typhoid fever" in history, in *Sec. 2.2.1*; the finite number of the categories of clincal waste, like "needles," "scalpels," "lancets," "identifiable body parts," etc., in *Sec. 2.2.2*; the finite number of the categories of litter, like "containers," "wrappers and paper product," "furniture," "appliances," "old electronics," "abandoned vehicles" and "construction materials," in *Sec. 2.3.1*; the finite number of spacecrafts which were lanuched by humans so far, as of 2012, in *Sec. 2.3.2*; the finite number of the properties of hazardous waste, like "ignitability," "reactivity," "corrosivity," and "toxicity," in *Sec. 2.4.1*; the finite number of "different kinds of sites," where environmental discrimination has historically occurred, like "waste disposal," "manufacturing," "energy production," "highways," "ports," and "airports," in *Sec. 2.4.2*; the finite number of the categories of moral concerns by ethical consumers, like "the environment," "people," "animals," "politics," and "product," in *Sec. 2.5.1*; the finite number of the categories of those people who "eschew all technology" or technological products, like the 2 categories known as "Luddites" and "neo-Luddites," in *Sec. 2.5.2*; the finite number of the categories of waste to fall under the jurisdiction of EPR, like "industrial waste," "electronic waste," etc., in *Sec. 3.2.1*; the finite number of the categories of discarded items in dumpsters, like "business waste," "chemical waste" "clinical waste," "commercial waste," etc., in *Sec. 3.2.2*; the finite number of the categories of pollutants by incinerators, like "fine particulate, heavy metals, trace dioxin and acid gas emissions," in *Sec.*

3.3.1; the finite number of the categories of foodstuffs and non-foodstuffs from animal products and animal by-products as listed earlier, in Sec. *3.3.2*; the finite number of the categories of "highly dispersed wastes" to be recycled, like "chlorinated organic hydrocarbons that have bioaccumulated in animal and human tissues across the globe, the copper dispersed in fungicides, the lead in widely applied paints, or the zinc oxides present in the finely dispersed rubber powder that is abraded from automobile tires," in *Sec. 3.4.1*; the finite number of the categories of waste to be treatable by waste converters, like "solid waste," "hazardous waste," etc., in *Sec. 3.4.2*; the finite number of the categories of materials used in products which can be targeted by Zero Waste, like "steel," "copper," "glass," and "plastics" in bottles, computers, etc., in *Sec. 3.5.1*; and the finite number of the categories of food suppliers like "supermarkets, grocery stores, and restaurants" targeted by dumpster divers for food wastes, in *Sec. 3.5.2*).

On the other hand, there is transfinitty (e.g., the transfinite number of all the individiual rats and fleas which have ever existed in history, in *Sec. 2.2.1*; the transfinite number of all the individual "needles," "scalpels," "lancets," "identifiable body parts," etc. that have ever existed or produced in history, in *Sec. 2.2.2*; the transfinite number of all the individual pieces of "containers," "wrappers and paper product," "furniture," "appliances," "old electronics," "abandoned vehicles" and "construction materials" in history that have ever been produced, in *Sec. 2.3.1*; the transfinite number of all the individual space particles, both natural and man-made, which have ever traveled near or to the Earth in the history of the universe and could hit any spacecraft or any object in the way, in *Sec. 2.3.2*; the transfinite number of all the individual pieces of hazardous waste in history that have ever existed, in *Sec. 2.4.1*; the transfinite number of all the individual waste items in all the sites of "waste disposal," "manufacturing," "energy production," "highways," "ports," and "airports" in history that have affected minorities, in *Sec. 2.4.2*; the transfinite number of all the individual things in "the environment," "people," "animals," "politics," and "product" that have ever existed in history, in *Sec. 2.5.1*; the transfinite number of all the individual pieces of technological products in history that have ever been produced, in *Sec. 2.5.2*; the transfinite number of all the individual pieces of products in history that have ever existed and can be classified under EPR, in *Sec. 3.2.1*; the transfinite number of all the individual items of wastes in history that can be found in dumpsters, in *Sec. 3.2.2*; the transfinite number of all the individual pieces of "fine particulate, heavy metals, trace dioxin and acid gas emissions," and so on in history that have ever been in the environment, in *Sec. 3.3.1*; the transfinite number of all the individual pieces of foodstuffs and non-food stuffs from animal products and animal by-products in history that have ever existed, in *Sec. 3.3.2*; the transfinite number of all the individual pieces of "highly dispersed wastes" in history that have ever existed, like "chlorinated organic hydrocarbons that have bioaccumulated in animal and human tissues across the globe, the copper dispersed in fungicides, the lead in widely applied paints, or the zinc oxides present in the finely dispersed rubber powder that is abraded from automobile tires," in *Sec. 3.4.1*; the transfinite number of all the indivdiual pieces of "solid waste," "hazardous waste," etc. in history that have ever existed, in *Sec. 3.4.2*; the transfinite number of all the individual pieces of "steel," "copper," "glass," and "plastics" in bottles, computers, etc. in history that can be targeted by Zero Waste, in *Sec. 3.5.1*; and the transfinite number of all the individual items of food wastes in history that have ever been discarded by "supermarkets, grocery stores, and restaurants," in *Sec. 3.5.2*).

And the reverse direction also holds true.

Once more, it should be stressed, however, that there are different shades of gray (or different degrees of truth) in the two opposites, and the classification is not necessarily mutually exclusive either. In fact, even when some combinations of the two occur, they only end up sharing the same dialectic relationship, but in a different degree.

Also, there are some other relationships (as a kind of family resemblance) which have something in common with the principle, but they are not exactly the same but only more or less comparable, which varies from case to case.

Good examples of family resemblance in relation to the principle are boundedness-quasiunboundedness, smallness-largeness, microness-macroness, boundedness-unboundedness, and the like.

8th Thesis: The Preciseness-Vagueness Principle

The eighth thesis is called the preciseness-vagueness principle (on the refinement of things) in the category of "structure" in existential dialectics (which was first out worked out in *FIA*), in that both preciseness and vagueness are important, not that one is better than the other, but that both are used, in different degrees of preference, in accordance to the contextual application from the perspectives of nature, the mind, culture, and society. There is no preciseness without vagueness—and vice versa.

For instance, on the one hand, there is preciseness (e.g., the precise identification of the three epidemics like "the Bubonic Plague, cholera and typhoid fever" in the article aforecited, in *Sec. 2.2.1*; the precise identification of three methods of clinical waste disposal in the article aforecited, in *Sec. 2.2.2*; the precise identification of 6 solutions to litter in the article aforecited, in *Sec. 2.3.1*; the precise identification of 3 solutions to space debris in the article aforecited, in *Sec. 2.3.2*; the precise identification of 4 properties of hazardous waste in the Basel Convention, namely, "explosive, flammable, toxic, or corrosive," in *Sec. 2.4.1*; the precise identification of 2 barriers to minority participation in the environmental justice movement in the article aforecited, in *Sec. 2.4.2*; the precise identification of 4 types of "products" that are targeted by ehtical consumers in the article aforecited, in *Sec. 2.5.1*; the precise identification of 4 major traditions of simple lifestyle in the article aforecited, in *Sec. 2.5.2*; the precise identification of 3 approaches to EPR, like "mandatory," "negotiated," and "voluntary," in the article aforecited, in *Sec. 3.2.1*; the precise identification of 8 reasons for dumpster diving in the article aforecited, in *Sec. 3.2.2*; the precise identification of 8 WtE methods, including incineration, in the article aforecited, in *Sec. 3.3.1*; the precise identification of 2 major uses of animal products and animal by-products, namely, pet food and soil amendment, in the article aforecited, in *Sec. 3.3.2*; the precise identification of 8 major disagreements on the cost-benefit analysis of waste recycle in the article aforecited, in *Sec. 3.4.1*; the precise identification of 5 applications of the waste converter in the article aforecited, in *Sec. 3.4.2*; the precise identification of 6 disputes in the debate on zero waste in the article aforecited, in *Sec. 3.5.1*; and the precise identification of 6 characteristics of freeganism in the article aforecited, in *Sec. 3.5.2*).

On the other hand, there is vagueness (e.g., the vagueness in the identification of the three epidemics like "the Bubonic Plague, cholera and typhoid fever" in the article aforecited, since it is not clear why there must be only three, not 5, 7, 8, 10, 12, and so on, in *Sec. 2.2.1*; the vagueness in the identification of the three methods of clinical waste disposal in the article aforecited, since it is not clear why there must be only 3, not 4, 5, 6, 7, and so on, in *Sec. 2.2.2*; the vagueness in the identification of the 6 solutions to litter in the article aforecited, since it is not clear why there must be only 6, not 7, 8, 9, and so on, in *Sec. 2.3.1*; the vagueness in the identification of the 3 solutions to space debris in the article aforecited, since it is not clear why there must be only 3, not 4, 5, 6, and so on, in *Sec. 2.3.2*; the vagueness in the identification of the 4 properties of hazardous waste in the Basel Convention, namely, "explosive, flammable, toxic, or corrosive," since it is not clear why there must be only 4, not 5, 6, 7, and so on, because, for the critics, there are other properties like "radioactive," etc., in *Sec. 2.4.1*; the vagueness in the identification of the 2 barriers to minority participation in the environmental justice movement in the article aforecited, since it is not clear why there must be only 2, not 3, 4, 5, and so on, in *Sec. 2.4.2*; the vagueness in the identification of the 4 types of "products" that are targeted by ehtical consumers in the article aforecited, since it is not clear why there must be only 4, not 5, 6, 7, and so on, in *Sec. 2.5.1*; the vagueness in the identification of the 4 major traditions of simple lifestyle in the article aforecited, since it is not clear why there must be only 4, not 5, 6, 7, and so on, in *Sec. 2.5.2*; the vagueness in the identification of the 3 approaches to EPR, like "mandatory," "negotiated," and "voluntary," in the article aforecited, since it is not clear why there must be only 3, not 4, 5, 6, and so on, in *Sec. 3.2.1*; the vagueness in the identification of the 8 reasons for dumpster diving in the article aforecited, since it is not clear why there must be only 8, not 9, 10, 11, and so on, in *Sec. 3.2.2*; the vagueness in the identification of the 8 WtE methods, including incineration, in the article aforecited, since it is not clear why there must be only 8, not 9, 10, 11, and so on, in *Sec. 3.3.1*; the vagueness in the identification of the 2 major uses of animal products and animal by-products, namely, pet food and soil amendment, in the article aforecited, since it is not clear why there must be only 2, not 3, 4, 5, and so on, in *Sec. 3.3.2*; the vagueness in the identification of the 8 major disagreements on the cost-benefit analysis of waste recycle in the article aforecited, since it is not clear why there must be only 8, not 10, 13, 15, and so on, in *Sec. 3.4.1*; the vagueness in the identification of the 5 applications of the waste converter in the article aforecited, since it is not clear why there must be only 5, not 8, 9, 10, and so on, in *Sec. 3.4.2*; the vagueness in the identification of the 6 disputes in the debate on zero waste in the article aforecited, since it is not clear why there must be only 6, not 8, 9,10, 11, and so on, in *Sec. 3.5.1*; and the vagueness in the identification of the 6 characteristics of freeganism in the article aforecited, since it is not clear why there must be only 6, not 8, 9, 10, and so on, in *Sec. 3.5.2*).

And the reverse direction also holds true.

Once more, it should be stressed, however, that there are different shades of gray (or different degrees of truth) in the two opposites, and the classification is not necessarily mutually exclusive either. In fact, even when some combinations of the two occur, they only end up sharing the same dialectic relationship, but in a different degree.

Also, there are some other relationships (as a kind of family resemblance) which have something in common with the principle, but they are not exactly the same but only more or less comparable, which varies from case to case.

Good examples of family resemblance in relation to the principle are clarity-ambiguity, directness-indirectness, quantitativeness-qualitativeness, specificity-obscurity, thickness-thinness, describability-nondescribability, concretness-abstractness, translatability-untranslatability, specificity-generality, clearcutness-transitionalness, and the like.

9th Thesis: The Simpleness-Complicatedness Principle

The ninth thesis refers to the simpleness-complicatedness principle (on the interconnection among things) in the category of "structure" in existential dialectics (which was first out worked out in *FIA*), in that both simpleness and complicatedness are vital, without favoring one over the other, and each is utilized, depending on the basis of the perspectives of nature, the mind, culture, and society. There is no simpleness without complicatedness—and vice versa.

For instance, on the one hand, there is simpleness (e.g., the relatively simple analysis of the bubonic plague as a desirable biological weapon by the Imperial Japanese Army Air Service in 1940, in *Sec. 2.2.1*; the relatively simple analysis of proper clinical waste disposal by the advocates, in *Sec. 2.2.2*; the relatively simple analysis of the prevention of litter by the advocates, in *Sec. 2.3.1*; the relatively simple analysis of the solutions to space debris by the advocates, in *Sec. 2.3.2*; the relatively simple analysis of the Basel Convention by the advocates, in *Sec. 2.4.1*; the relatively simple analysis of environmental justice movement by the advocates, in *Sec. 2.4.2*; the relatively simple analysis of ethical consumerism by its advocates, in *Sec. 2.5.1*; the relatively simple analysis of simple living by the advocates, in *Sec. 2.5.2*; the relatively simple analysis of EPR by the proponents, in *Sec. 3.2.1*; the relatively simple analysis of dumpster diving by the advocates, in *Sec. 3.2.2*; the relatively simple analysis of each of the WtE methods by its advocates, in *Sec. 3.3.1*; the relatively simple analysis of animal products and animal by-products to be used for foodstuffs and non-foodstuffs by the advocates, in *Sec. 3.3.2*; the relatively simple analysis of waste recycle with the tool of cost-benefit analysis by the advocates, in *Sec. 3.4.1*; "applications outside of waste treatment centers are increasingly common due to the…simplicity of modern converters," in *Sec. 3.4.2*; the relatively simple analysis of Zero Waste by the advocates, in *Sec. 3.5.1*; and the relatively simple analysis of the ideology of freeganism by its advocates, in *Sec. 3.5.2*).

On the other hand, there is complicatedness (e.g., the relatively more complicated analysis of the bubonic plague as a biological weapon, by challenging the claims and assumptions, as shown in the problems pointed out by the critics as cited, in *Sec. 2.2.1*; the relatively more complicated analysis of proper clinical waste disposal, by challenging the claims and assumptions, as shown in the problems pointed out by the critics as cited, in *Sec. 2.2.2*; the relatively more complicated analysis of the prevention of litter, by challenging the claims and assumptions, as shown in the problems with complicated causes and other issues pointed out by the critics as cited, in *Sec. 2.3.1*; the relatively more complicated analysis of the solutions to space debris, by challenging the claims and assumptions, as shown in the problems pointed out by the critics as cited, in *Sec. 2.3.2*; the relatively more complicated analysis of the Basel Convention, by challenging the claims and assumptions, as shown in the problems pointed out by the critics as cited, in *Sec. 2.4.1*; the relatively more complicated analysis of the environmental justice movement, by challenging the claims and assumptions, as shown in the problems pointed out by the critics as cited, in *Sec. 2.4.2*; the relatively more complicated analysis of ethical consumerism, by challenging the claims and assumptions, as shown in the problems pointed out by the critics as cited, in *Sec. 2.5.1*; the relatively more complicated analysis of simple living, by challenging the claims and assumptions, as shown in the problems pointed out by the critics as cited, in *Sec. 2.5.2*; the relatively more complicated analysis of EPR, by challenging the claims and assumptions, as shown in the problems pointed out by the critics as cited, in *Sec. 3.2.1*; the relatively more complicated analysis of dumpster diving, by challenging the claims and assumptions, as shown in the problems pointed out by the critics as cited, in *Sec. 3.2.2*; the relatively more complicated analysis of WtE methods, by challenging the claims and assumptions, as shown in the problems pointed out by the critics as cited, in *Sec. 3.3.1*; the relatively more complicated analysis of animal products and animal by-products to be used for foodstuffs and non-foodstuffs, by challenging the claims and assumptions, as shown in the problems pointed out by the critics as cited, in *Sec. 3.3.2*; the relatively more complicated analysis of waste recycle with the tool of cost-benefit analysis, by challenging the claims and assumptions, as shown in the problems pointed out by the critics as cited, in *Sec. 3.4.1*; nowadays "autoclaves employ the same method of sterilization as converters, but do so using larger and more complex equipment," in *Sec. 3.4.2*; the relatively more complicated analysis of Zero Waste, by challenging the claims and assumptions, as shown in the problems pointed out by the critics as cited, in *Sec. 3.5.1*; and the relatively more complicated analysis of the ideology of freeganism, by challenging the claims and assumptions, as shown in the problems pointed out by the critics as cited, in *Sec. 3.5.2*).

And the reverse direction also holds true.

Again, it should be stressed, however, that there are different shades of gray (or different degrees of truth) in the two opposites, and the classification is not necessarily mutually exclusive either. In fact, even when some combinations of the two occur, they only end up sharing the same dialectic relationship, but in a different degree.

Also, there are some other relationships (as a kind of family resemblance) which have something in common with the principle, but they are not exactly the same but only more or less comparable, which varies from case to case.

Good examples of family resemblance in relation to the principle are inflexibility-flexibility, standardization-specialization, imperfectness-perfectness, superficiality-depth, shallowness-deepness, economicalness-elaboratedness, plainness-circumspection, onesidedness-multisidedness, succinctness-duplicatedness, and the like.

10th Thesis: The Openness-Hiddenness Principle

The tenth thesis refers to the openness-hiddenenss principle (on the detection of things) in the category of "structure" in existential dialectics (which was already worked out in my previous books, especially in *FPHU*), in that reality has its hidden face, just as it is open to outside view in some other ways. There is no openness without hiddenness—and vice versa.

For instance, on the one hand, there is openness (e.g., the open exploration of the use of the bubonic plague as a biological weapon by the Imperial Japanese Army Air Service in 1940, in *Sec. 2.2.1*; the open exploration of what constitutes clinical waste by a given definition, in *Sec. 2.2.2*; the open exploration of using legislation to enforce proper litter disposal, in *Sec. 2.3.1*; the open exploration of the so-called "Kessler Syndrome" by Donald Kessler and later other researchers, in that, "instead of the density of debris being a measure of the number of items launched into orbit, it was that number plus any new debris caused when they collided," so "if the new debris did not decay from orbit before impacting another object, the number of debris items would continue to grow even if there were no new launches," in *Sec. 2.3.2*; the open exploration of "a general prohibition on the exportation or importation of wastes between Parties and non-Parties" in the Basel Convention, in *Sec. 2.4.1*; the open exploration of using the environmental justice movement to end environmental discrimination against minorities, in *Sec. 2.4.2*; the open exploration, by ethical consumers, of buying "ethical products," in *Sec. 2.5.1*; the open exploration of simple living by those in the Judeo-Christian tradition, in *Sec. 2.5.2*; the open exploration of producer responsibility by the advocates of EPR, in *Sec. 3.2.1*; the open exploration of collecting discarded items in dumpsters for reuse, in *Sec. 3.2.2*; the open exploration of using incinerators to convert waste into energy, in *Sec. 3.3.1*; the open exploration of selling animal products and animal by-products by the meat industry, in *Sec. 3.3.2*; the open exploration of the costs and benefits of waste recycle for an individual business, by way of fiscal measurement, in *Sec. 3.4.1*; the open exploration of using the waste converter for waste recycle, in *Sec. 3.4.2*; the open exploration by "many commercial or industrial companies" like "Staples, Home Depot, Toyota, General Motors and computer take-back campaigns" to "embrace Zero Waste" in their market-based campaigns, in *Sec. 3.5.1*; and the open exploration of the possibility of "reclaiming and eating food that has been discarded" by the freegans, in *Sec. 3.5.2*).

On the other hand, there is hiddenness (e.g., the hidden bias in the view of using the bubonic plague as a biological weapon by the Imperial Japanese Army Air Service in 1940, because the Chinese victims and their family members considered its use as immoral and inhumane, in *Sec. 2.2.1*; the hidden bias in each definition of clinical waste, because "defining what objects were to be classified as medical wastes was crucial to ensure that all municipalities…would be monitoring similar materials," so different definitions end up monitoring different materials, in *Sec. 2.2.2*; the hidden bias in using legislation to enforce proper litter disposal, since there is no universal regulation of littering behavior and, in addition, different countries and localities have different customs and legislations concerning littering behavior, in *Sec. 2.3.1*; the hidden nature of the "Kessler Syndrome," because, as explained by Kessler, "the start of a cascade would not be obvious until the situation was well advanced, which might take years," in *Sec. 2.3.2*; the hidden bias in the Basel Convention, because, for the critics, it does not go far enough to protect less developed countries from the spread of hazardous wastes and often favors major powers with those notorious exceptional provisions, in *Sec. 2.4.1*; the hidden bias in the Environmental Justice Movement, because, for the critics, it has been plagued with "three different forms" of elitism, with opposite effects, in *Sec. 2.4.2*; the hidden bias in ethical consumerism, because, for the critics, it can be counter-productive in encouraging immoral behavior instead, that ethical consumerism is a mistake in thinking, etc., in *Sec. 2.5.1*; the hidden bias in the simple living by those in the Judeo-Christian tradition, because, for the critics like Friedrich Nietzsche, asceticism in the Judeo-Christian tradition is essentially life-denying, in *Sec. 2.5.2*; the hidden bias in EPR, because it does not focus on consumer responsibility, in *Sec. 3.2.1*; the hidden nature of the discarded items in dumpsters, because they "may have faults which are not apparent even if examined," or "furniture may have not easily seen bedbugs which may cause an infestation," or "faulty electrical and mechanical goods can be hazardous and dangerous," etc., in *Sec. 3.2.2*; the hidden bias in the use of incinerators to convert waste into energy," because of "the potential for pollutants to enter the atmosphere with the flue gases from the boiler," and "these pollutants can be acidic and in the 1980s were reported to cause environmental damage by turning rain into acid rain," together with other problems over the years, in *Sec. 3.3.1*; the hidden bias in the sale of animal products and animal by-products by the meat industry, because their sale by the meat industry gives it an advantage "to compete economically with industries selling sources of vegetable protein," in *Sec. 3.3.2*; the hidden bias in fiscal measurement of waste recycle, because it ignores negative externalities, in *Sec. 3.4.1*; the hidden bias in the use of the waste converter, because "most products are not designed with recycling in mind," in *Sec. 3.4.2*; the hidden bias in the market-based campaigns by "many commercial or industrial companies" like "Staples, Home Depot, Toyota, General Motors and computer take-back campaigns" to "embrace Zero Waste," because they "usually mean no more than a major materials recycling effort, having no bearing on product redesign." in *Sec. 3.5.1*; and the hidden bias in freeganism, because of its hidden extremist roots in religion, in *Sec. 3.5.2*).

And the reverse direction also holds true.

Again, it should be stressed, however, that there are different shades of gray (or different degrees of truth) in the two opposites, and the classification is not necessarily mutually exclusive either. In fact, even when some combinations of the two occur, they only end up sharing the same dialectic relationship, but in a different degree.

Also, there are some other relationships (as a kind of family resemblance) which have something in common with the principle, but they are not exactly the same but only more or less comparable, which varies from case to case.

Good examples of family resemblance in relation to the principle are overtness-covertness, participation-obstruction, publicness-privateness, openness-closedness, surface-depth, transparency-secrecy, openness-biasedness, , exposition-illusion, and so on.

11th Thesis: The Denseness-Emptiness Principle

The eleventh thesis concerns the denseness-emptiness principle (on the distribution of entities in space) in the category of "structure" in existential dialectics, which is first proposed in *FPHUP*, in that both density and void are needed, in relation to the mind, nature, culture, and society, albeit in different ways. There is no denseness without emptiness—and vice versa.

Lest any misunderstanding occurs, the term "void" is used here only as an approximation of emptiness (depending on the degree of the lack of density), since, in physics, it is well known that "empty" space is not really empty all the way, because it can be full of energy (e.g., random quantum fluctuations at the sub-atomic level, and, for that matter, dark energy in the universe) and matter (e.g., different versions of sub-atomic particles, and, for that matter, dark matter in the universe). (F. Wilczek 2008)

For instance, on the one hand, there is denseness (e.g., the relatively denser concentration of patients who died from the bubonic plague when untreated with antibiotics like "streptomycin and gentamicin, tetracyclines," and so on, as shown in the "mortality rate of 50-90% in untreated cases," in *Sec. 2.2.1*; the relatively denser concentration of items like "needles," "scalpels," "lancets," "identifiable body parts," etc. to fall under the jursidiction of the MWTA, in *Sec. 2.2.2*; the relatively denser concentration of litter in litter traps, in *Sec. 2.3.1*; the relatively denser concentration of "collisions…at very high relative velocities, typically several kilometres per second," in the LEO region, in *Sec. 2.3.2*; the relatively denser concentration of hazardous wastes from developed countries to less developed countries prior to the creation of the Basel Convention, in *Sec. 2.4.1*; the relatively denser concentration of minorities and working class people in the environmental justice movement nowadays, in *Sec. 2.4.2*; the relatively denser concentration of concerns with moral issues by ethical consumers, in *Sec. 2.5.1*; the relatively denser concentration of "vegan diets" in the Movement for Compassionate Living, in *Sec. 2.5.2*; the relatively denser concentration of concerns, by the proponents of EPR, with producer responsibility in EPR, in *Sec. 3.2.1*; the relatively denser concentration of discarded items in dumpsters, in *Sec. 3.2.2*; the relatively denser concentration of using incinerators to convert waste into energy nowadays, in *Sec. 3.3.1*; the relatively denser concentration of foodstuffs from animal products among those in mainstream society who do not mind to consume non-human animal diets, in *Sec. 3.3.2*; the relatively denser concentration of "the traditional job of recycling" among "the entrepreneurial poor such as the karung guni, Zabaleen, the rag-and-bone man, waste picker, and junk man," in *Sec. 3.4.1*; the relatively denser concentration of concerns with using "external water input" in autoclaves, in *Sec. 3.4.2*; the relatively denser concentration of concerns with the use of "governmental and non-governmental institutions" by those who favor governance for Zero Waste, in *Sec. 3.5.1*; and the relatively denser concentration of concerns with working and consuming less among the freegans, in *Sec. 3.5.2*).

On the other hand, there is emptiness (e.g., the relatively less dense, or more empty, concentration of patients who died from the bubonic plague when treated with antibiotics like "streptomycin and gentamicin, tetracyclines," and so on, as shown in the mortality rate of only "about 1-15%," in *Sec. 2.2.1*; the relatively less dense, or more empty, concentration of space debris to fall under the jursidiction of the MWTA, in *Sec. 2.2.2*; the relatively less dense, or more empty, concentration of litter in areas benefited from regular clean-up activities, in *Sec. 2.3.1*; the relatively less dense, or more empty, concentration of major collisions at higher altitudes, unlike those in LEO, and "this means that the debris field from such a collision is not the same as a LEO collision and does not pose the same sort of risks," in *Sec. 2.3.2*; the relatively less dense, or more empty, concentration of hazardous wastes from developed countries to less developed countries, under the Basel Ban Amendment to the Basel Convention, at least on paper, in *Sec. 2.4.1*; the relatively less dense, or more empty, concentration of minorities and working class people in the early environmental justice movement in the older days, in *Sec. 2.4.2*; the relatively less dense, or more empty, concentration of concerns with moral issues by market-oriented capitalists, who are not environment-friendly, for instance, in *Sec. 2.5.1*; the relatively less dense, or more empty, concentration of "vegan diets" among those in mainstream society who consume animal products instead, in *Sec. 2.5.2*; the relatively less dense, or more empty, concentration of concerns, by the proponents of EPR, with the subsequent consequences that "producers would be adding recycling costs into the initial price tag" or that "manufacturers may use takeback programs to take secondhand electronics off the reuse market, by shredding rather than reusing or repairing goods that come in for recycling," and other problems aforecited,in *Sec. 3.2.1*; the relatively less dense, or more empty, concentration of discarded items in refrigerators, in *Sec. 3.2.2*; the relatively less dense, or more empty, concentration of using alternative WtE methods other than incineration nowadays, in *Sec. 3.3.1*; the relatively less dense, or more empty, concentration of foodstuffs from animal products among those for vegetarian, kosher, and halaal diets only, in *Sec. 3.3.2*; the relatively less dense, or more empty, concentration of "the traditional job of recycling" among the rich, in *Sec. 3.4.1*; the relatively less dense, or more empty, concentration of concerns with using "external water input" in modern converters, which instead "utilize the moisture content already present in the conversion cell to generate steam sterilization conditions," in *Sec. 3.4.2*; the relatively less dense, or more empty, concentration of concerns with "formal organisations and procedures of the public sector" by those who favor governance for Zero Waste, in *Sec. 3.5.1*; and the relatively less dense, or more empty, concentration of concerns with working and consuming less among the capitalists, in *Sec. 3.5.2*).

And the reverse direction also holds true.

Again, it should be stressed, however, that there are different shades of gray (or different degrees of truth) in the two opposites, and the classification is not necessarily mutually exclusive either. In fact, even when some combinations of the two occur, they only end up sharing the same dialectic relationship, but in a different degree.

Also, there are some other relationships (as a kind of family resemblance) which have something in common with the principle, but they are not exactly the same but only more or less comparable, which varies from case to case.

Good examples of family resemblance in relation to the principle are fullness-voidness, nearness-farness, concentration-dispersion, denseness-nothingness, something-nothing, heaviness-lightness, composition-emptiness, solidness-void, presence-absence, focus-unfocus, emphasis-disemphasis, and the like.

In Relation to Process

The next four theses are in relation to process in existential dialectics, as shown below.

12th Thesis: The Change-Constancy Principle

The twelfth thesis concerns the change-constancy principle (on the alteration of things) in the category of "process" in existential dialectics, which is first named in *BCPC* (and were worked out in my books like *FHC, FCD,* and *FPHC),* in that change occurs over time, although constancy is also allowed.

Old players fade away, and new ones emerges, with ever new causes and ever new forms—and asymmetry undergoes changes over time too, so does symmetry. There is no change without constancy—and vice versa.

For instance, on the one hand, there is change (e.g., the ever new ways to understand bacteria and diseases, as shown in the study of biodegradable waste and epidemics, in *Sec. 2.2.1*; the ever new ways to understand clinical waste, as shown in the study on clinical waste and diseases, in *Sec. 2.2.2*; the ever new ways to deal with waste, as shown in the study on litter and the environment, in *Sec. 2.3.1*; the ever new ways to use technology to handle space problems, as shown in the study on space debris, in *Sec. 2.3.2*; the ever new ways to manage waste, as shown in the creation of the Basel Convention, in *Sec. 2.4.1*; the ever new ways to fight for environmental justice, as shown in the history of the environmental justice movement, in *Sec. 2.4.2*; the ever new ways to study consumption, as shown in ethical consumerism, in *Sec. 2.5.1*; the ever new ways to deal with human needs, as shown in the simple living movement, in *Sec. 2.5.2*; the ever new ways to deal with waste, as shown in EPR, in *Sec. 3.2.1*; the ever new ways to handle waste, as shown in dumpster diving, in *Sec. 3.2.2*; the ever new ways to deal with wastes, as shown in the use of WtE methods, in *Sec. 3.3.1*; the ever new ways to deal with waste, as shown in the use of animal products and animal by-products, in *Sec. 3.3.2*; the ever new ways to deal with waste, as shown in the cost-benefit analysis of waste recycle, in *Sec. 3.4.1*; the ever new ways to recycle waste, as shown in the use of the waste converter, in *Sec. 3.4.2*; the ever new ways to treat waste, as shown in the idea of Zero Waste, in *Sec. 3.5.1*; and the ever new ways to reclaim wastes, as shown in freeganism, in *Sec. 3.5.2*).

On the other hand, there is constancy (e.g., the ever constant existence of different problems in the ever new ways to understand bacteria and diseases, as shown in the problems with the study of biodegradable waste and epidemics pointed out by the critics, in *Sec. 2.2.1*; the ever constant existence of different problems in the ever new ways to understand clinical waste, as shown in the problems with the study on clinical waste and diseases pointed out by the critics, in *Sec. 2.2.2*; the ever constant existence of different problems in the ever new ways to deal with waste, as shown in the problems with the study on litter and the environment pointed out by the critics, in *Sec. 2.3.1*; the ever constant existence of different problems with the ever new ways to use technology to handle space problems, as shown in the problems with the study on space debris pointed out by the critics, in *Sec. 2.3.2*; the ever constant existence of different problems with the ever new ways to manage waste, as shown in the problems with the Basel Convention pointed out by the critics, in *Sec.*

2.4.1; the ever constant existence of different problems in the ever new ways to fight for environmental justice, as shown in the problems with the history of the environmental justice movement pointed out by the critics, in *Sec. 2.4.2*; the ever constant existence of different problems in the ever new ways to study consumption, as shown in the problems with ethical consumerism pointed out by the critics, in *Sec. 2.5.1*; the ever constant existence of different problems in the ever new ways to deal with human needs, as shown in the problems with the simple living movement pointed out by the critics, in *Sec. 2.5.2*; the ever constant existence of different problems with the ever new ways to deal with waste, as shown in the problems with EPR pointed out by the critics, in *Sec. 3.2.1*; the ever constant existence of different problems in the ever new ways to handle waste, as shown in the problems with dumpster diving pointed out by the critics, in *Sec. 3.2.2*; the ever constant existence of different problems in the ever new ways to deal with wastes, as shown in the problems with WtE methods pointed out by the critics, in *Sec. 3.3.1*; the ever constant existence of different problems in the ever new ways to deal waste, as shown in the problems with the use of animal products and animal by-products pointed out by the critics aforecited, in *Sec. 3.3.2*; the ever constant existence of different problems in the ever new ways to deal with waste, as shown in the problems with the cost-benefit analysis of waste recycle pointed out by the critics, in *Sec. 3.4.1*; the ever constant existence of different problems in the ever new ways to recycle waste, as shown in the problems with the use of the waste converter pointed out by the critics, in *Sec. 3.4.2*; the ever constant existence of different problems in the ever new ways to treat waste, as shown in the problems with Zero Waste pointed out by the critics, in *Sec. 3.5.1*; and the ever constant existence of different problems in the ever new ways to reclaim wastes, as shown in the problems with freeganism pointed out by the critics, in *Sec. 3.5.2*).

And the reverse direction also holds true.

Again, it should be stressed, however, that there are different shades of gray (or different degrees of truth) in the two opposites, and the classification is not necessarily mutually exclusive either. In fact, even when some combinations of the two occur, they only end up sharing the same dialectic relationship, but in a different degree.

Also, there are some other relationships (as a kind of family resemblance) which have something in common with the principle, but they are not exactly the same but only more or less comparable, which varies from case to case.

Good examples of family resemblance in relation to the principle are dynamicness-staticness, instability-stability, and the like.

13th Thesis: The Order-Chaos Principle

The thirteenth thesis is called the the order-chaos principle (on the pattern of things) in the category of "process" in existential dialectics (as first worked out in *FC*), in that both order and chaos are vital in the process of change in the world.

The preference for order is biased, since it does not give sufficient attention to the vital role of chaos in the transformation of the world (without somehow reducing it for the understanding of order). There is no order without chaos—and vice versa.

The scientific search for order in the world is often a hidden bias in its ontological obsession with order, since chaos is often treated as the "bad" guy, with order as the "good" guy (for the end goal of science).

Neither order nor chaos is the final end of the world, and one is not to be treated as the means for the other in the transformation of things. Both are fundamental in their recurrent dialectical interactions with each other over time, without reducing one for the other.

For instance, on the one hand, there is order (e.g., the more or less orderly view of the streets in contemporary French cities, which had been installed with sewers and other sanitary systems since the 1950, in *Sec. 2.2.1*; the more or less orderly appearance of clinical waste, if looked from the vantage point of the MWTA between 1988 and 1991, when it enforced proper clinical waste disposal, in *Sec. 2.2.2*; the more or less orderly appearance of clean-up areas in the aftermath of intensive anti-litter campaigns, in *Sec. 2.3.1*; the more or less orderly appearance of those orbits "with a low enough density" of space debris, so that "the addition of debris through impacts is slower than their rate of decay, and the problem does not become significant," so things do not look so chaotic, with so many fewer collisions, in *Sec. 2.3.2*; the more or less orderly appearance of hazardous waste management nowadays, with the creation of the Basel Convention, which "places a general prohibition on the exportation or importation of wastes between Parties and non-Parties," in *Sec. 2.4.1*; the more or less orderly view of the environmental justice movement, if looked from the vantage point of the three clear contributions of the Civil Rights Movement to the movement, in *Sec. 2.4.2*; the more or less orderly appearance of ethical consumerism, if looked from the vantage point of a single definition of what constitutes "moral," as in the "Judeo-Christian scriptures," which "direct followers towards practising good stewardship of the Earth, under an obligation to a God who is believed to have created the planet for us to share with other creatures," in *Sec. 2.5.1*; the more or less orderly view of the simple living movement, if looked from the vantage point of its focus on "voluntary practices to simplify one's lifestyle," in *Sec. 2.5.2*; the more or less orderly reality of waste recycling if looked from the vantage point of the the advocates of EPR to solely focus on producer responsibility, in *Sec. 3.2.1*; the more or less orderly appearance of the permissability of dumpster diving if looked only from the vantage point of "a law issued in 2000" in Italy which "declared dumpster diving to be legal," in *Sec. 3.2.2*; the more or less orderly appearance of WtE methods, if looked only from the vantage point of incineration as the most common implementation, in *Sec. 3.3.1*; the more or less orderly appearance of animal products and animal by-products, if looked only from the vantage point of their advocates for food consumption, in *Sec. 3.3.2*; the more or less orderly appearance of wastes, if looked from the vantage point of a landfill tightly packed with wastes in a small area, in *Sec. 3.4.1*; the more or less orderly appearance of waste treatment, if looked from the vantage point of the neatly organized conversion cycle of the waste converter, in *Sec. 3.4.2*; the more or less orderly view of waste treatment, if looked from the vantage point of "construction," which favors "the design of buildings as assemblages of high level components, not their creation from rough materials such as lumber, cement or plaster," so that "these re-designed, resource-efficient systems will be reused many times as the products that incorporate them are reused," in *Sec. 3.5.1*; and the more or less orderly appearance of food wastes, if looked from the vantage point of them being contained inside dumpsters of food suppliers like "supermarkets, grocery stores, and restaurants," in *Sec. 3.5.2*).

On the other hand, there is chaos (e.g., the more or less chaotic view of the streets in French cities in the 19th century and before, which were often filled with "great piles of refuse and excrement down the centre of the streets," and "cholera outbreaks were common and ravaged the population," in *Sec. 2.2.1*; the more or less chaotic appearance of clinical waste, if looked from the vantage point of "the '30-mile garbage slick' composed primarily of medical and household wastes" spreading in "numerous New Jersey and New York beaches" in 1987, before the enactment of the MWTA, in *Sec. 2.2.2*; the more or less chaotic appearance of abandoned areas suffering from massive litter, in *Sec. 2.3.1*; the more or less chaotic appearance of space debris in those orbits characterized by a "Kessler Syndrome," in that, "instead of the density of debris being a measure of the number of items launched into orbit, it was that number plus any new debris caused when they collided," so "if the new debris did not decay from orbit before impacting another object, the number of debris items would continue to grow even if there were no new launches," and more collisions would continue, in *Sec. 2.3.2*; the more or less chaotic appearance of hazardous waste management in the older days before the creation of the Basel Convention, as shown in "the Khian Sea waste disposal incident, in which a ship carrying incinerator ash from the city of Philadelphia in the United States after having dumped half of its load on a beach in Haiti, was forced away where it sailed for many months, changing its name several times," and "unable to unload the cargo in any port, the crew was believed to have dumped much of it at sea," in *Sec. 2.4.1*; the more or less chaotic view of the environmental justice movement, if looked from the myriad conflicting vantage points of its different forms of elitism by some groups against other groups, further compounded by the different conflicts between different "supporters of economic growth" and different proponents of environmental justice, in *Sec. 2.4.2*; the more or less chaotic appearance of ethical consumerism, if looked from the different vantage points of competing inconsistent definitions of what constitutes something "moral" in the first place, because "the definition has become wider and means different things to different groups of people" over time, in *Sec. 2.5.1*; the more or less chaotic view of the simple living movement, if looked from the vantage point of the myraid reasons behind those who choose simple living, like "spirituality, health, increase in 'quality time' for family and friends, work-life balance, personal taste, frugality, or reducing personal ecological footprint and stress," as well as "anti-war movements," "conservation, degrowth, social justice, ethnic diversity, tax resistance and sustainable development," just to cite some of them, in *Sec. 2.5.2*; the more or less chaotic reality of waste recycling if looked from the multiple vantage points of the critics of EPR, as shown in the numerous problems pointed out by the critics, like the examples that its measurements are not consistent over time, that it can be counter-productive, that it can be misused for something else, etc., in *Sec. 3.2.1*; the more or less chaotic appearance of the permissability of dumpster diving if looked from the myriad often conflicting vantage points of numerous laws in different countries and localities about dumpster diving, in *Sec. 3.2.2*; the more or less chaotic appearance of WtE methods, if looked from the multiple vantage points of different competing, or even conflicting, WtE methods, in *Sec. 3.3.1*; the more or less chaotic appearance of animal products and animal by-products, if looked from the multiple often conflicting vantage points of different groups in regard to the pros and cons of using them for food consumption, as well as the different uses for foodstuffs and non-foodstuffs aforediscussed, in *Sec. 3.3.2*; the more or less chaotic appearance of wastes, if looked from the vantage point of "highly dispersed wastes" like "chlorinated organic hydrocarbons that have bioaccumulated in animal and human tissues across the globe," etc., in *Sec. 3.4.1*; the more or less chaotic appearance of waste treatment, if looked from the vantage point of the disorderly, unhealthy piles of waste in overused landfills in underdeveloped countries, in *Sec. 3.4.2*; the more or less chaotic view of waste treatment, if looked from the vantage point of "deconstruction," as shown in the "demolition of traditional buildings" which "has been long done by wrecking ball or bulldozer" into myriad pieces, in *Sec. 3.5.1*; and the more or less chaotic appearance of food wastes, if looked from the vantage point of them being thrown away on the streets and in different dump sites all over the world, in *Sec. 3.5.2*).

And the reverse direction also holds true.

Once more, it should be stressed, however, that there are different shades of gray (or different degrees of truth) in the two opposites, and the classification is not necessarily mutually exclusive either. In fact, even when some combinations of the two occur, they only end up sharing the same dialectic relationship, but in a different degree.

Also, there are some other relationships (as a kind of family resemblance) which have something in common with the principle, but they are not exactly the same but only more or less comparable, which varies from case to case.

Good examples of family resemblance in relation to the principle are lawfulness-disorder, order-disorder, pattern-chaos, linearity-nonlinearity, unity-disunity, linearity-irreversalness, etc.

14th Thesis: The Slowness-Quickness Principle

The fourteenth thesis concerns the slowness-quickness principle (on the speed of change) in the category of "process" in existential dialectics (which was first worked out in *FIA*), in that both slowness and quickness co-exist, with their own internal tension, to the extent that each fights for its own relevance with the other, in accordance to the perspectives of nature, the mind, culture, and society, without one being the victor and the other being the vanquished in the long haul. There is no slowness without quickness—and vice versa.

For instance, on the one hand, there is slowness (e.g., the relatively slower speed of waste to become a critical issue in antiquity, "due to low population density, coupled with insignificant exploitation of natural resources," in *Sec. 2.2.1*; "hot air at 160 °C takes two hours to achieve" the same sterility for clinical waste that steam at 134 °C takes in three minutes, in *Sec. 2.2.2*; the relatively slower speed of some items of litter like "condensed glass, styrofoam or plastic" waste to biodegrade, because they can "possibly remain…in the environment for over a million years," in *Sec. 2.3.1*; "relative velocities in GEO are low, compared with those between objects in largely random low earth orbits," in *Sec. 2.3.2*; the relatively slower speed of the "Basel Ban Amendment to the Basel Convention" to be ratified, if the requirement is that "amendments to the convention enter into force after ratification of 'three-fourths of the Parties who accepted them,'" in *Sec. 2.4.1*; the relatively slower speed of the early environmental justice movement in the older days to fight for environmental justice, when "supporters of economic growth have taken advantage of environmentalists' neglect of minorities," in *Sec. 2.4.2*; the relatively slower speed of ethical consumerism to be influential in less developed/poor countries, because it has spread primarily in developed countries, in *Sec. 2.5.1*; the relatively slower speed of getting broccoli if having to grow it onself, since, on average, "broccoli takes 105 days to grow a head," as ponted out by farmer Matt Moore, in *Sec. 2.5.2*; the relatively slower speed of companies to engage in technical innovations if regulated by EPR, according to "organizations and researchers against EPR," as they claim that "the mandate would slow technical innovation and impede technological process," because it adds extra costs to producers and contributes to other problems, in *Sec. 3.2.1*; the relatively slower speed of a discarded item left on private premises to be the target of collection by dumpster divers, in *Sec. 3.2.2*; the relatively slower speed of "municipal solid wastes" to decompose naturally, in *Sec. 3.3.1*; the relatively slower speed of animal by-products to be used for foodstuffs because they have to go through the process of "rendering," which is time-consuming, in *Sec. 3.3.2*; the relatively slower speed of wastes to be recycled if they spread out all over the globe in "highly dispersed" and "diluted" forms, in *Sec. 3.4.1*; the relatively slower speed of the "older models" of the waste converter, which "required up to several hours to saturate and sterilize the same load" of waste, in *Sec. 3.4.2*; the relatively slower speed of Zero Waste to be achieved if done by the government alone, because "the government alone does not have the cognitive breadth to determine how to reach the ultimate target of Zero Waste which requires the involvement of businesses, NGOs, the public and the state in governing," in *Sec. 3.5.1*; and the relatively slower speed of unused buildings to be occupied if based on a free market economy, in *Sec. 3.5.2*).

On the other hand, there is quickness (e.g., the relatively quicker speed of waste to become a critical issue nowadays, "due to the increase in population and the massive migration of people to industrial towns and cities from rural areas," in *Sec. 2.2.1*; "steam at 134 °C can achieve in three minutes the same sterility [for clinical waste] that hot air at 160 °C takes two hours to achieve," in *Sec. 2.2.2*; the relatively quicker speed of some biodegradable items of litter to eventually decompose, in *Sec. 2.3.1*; relative velocities of "objects in largely random low earth orbits" like LEO are higher, in *Sec. 2.3.2*; the relatively quicker speed of the "Basel Ban Amendment to the Basel Convention" to be ratified, if the requirement is that "amendments to the convention enter into force after ratification" of only a small number of the Parties who accepted them, in *Sec. 2.4.1*; the relatively quicker speed of the environmental justice movement nowadays to fight for environmental justice, when it receives help from the Civil Rights Movement, in *Sec. 2.4.2*; the relatively quicker speed of the fall of the Apartheid regime, because of the contribution by "consumer boycotts of South Africa over apartheid," in *Sec. 2.5.1*; the relatively quicker speed of getting broccoli if one buys broccoli at a supermarket and gets it instantly, without waiting for 105 days, in *Sec. 2.5.2*; the relatively quicker speed of companies to engage in technical innovations if not burdened by EPR at all, because "organizations and researchers against EPR" claim that "the mandate [of EPR] would slow technical innovation and impede technological process," because it adds extra costs to producers and contributes to other problems, in *Sec. 3.2.1*; the relatively quicker speed of a discarded item left in public dumpsters to be the target of collection by dumpster divers, in *Sec. 3.2.2*; the relatively quicker speed of "municipal solid wastes" to be dealt with by incinerators instead, in *Sec. 3.3.1*; the relatively quicker speed of animal products to be used for foodstuffs because they do not need to go through the process of "rendering" as much, in *Sec. 3.3.2*; the relatively quicker speed of wastes to be recycled if they concentrate in a landfill, in *Sec. 3.4.1*; the relatively quicker speed of "the modern converter," which "achieves saturation within 10–15 minutes due to the high degree of pulverization preceding the sterilization phase," in *Sec. 3.4.2*; the relatively quicker speed of Zero Waste to be achieved with "the involvement of businesses, NGOs, the public and the state in governing," etc., in *Sec. 3.5.1*; and the relatively quicker speed of unused buildings to be occupied if based on an alternative practice known as squatting, in *Sec. 3.5.2*).

And the reverse direction also holds true.

Again, it should be stressed, however, that there are different shades of gray (or different degrees of truth) in the two opposites, and the classification is not necessarily mutually exclusive either. In fact, even when some combinations of the two occur, they only end up sharing the same dialectic relationship, but in a different degree.

Also, there are some other relationships (as a kind of family resemblance) which have something in common with the principle, but they are not exactly the same but only more or less comparable, which varies from case to case.

Good examples of family resemblance in relation to the principle are inconvenience-convenience, passiveness-activenessness, gradualness-abruptness, deceleration-acceleration, difficulty-ease, and the like.

15th Thesis: The Expansion-Contraction Principle

The fifteenth thesis is called the expansion-contraction principle in the category of "process" in existential dialectics, in that entities in the world can both expand in some ways and contract in other ones, as part of their nature. There is no expansion without contraction—and vice versa.

This principle, although not so explicitly called, was already used in my previous works on different topics (e.g., the theory of floating consciousness in *FCD* and *FPHC*, the union of the unions in *BWT*, and the cyclical progression of system fragmentation and integration in *FCD*).

For instance, on the one hand, there is expansion (e.g., the relative spread of cities with sewer systems nowadays, in *Sec. 2.2.1*; the relative spread of the use of "many surgical instruments [such as forceps, needle holders, and scalpel handles]" which are "single-use...items" nowadays, in *Sec. 2.2.2*; the relative spread of the practice of using litter bins on the streets nowadays, in *Sec. 2.3.1*; the relative expansion of "the testing of anti-satellite weapons carried out by both the U.S. and Soviet Union in the 1960s and 1970s," in *Sec. 2.3.2*; the relative expansion of the types of hazardous waste to be included for "the prohibition on the exportation or importation of wastes between Parties and non-Parties" under the Basel Convention and the "Basel Ban Amendment nowadays, in *Sec. 2.4.1*; the relative spread of the environmental justice movement nowadays, in *Sec. 2.4.2*; the relative spread of ethical consumerism in developed countries nowadays, in *Sec. 2.5.1*; the relative rise of materialist capitalist culture in our time, in *Sec. 2.5.2*; the relative expansion of "more sustainable, less toxic, and easily recyclable" products nowadays among those who comply with EPR, in *Sec. 3.2.1*; the relative expansion of dumpsters nowadays in developed countries, since "waste management became a critical issue," due to "the increase in population and the massive migration of people to industrial towns and cities from rural areas," in *Sec. 3.2.2*; the relative expansion of the use of some new types of incinerators nowadays, which "reduce the volume of the original waste by 95-96%," in *Sec. 3.3.1*; the relative spread of the use of food packaging nowadays, as shown by the

AAFCO regulations in regard to pet food packaging, in *Sec. 3.3.2*; the relative spread of of "large recycling organizations" to do waste recycle in developed countries nowadays, in *Sec. 3.4.1*; the relative expansion of the use of waste converters nowadays, like "upcycling," which is "the process of converting waste materials or useless products into new materials or products of better quality" and "has shown significant growth across the United States" in our time, in *Sec. 3.4.2*; the relative spread of the Zero Waste movement, especially in 1998-2002, in *Sec. 3.5.1*; and the relative spread of the practice of freeganism in the developed world nowadays, in *Sec. 3.5.2*).

On the other hand, there is contraction (e.g., the relative decrease of cities without any sewer systems nowadays, where the streets can be filled with "great piles of refuse and excrement down the centre of the streets," in *Sec. 2.2.1*; the relative decline of the use of "surgical instruments [such as forceps, needle holders, and scalpel handles]" which are "reusable items" nowadays, in *Sec. 2.2.2*; the relative decline of the practice of throwing "waste and human wastes out of the window which would decompose in the street" nowadays, although this was common for Europeans in the older days, in *Sec. 2.3.1*; the relative decline of ASAT testing in late 1970s, when "the U.S.'s only active weapon, Program 437, was shut down in 1975," in *Sec. 2.3.2*; the relative decline or decrease of the incidents like the *Khian* Sea waste disposal incident and the 1988 Koko case nowadays, because of the Basel Convention, in *Sec. 2.4.1*; the relative decrease of market-oriented waste management nowadays, in *Sec. 2.4.2*; the relative decline of corporations which still engage in unethical behavior nowadays in developed countries, in *Sec. 2.5.1*; the relative decline of Judeo-Christian asceticism in our time, in *Sec. 2.5.2*; the relative decline of electronics nowadays which cannot be safely recycled, such as the old-fashioned "Lithium-ion polymer batteries" and others, in *Sec. 3.2.1*; the relative decline of people nowadays in developed countries to simply "throw their...wastes out of the window which would decompose in the street," as was the case for the Europeans in the older days; in *Sec. 3.2.2*; the relative decline nowadays of the use of the older types of "incineration plants, which neither recovered energy nor materials," in *Sec. 3.3.1*; the relative decline of the practice of producing products without a lot of detailed info about them in antiquity, in *Sec. 3.3.2*; the relative decline of "the entrepreneurial poor such as the karung guni, Zabaleen, the rag-and-bone man, waste picker, and junk man" to do waste recycle in developed countries nowadays, in *Sec. 3.4.1*; the relative decline of the demand on landfills nowadays, since "waste conversion...now allows nearly 99% of all MSW to be reused in some way, thus sharply reducing the demand on landfills" in the U.S., in *Sec. 3.4.2*; the relative decline of the idea of "recycling" as simply "placing bottles and cans in a recycle bin," especially after the rise of the Zero Waste movement, in *Sec. 3.5.1*; and the relative decline of environment-unfriendly practices in the developed world nowadays, in *Sec. 3.5.2*).

And the reverse direction also holds true.

Again, it should be stressed, however, that there are different shades of gray (or different degrees of truth) in the two opposites, and the classification is not necessarily mutually exclusive either. In fact, even when some combinations of the two occur, they only end up sharing the same dialectic relationship, but in a different degree.

Also, there are some other relationships (as a kind of family resemblance) which have something in common with the principle, but they are not exactly the same but only more or less comparable, which varies from case to case.

Good examples of family resemblance in relation to the principle are conquest-autarky, rise-fall, spread-shrink, extendingness-shorteningness, widening-narrowing, construction-deconstruction, expansion-destruction, expansion-decline, construction-demolition, creation-destruction, and so forth.

In Relation to Agency

The next eight theses are in relation to agency in existential dialectics, as shown below.

16th Thesis: The Theory-Praxis Principle

The sixteenth thesis is the theory-praxis principle (on the duality of knowledge) in the category of "agency" in existential dialectics (which was first worked out in *FPHE*), in that, if there is theoretical construction, there is likewise its practical application, both technical and normative. There is no theory without praxis—and vice versa.

For instance, on the one hand, there is theory (e.g., the theoretical construction of different antibacterial compounds in chemistry, in *Sec. 2.2.1*; the theoretical understanding of "infectious or biohazardous" disease in medicine, in *Sec. 2.2.2*; the theoretical argument about the harmful effects of litter on wildlife, in *Sec. 2.3.1*; the theoretical understanding of "micro-meteorites" by Fred Whipple and Fletcher Watson for the study of "the interplanetary dust cloud" in cosmology after WWII, in *Sec. 2.3.2*; the theoretical understanding of the properties of hazardous waste, like "ignitability," "reactivity," "corrosivity," and "toxicity," in chemistry, in *Sec. 2.4.1*; the theoretical indoctrination of the Civil Rights Movement in regard to "social justice, equal protection, and an end to institutional discrimination," in *Sec. 2.4.2*; the theoretical construction of the "good stewardship of the Earth" in "Judeo-Christian scriptures," in *Sec. 2.5.1*; the theoretical construction of environmentalism in environmental studies, in *Sec. 2.5.2*; the theoretical construction of "sustainability" in environmental studies, in *Sec. 3.2.1*; the theoretical construction of "sustainability" in environmental studies, in *Sec. 3.2.2*; the theoretical construction of "thermal efficiency" in thermodynamics, in *Sec. 3.3.1*; the theoretical understanding of "heating" and "drying" in physics and chemistry, in *Sec. 3.3.2*; the theoretical understanding of "environmental design" to resolve wastes, like waste recycle, in *Sec. 3.4.1*; the theoretical understanding of "pasteurization" in microbiology, in *Sec. 3.4.2*; the theoretical construction of "Design Efficiency" and "Design Effectiveness" in the Zero Waste Institute, in *Sec. 3.5.1*; and the theoretical construction of "anti-globalization" worldviews against corporate capitalism in contemporary times, in *Sec. 3.5.2*).

On the other hand, there is praxis (e.g., the practical application of different antibacterial compounds in chemistry to the field of medicine, like the use of "streptomycin," "gentamicin," "tetracyclines," etc. for the treatment of bubonic plague, in *Sec. 2.2.1*; the practical application of the knowledge of "infectious or biohazardous" disease in medicine to the field of waste management, like the enforcement by the MWTA for proper clinical waste disposal to avoid the spread of infectious disease, in *Sec. 2.2.2*; the practical application of the knowledge about the harmful effects of litter on wildlife to the field of litter treatment, like different legislations in the United States, where "all 50 states have anti-litter laws," in *Sec. 2.3.1*; the practical application of the theoretical understanding of "micrometeorites" in cosmology to the field of space race for the measurement of space debris, in *Sec. 2.3.2*; the practical application of the theoretical understanding of the properties of hazardous waste, like "ignitability," "reactivity," "corrosivity," and "toxicity," in chemistry to the field of hazardous waste management, like the creation of the Basel Convention, in *Sec. 2.4.1*; the practical application of the theoretical indoctrination of the Civil Rights Movement in regard to "social justice, equal protection, and an end to institutional discrimination" to the field of environmental justice, like its practical influence on the environmental justice movement, in *Sec. 2.4.2*; the practical application of the theoretical construction of the "good stewardship of the Earth" in "Judeo-Christian scriptures" to the field of ethics for ethical consumerism, in *Sec. 2.5.1*; the practical application of environmentalism in environmental studies to the field of spiritual studies, like the simple living movement for waste reduction and a more sustainable environment, in *Sec. 2.5.2*; the practical application of the knowledge of "sustainability" in environmental studies to the field of economics, like the call for EPR in manufacturing, in *Sec. 3.2.1*; the practical application of the knowledge of "sustainability" in environmental studies to the field of waste management, like "dumpster diving" which "becomes an environmentalist endeavor…and is…practiced by many pro-green communities," in *Sec. 3.2.2*; the practical application of "thermal efficiency" in thermodynamics to the field of waste treatment for WtE, but "thermal efficiency" is limited in real life, because "real engines have many departures from ideal behavior that waste energy, reducing actual efficiencies far below the theoretical values" which are possible only in ideal situations, in *Sec. 3.3.1*; the practical application of the understanding of "heating" and "drying" in physics and chemistry to the field of meat industry, like the process of "rendering," in *Sec. 3.3.2*; what is conceptual and theoretical is not necessarily practical, because "complete recycling is impracticable as highly dispersed wastes become so diluted that the energy needed for their recovery becomes increasingly excessive, in *Sec. 3.4.1*; the practical application of the understanding of "pasteurization" in microbiology to the field of waste treatment, as shown in the use of the waster converter, in *Sec. 3.4.2*; the practical application of "Design Efficiency" and "Design Effectiveness" in the Zero Waste Institute to the realm of practice, as shown by the Zero Waste International Alliance, which helps a lot of "communities across the globe that have created public policy to promote zero-waste practices," in *Sec. 3.5.1*; and the practical application of "anti-globalization" worldviews against corporate capitalism in contemporary times to the field of waste treatment, like freeganism, in *Sec. 3.5.2*).

And the reverse direction also holds true.

Again, it should be stressed, however, that there are different shades of gray (or different degrees of truth) in the two opposites, and the classification is not necessarily mutually exclusive either. In fact, even when some combinations of the two occur, they only end up sharing the same dialectic relationship, but in a different degree.

Also, there are some other relationships (as a kind of family resemblance) which have something in common with the principle, but they are not exactly the same but only more or less comparable, which varies from case to case.

Good examples of family resemblance in relation to the principle are discovery-application, knowledge-action, invention-innovation, theory-intentionality, discourse-action, knowledge-interest, knowledge-motivation, universality-particularity, and so forth.

17th Thesis: The Convention-Novelty Principle

The seventeenth thesis is the convention-novelty principle (on the nature of creative thinking) in the category of "agency" in existential dialectics (which was first worked out in *FPHCT*), in that, if there is conventional wisdom, there is likewise novel challenge, to the extent that both convergent and divergent thinking are part of life. There is no convention without novelty—and vice versa.

As srummarized from *FPHCT* (almost verbatim here), there are (a) "creative techniques" and (b) "creative traits," which, when satisfied—in relation to the larger context of the mind, nature, society, and culture—can be used to enhance creative works.

In addition, creative thinking has its own possibilities and limits (in relation to invention), just as it has its own promises and pitfalls (in relation to innovation)— as already analyzed in *FPHCT*).

In the end, creative thinking has its own desirability and dark sides (as also already analyzed in *FPHCT*).

With these clarifications in mind—there are good empirical examples for the convention-novelty princple.

For instance, on the one hand, there is convention (e.g., the conventional wisdom of using antibiotics to fight the bubonic plague, in *Sec. 2.2.1*; the conventional wisdom of using "surgical instruments [such as forceps, needle holders, and scalpel handles]" which are "reusable" and "sterilizable" in the older days, in *Sec. 2.2.2*; the conventional practice of the Europeans in the older days when they threw "waste and human wastes out of the window which would decompose in the street," in *Sec. 2.3.1*; the conventional wisdom about space debris in the older days when it was thought that "the density of debris" is "a measure of the number of items launched into orbit," in *Sec. 2.3.2*; the conventional wisdom of the Basel Convention in the older days in regard to its exclusion of radioactive waste in the original definition of hazaroud swaste, in *Sec. 2.4.1*; the conventional wisdom of the early environmental movement to focus on "wilderness protection and wildlife preservation," in *Sec. 2.4.2*; the conventional custom of some corporations which invest "a portion of their profits in, for example, the arms industry" in the older days, in *Sec. 2.5.1*; the conventional wisdom about "consumerist mentality" in modern capitalist culture, in *Sec. 2.5.2*; the conventional wisdom about making "complex electronics that can be difficult to safely recycle, such as

Lithium-ion polymer batteries" in the older days, in *Sec. 3.2.1*; the conventional wisdom that "most people who resort to dumpster-diving are forced to do so out of economic necessity," in *Sec. 3.2.2*; the conventional wisdom of using incineration for WtE, in *Sec. 3.3.1*; the conventional wisdom of using animal products and animal by-products as pet food, in *Sec. 3.3.2*; the conventional wisdom about waste recycle to be good for the environment, in *Sec. 3.4.1*; the conventional wisdom of using autoclaves for waste treatment, in *Sec. 3.4.2*; the conventional wisdom about waste recycle, in *Sec. 3.5.1*; and the conventional wisdom about work ethics for working more to earn a good living, in *Sec. 3.5.2*).

On the other hand, there is novelty (e.g., the alternative novel challenge to the conventional wisdom about using antibiotics to fight the bubonic plague, by the new idea of the Imperial Japanese Army Air Service in 1940 to use the bubonic plague as a biological weapon instead, in *Sec. 2.2.1*; the alternative novel challenge to the conventional wisdom of using "reusable" and "sterilizable" surgical instruments in the older days, by the new idea of using "surgical instruments [such as forceps, needle holders, and scalpel handles]" which are "single-use" nowadays, in *Sec. 2.2.2*; the alternative novel challenge to the conventional practice of the Europeans in the older days to throw waste on the street, by the new idea of different solutions to litter nowadays, like the use of litter bins, litter traps, etc., in *Sec. 2.3.1*; the alternative novel challenge to the conventional wisdom about space debris in the older days, by the new idea of the so-called "Kessler Syndrome" by Donald Kessler and later other researchers, when they argued that, "instead of the density of debris being a measure of the number of items launched into orbit, it was that number plus any new debris caused when they collided," so "if the new debris did not decay from orbit before impacting another object, the number of debris items would continue to grow even if there were no new launches," in *Sec. 2.3.2*; the alternative novel challenge to the conventional wisdom about the exclusion of radioactive waste in the original definition of hazarouds waste in the Basel Convention in the older days, by the new idea of the "Basel Ban Amendment to the Basel Convention" nowadays, which seeks a broader definition, in *Sec. 2.4.1*; the alternative novel challenge to the conventional wisdom of the early environmental movement to focus on "wilderness protection and wildlife preservation, by the new idea nowadays of the environmental movement to focus more on "low-income minorities" to ban environmental discrimination, in *Sec. 2.4.2*; the alternative novel challenge to the conventional custom of some corporations which invest "a portion of their profits in, for example, the arms industry" in the older days, by the new idea of ethical consumerism to boycott these corporations, in *Sec. 2.5.1*; the alternative novel challenge to the conventional wisdom about "consumerist mentality" in modern capitalist culture, by the new idea of a new economics based on the simple living movement for "simple prosperity" by David Wann, in *Sec. 2.5.2*; the alternative novel challenge to the conventional wisdom about making "complex electronics that can be difficult to safely recycle," by the new idea of EPR, which suggests that "it financially encourages manufacturers to design for recycling and make products last longer," in *Sec. 3.2.1*; the alternative novel challenge to the conventional wisdom that "most people who resort to dumpster-diving are forced to do so out of economic necessity," by the new idea nowadays that discarded items in dumpster can be used for police investigation, the creation of art works, academic research, etc., in *Sec. 3.2.2*; the alternative novel challenge to the conventional wisdom of using incineration for WtE, by the new idea of using alternative technologies for WtE, like "thermal depolymerization" and "plasma arc gasification," in *Sec. 3.3.1*; the alternative novel challenge to the conventional wisdom of using animal products and animal by-products as pet food, by the new idea of making a new category of pet food which "emphasizes the use of human-grade meat sources only, with no animal meat by-products," although AAFCO found this claim misleading, in *Sec. 3.3.2*; the alternative novel challenge to the conventional wisdom about waste recycle to be good for the environment, by the new idea of Steven Landsburg, who argued instead that "paper recycling actually reduces tree populations" and is thus bad for the environment, in *Sec. 3.4.1*; the alternative novel challenge to the conventional wisdom of using autoclaves for waste treatment, by the new idea of the use of the waste converter instead, which differs from the use of autoclaves, in *Sec. 3.4.2*; the alternative novel challenge to the conventional wisdom about waste recycle, by the new idea of Zero Waste, which is more than just waste recycle but also waste reuse, in *Sec. 3.5.1*; and the alternative novel challenge to the conventional wisdom about work ethics for working more to earn a good living, by the new idea of "working less" instead in freeganism for the anti-capitalist movement, in *Sec. 3.5.2*).

And the reverse direction also holds true.

Again, it should be stressed, however, that there are different shades of gray (or different degrees of truth) in the two opposites, and the classification is not necessarily mutually exclusive either. In fact, even when some combinations of the two occur, they only end up sharing the same dialectic relationship, but in a different degree.

Also, there are some other relationships (as a kind of family resemblance) which have something in common with the principle, but they are not exactly the same but only more or less comparable, which varies from case to case.

Good examples of family resemblance in relation to the principle are convergence-divergence (or convergent thinking vs. divergent thinking), normalness-nonnormalness, conformity-nonconformity, familiarity-freshness, inertia-newness, and so on.

18th Thesis: The Evolution-Transformation Principle

The eighteenth thesis refers to the evolution-transformation principle (on the multiple kinds of agency). in the category of "agency" in existential dialectics (which was first worked out in *FAE* and then in other books of mine).

This principle—and the symmetry-asymmetry principle, for instance—are both about agency. More precisely, the word "agency," in a formal definition, refers to "a person or thing through which power is exerted or an end is achieved." (MWD 2007) It therefore does not have to necessarily involve an intelligent lifeform.

Because of this dual meaning in agency, the evolution-transformation principle is more concerned with *the kind of agency*, that is, both about the *evolution* in the state of nature (e.g., an object of natural beauty) and the *transformation* in the world of intelligent lifeforms (e.g., a work of art). There is no evolution without transformation—and vice versa.

In classical Darwinian evolutionary theory (as more detailedly analyzed in *BNN*), evolution is "blind." But in the human world, change often takes place because of the conscious intervention of humans in transforming society and culture, just to cite two instances.

And the transformative part of the principle precisely refers to the other dimension in the dual meaning of agency, in giving technology (as invented by intelligent lifeforms like humans and, soon, post-humans) a major role to play in the change of the world, which is something that I extensively analyzed in *FHC* in the context of the technophilic lifeworld, espeically though not exclusively since modern times. The same theme is also elaborated in many other books of mine.

For instance, on the one hand, there is evolution (e.g., the natural evolution of humans to get infected with diseases from time to time in the state of nature, in *Sec. 2.2.1*; the natural evolution of humans to throw away clinical waste like "identifiable body parts" in the state of nature in a primitive way, in *Sec. 2.2.2*; the natural evolution of humans to throw away waste in the state of nature in a primitive way, in *Sec. 2.3.1*; the natural evolution of humans to live on Earth in the state of nature in a primitive way, in *Sec. 2.3.2*; the natural evolution of humans to throw away hazardous wastes in the state of nature in a primitive way, in *Sec. 2.4.1*; the natural evolution of humans to throw away wastes in the state of nature in a primitive way, in *Sec. 2.4.2*; the natural evolution of humans to throw wastes or to pollute the environment in the state of nature in a primitive way, in *Sec. 2.5.1*; the natural evolution of humans to live by foraging in the state of nature in a primitive way, in *Sec. 2.5.2*; the natural evolution of humans to throw away wastes in the state of nature in a primitive way, in *Sec. 3.2.1*; the natural evolution of humans to throw away wastes in the state of nature in a primitive way, in *Sec. 3.2.2*; the natural evolution of organic wastes to decompose naturally in the state of nature, in *Sec. 3.3.1*; the natural evolution of humans to consume animal products and animal by-products in the state of nature in a primitive way, like foraging, in *Sec. 3.3.2*; the natural evolution of humans, in that "humans have evolved to be sensitive to short term survival goals," in *Sec. 3.4.1*; the natural evolution of humans to throw away wastes in the state of nature in a primitive way," in *Sec. 3.4.2*; the natural evolution of humans to live in caves in the state of nature in a primitive way, in *Sec. 3.5.1*; and the natural evolution of humans to live in caves in the state of nature in a primitive way, in *Sec. 3.5.2*).

On the other hand, there is transformation (e.g., the technical transformation of human ability to deal with infection by the invention of antibiotics like "streptomycin," "gentamicin," "tetracyclines," etc. for the treatment of bubonic plague, in *Sec. 2.2.1*; the technical transformation of human ability to dispose clinical waste by the invention of autoclaves, in *Sec. 2.2.2*; the technical transformation of human ability to throw away waste by the invention of picking-up, container deposit schemes, monitering sites, etc., *Sec. 2.3.1*; the technical transformation of human ability to live by the invention of spacecrafts for space exploration, in *Sec. 2.3.2*; the technical transformation of human ability to manage hazardous wastes by the invention of legal institutions like the Basel Convention, in *Sec. 2.4.1*; the technical transformation of the human treatment of wastes by the invention of the environmental justice movement to ban environmental discrimination against minorities, in *Sec. 2.4.2*; the technical transformation of human practice of dealing with wastes and the environment by the invention of ethics, like ethical consumerism for a "carbon-neutral zone," in *Sec. 2.5.1*; the technical transformation of human ability to live by the invention of the ideology of "simple living" for a more sustainable environment, in *Sec. 2.5.2*; the technical transformation of human ability to handle waste by the invention of EPR for a systematic "integration of environmental costs associated with goods throughout their life cycles into the market price of the products," in *Sec. 3.2.1*; the technical transformation of human ability to deal with wastes by the invention of dumpsters nowadays, in *Sec. 3.2.2*; the technical transformation of wastes by the man-made invention of WtE methods to decompose wastes for energy creation, in *Sec. 3.3.1*; the technical transformation of human ability to deal with animal products and animal by-products by the invention of "rendering," in *Sec. 3.3.2*; the technical transformation of human ability to deal with survival goals by the invention of "recycling behaviour," which "is not natural because it requires a focus and appreciation for long term planning," in *Sec. 3.4.1*; the technical transformation of human ability to deal with wastes by the invention of the modern waste converter, in *Sec. 3.4.2*; the technical transformation of human ability to cope with shelters by the invention of Zero Waste, which invents "the design of buildings as assemblages of high level components," so that "these re-designed, resource-efficient systems will be reused many times as the products that incorporate them are reused, in *Sec. 3.5.1*; and the technical transformation of human ability to live by the invention of squatting in freeganism, in *Sec. 3.5.2*).

And the reverse direction also holds true.

Again, it should be stressed, however, that there are different shades of gray (or different degrees of truth) in the two opposites, and the classification is not necessarily mutually exclusive either. In fact, even when some combinations of the two occur, they only end up sharing the same dialectic relationship, but in a different degree.

Also, there are some other relationships (as a kind of family resemblance) which have something in common with the principle, but they are not exactly the same but only more or less comparable, which varies from case to case.

Good examples of family resemblance in relation to the principle are inorganicness-volition, naturalness-technologicalness, naturalness-nonnaturalness, nonwillingness-willingness, inorganicness-motivation, evolution-creation, naturalness-artificialness, and so on.

19th Thesis: The Symmetry-Asymmetry Principle

The nineteenth thesis is labeled as the symmetry-asymmetry principle (on the relationships among existents) in the category of "agency" in existential dialectics (which was also already worked out in my previous books), in that there is no asymmetry without symmetry—and vice versa.

For instance, the Same can be symmetric and asymmetric towards the Others. But in case of asymmetry (as analyzed in *BDPD*), oppression and self-oppression can occur. So, when the Same is asymmetric towards the Others, the Same can also be relatively asymmetric towards itself in self-oppression, just as the Others can be likewise towards themselves. The subsequent oppressiveness is dualistic, as much by the Same against the Others and itself, as by the Others against the Same and themselves. Both oppression and self-oppression can be achieved by way of downgrading differences between the Same and the Others and of accentuating them.

This is true, even though not all forms of asymmetry have to be about oppression and self-oppression.

In addition, from Chapter Three of *FPHG*, symmetry is not perfect, to be understood in an approximate sense under many life circumstances. With this clarification in mind, hereafter are some empirical examples.

For instance, on the one hand, there is symmetry (e.g., the existence of different lifeforms like humans and bacteria, in *Sec. 2.2.1*; the existence of different actors in the field of clinical waste disposal, like the federal government and "private waste contractors," in *Sec. 2.2.2*; the existence of different actors on the street, like cops and citizens, in *Sec. 2.3.1*; the existence of different countries on Earth, in *Sec. 2.3.2*; the existence of different parties in the Basel Convention, in *Sec. 2.4.1*; the existence of different groups in the environmental justice movement, in *Sec. 2.4.2*; the co-existence of different social groups, in *Sec. 2.5.1*; the existence of different lifestyles, like simple living and conspicuous consumption, in *Sec. 2.5.2*; the existence of different forms of responsibility, like producer-based and consumer-based, in *Sec. 3.2.1*; the existence of different practices of dumpster diving in different countries, in *Sec. 3.2.2*; the existence of different WtE methods, in *Sec. 3.3.1*; the existence of different intelligent life forms, like animals and humans, in *Sec. 3.3.2*; the existence of different social classes, in *Sec. 3.4.1*; the existence of different methods of waste treatment, like "waste converters," "plasma-arc gasification," "micro-wave and other irradiation sterilization," and "autoclaves," in *Sec. 3.4.2*; the existence of different entities in nature, like animals, humans, materials, etc., in *Sec. 3.5.1*; and the existence of different social classes, like the poor and the rich, in *Sec. 3.5.2*).

On the other hand, there is asymmetry (e.g., humans can kill bacteria by antibiotics more than bacteria can kill humans by infection, at least in absolute numbers, in *Sec. 2.2.1*; the federal government empowered by the MWTA can punish those "private waste contractors" for dumping illegally "to avoid high fees," in *Sec. 2.2.2*; "police and local supervisors" like those known as "*buitengewoon opsporingsambtenaar*, or BOA" can "fine citizens for throwing away cans, bottles or wrappers onto the street," in *Sec. 2.3.1*; only some countries, like the USSR in the older days or Russia nowadays, the U.S., and now China, can send manned spacecrafts for space exploration, in *Sec. 2.3.2*; major powers can seek exceptional treatments within the Basel Convention, which then allows the United States, for instance, to ship "hazardous wastes to Basel Party countries," and "the OECD countries to continue trading in wastes with countries like the United States that have not ratified the Basel Convention," in *Sec. 2.4.1*; "minorities and working class people" are less represented than "the middle and upper class" in the environmental justice movement, in *Sec. 2.4.2*; "critics have...argued that the uneven distribution of wealth prevents consumerism, ethical or otherwise, from fulfilling its democratic potential," in *Sec. 2.5.1*; conspicuous consumption in modern capitalist mainstream culture is more dominant than simple living in the subculture of the simple living movement, in *Sec. 2.5.2*; producer responsibility is more dominant than consumer responsibility under the policy of EPR nowadays, in *Sec. 3.2.1*; "in many developing countries, food is rarely thrown away unless it is rotten as food is scarce in comparison to developed nations"—but in developed countries like "the United States where 40–50% of food is wasted, the trash contains a lot more food to gather," in *Sec. 3.2.2*; "incineration...is the most common WtE implementation," in *Sec. 3.3.1*; humans can use animal products and animal by-products for foodstuffs and non-foodstuffs in a way that animals cannot do the same to humans, at least not to the same extent, in *Sec. 3.3.2*; the poor like "the karung guni, Zabaleen, the rag-and-bone man, waste picker, and junk man" do more waste recycle on the streets than the rich, in *Sec. 3.4.1*; waste converters gain more popularity nowadays, when contrasted with"plasma-arc gasification," "micro-wave and other irradiation sterilization," and "autoclaves," in *Sec. 3.4.2*; humans can recycle and reuse materials found in nature in a way that materials cannot recycle and reuse humans, as shown in "the creation of industry-specific more efficient reuse modalities," such that "industrial products are inherently non-natural," in *Sec. 3.5.1*; and the freegans live by reclaiming wastes but the rich live by wastefully consuming resources, in *Sec. 3.5.2*).

And the reverse direction also holds true.

Again, for the last time, it should be stressed, however, that there are different shades of gray (or different degrees of truth) in the two opposites, and the classification is not necessarily mutually exclusive either. In fact, even when some combinations of the two occur, they only end up sharing the same dialectic relationship, but in a different degree.

Also, there are some other relationships (as a kind of family resemblance) which have something in common with the principle, but they are not exactly the same but only more or less comparable, which varies from case to case.

Good examples of family resemblance in relation to the principle are balance-extremity, reflexiveness-unreflexiveness, equality-inequality, harmoniousness-unharrmoniousness, center-periphery, compatibility-incompatibility, proportion-disproportion, balance-unbalance, and something like that.

20th Thesis: The Softness-Hardness Principle

The twentieth thesis refers to the softness-hardness principle (on the force of change) in the category of "agency" in existential dialectics (which was first worked out in *ALD*), in that any change by an agent, be it organic (like humans) or non-organic (like natural objects), can occur in a forceful (aggressive) or gentle (pacific) way, which can come in all shapes and sizes, of course. There is no softness without hardness—and vice versa.

For instance, on the one hand, there is softness (e.g., the use of antibiotics to save human life against the bubonic plague, in *Sec. 2.2.1*; the compliance of those "private waste contractors," when they dispose clinical waste in accordance to the regulations under the MWTA, in *Sec. 2.2.2*; the cooperation with the legal authorities for anti-litter enforcement, in *Sec. 2.3.1*; the cooperation by Russia and America for the International Space Station nowadays, in *Sec. 2.3.2*; the support of the Basel Convention by the advocates, in *Sec. 2.4.1*; the inclusion of "the middle and upper class" in the early environmental justice movement, in *Sec. 2.4.2*; the endorsement of ethical consumerism by its advocates, in *Sec. 2.5.1*; the appreciation of simple lifestyle by its advocates, in *Sec. 2.5.2*; the praise of EPR by its proponents, in *Sec. 3.2.1*; the appreciation of dumpster diving, as shown in the reuse of "resources destined for the landfill" by dumpster divers, in *Sec. 3.2.2*; appreciation of incineration by its proponents, in *Sec. 3.3.1*; the appreciation of animal products and animal by-products by the advocates of their use for foodstuffs and non-foodstuffs, in *Sec. 3.3.2*; the appreciation of the cost-benefit analysis of waste recycle by the advocates, in *Sec. 3.4.1*; the application of the waster converter in hospitals, in *Sec. 3.4.2*; the appreciation of Zero Waste by its advocates, in *Sec. 3.5.1*; and the appreciation of environmentalism by the freegans, in *Sec. 3.5.2*).

On the other hand, there is hardness (e.g., the use of the bubonic plague as a biological weapon by the Imperial Japanese Army Air Service, in *Sec. 2.2.1*; the violation of those "private waste contractors," when they engage in dumping illegally "to avoid high fees," in *Sec. 2.2.2*; the punishment of those who litter on the street, in *Sec. 2.3.1*; the rivalry between Russia and America in international politics nowadays, in *Sec. 2.3.2*; the critique of the Basel Convention, as shown in the problems pointed out by the critics as cited, in *Sec. 2.4.1*; the exclusion of "minorities and working class people" in the early environmental justice movement, in *Sec. 2.4.2*; the rejection of ethcial consumerism by its critics, in *Sec. 2.5.1*; the critique of simple lifestyle by its critics, as shown in the problems aforecited, in *Sec. 2.5.2*; the critique of EPR by its critics, as shown in the problems as cited, in *Sec. 3.2.1*; "in Germany, a dumpster's contents are regarded as the property of the dumpster's owner," and "therefore, taking items from a dumpster is viewed as theft," in *Sec. 3.2.2*; the critique of incineration, as shown in the problems pointed out by the critics as cited, in *Sec. 3.3.1*; the critique of the use of animal products and animal by-products for foodstuffs and non-foodstuffs, as shown in the problems pointed out by the critics as cited, in *Sec. 3.3.2*; the critique of the cost-benefit analysis of waste recycle by the advocates, as shown in the problems pointed out by the critics as cited, in *Sec. 3.4.1*; the critique of the use of the waste converter, as shown in the problems pointed out by the critics as cited, in *Sec. 3.4.2*; the critique of Zero Waste by its critics, in *Sec. 3.5.1*; and the critique of environmentalism by the market-oriented capitalists, in *Sec. 3.5.2*).

And the reverse direction also holds true.

Again, it should be stressed, however, that there are different shades of gray (or different degrees of truth) in the two opposites, and the classification is not necessarily mutually exclusive either. In fact, even when some combinations of the two occur, they only end up sharing the same dialectic relationship, but in a different degree.

Also, there are some other relationships (as a kind of family resemblance) which have something in common with the principle, but they are not exactly the same but only more or less comparable, which varies from case to case.

Good examples of family resemblance in relation to the principle are peacefulness-violence, cooperation-competition, pacificity-aggression, dovishness-hawkishness, reward-punishment, peace-war, friendliness-unfriendliness, kindness-hostility, love-hatred, acceptance-discrimination, admiration-belittlement, inclusion-exclusion, and the like.

21st Thesis: The Seriousness-Playfulness Principle

The twenty-first thesis refers to the seriousness-playfulness principle (on the extent of seriousness) in the category of "agency" in existential dialectics (which is first worked out in *FPHH*), in that, if there is seriousness, there is also playfulness, especially for any agent with some kind of intelligence life, be it about humans, animals, or, later, post-humans. There is no seriousness without playfulness—and vice versa.

Lest any misunderstanding occurs, the word "playfulness" here should not be confused with other terms like "laughter," "smile," "joke," "tickleness," and "laugh-like vocalization" (in some animals), for instance.

Being playful can come in all shapes and sizes and therefore is not necessarily tied up with any of these terms, alhough it can result from any of them, or a combination of them, or something else altogether.

This principle, together with some other principles (especially though not exclusively, like the formalness-informalness principle), have important implications for what constitutes rationality, be it about "rationality of application" at the practical level or "rationality of knowledge" at the meta-theoretical level (as summarized in *Sec. 4.20* of *FPHH*).

For instance, on the one hand, there is seriousness (e.g., the serious business of those in history who tried to find treatments against the bubonic plague, in *Sec. 2.2.1*; the serious business of those trying to deal with proper clinical waste disposal, in *Sec. 2.2.2*; the serious business of those who try to clean up litter, in *Sec. 2.3.1*; the serious business of those who try to propose solutions to space debris, in *Sec. 2.3.2*; the serious business of those involved in the Basel Convention to deal with hazardous waste management, in *Sec. 2.4.1*; the serious business of those fighting for environmental justice, in *Sec. 2.4.2*; the serious business of the advocates of ethical consumerism to stop unethical behavior, in *Sec. 2.5.1*; the serious business of those in the simple living movement to reject conspicuous consumption in modern capitalist culture, in *Sec. 2.5.2*; the serious business of the proponents of EPR about producer responsibility, in *Sec. 3.2.1*; the serious business of those who engage in dumpster diving, in *Sec. 3.2.2*; the serious business of those who treat wastes by WtE, in *Sec. 3.3.1*; the serious business of those who use animal products and animal by-products, in *Sec. 3.3.2*; the serious business of those who criticize the cost-benefit analysis of waste recycle, in *Sec. 3.4.1*; the serious business of those who make good use of the waste converter, in *Sec. 3.4.2*; the serious business of those participating in the Zero Waste movement, in *Sec. 3.5.1*; and the serious business of those who practice freeganism, in *Sec. 3.5.2*).

On the other hand, there is playfulness (e.g., the playful part of those in history who tried to find treatments against the bubonic plague, when they played around with different solutions over the ages, like some "ethno-medical beliefs of prevention methods" by "walking around with flowers in or around their nose," carving "the symbol of the cross onto the front door of a house with the words 'Lord have mercy on us' near it," etc., in *Sec. 2.2.1*; the playful part of those trying to deal with proper clinical waste disposal, when they play around with different methods over time, like "on-site, truck service, and mail-back disposal," etc., in *Sec. 2.2.2*; the playful part of those who try to clean up litter, when they play around with different solutions over time, like the use of litter bins, litter traps, picking up, monitering sites, legislation and fines, etc., in *Sec. 2.3.1*; the playful part of those who try to propose solutions to space debris, when they play around with different proposals over time, like "foamy ball of aerogel or spray of water, inflatable balloons, electrodynamic tethers, boom electroadhesion, or dedicated 'interceptor satellites,'" in *Sec. 2.3.2*; the playful part of those involved in the Basel Convention to deal with hazardous waste management, when they play around with different definitions of hazardous waste over time, in *Sec. 2.4.1*; the playful part of those fighting for environmental justice, when they play around with different goals and tactics over time, as shown in the controversial history of its evolution, in *Sec. 2.4.2*; the playful part of the advocates of ethical consumerism, when they play around with different definitions of what constitutes something "moral" in the first place, as "the definition has become wider and means different things to different groups of people" over time, in *Sec. 2.5.1*; the playful part of those in the simple living movement, when they play around with differnet solutions, as different groups in the movement advocate different versions of simple living over time, like reducing consumption, reconsidering technology, increasing self-sufficiency, and simplifying diets, in *Sec. 2.5.2*; the playful part of the proponents of EPR about producer responsibility, when they play aroumd with different forms of enforcement over time, like "reuse," "buy-back," "recycling," or even the delegation of this responsibility to a third party known as "producer responsibility organization" instead, in *Sec. 3.2.1*; the playful part of those who engage in dumpster diving, when they play around with discaded items for different purposes over time, like the creation of art works, criminal investigation, waste reuse, academic research, etc., in *Sec. 3.2.2*; the playful part of those who treat wastes by WtE, when they play around with different options over time, as shown in the list of WtE technologies aforecited, in *Sec. 3.3.1*; the playful part of those who use animal products and animal by-products, when they play around with numerous ways to use them for foodstuffs and non-foodstuffs over time, as shown earlier, in *Sec. 3.3.2*; the playful part of those who criticize the cost-benefit analysis of waste recycle, when they play around with different ways to undermine the cost-benefit analysis over time, as shown in the 6 disagreements over the years aforecited, in *Sec. 3.4.1*; the playful part of of those who make good use of the waste converter, when they play around with different areas in which it can be used over the years, like hospitals, ships, farms, slaughterhouses, etc., in *Sec. 3.4.2*; the playful part of those participating in the Zero Waste movement, when they play around with different modalities of waste reuse over the years, like "the literal application of natural processes," "the creation of industry-specific more efficient reuse modalities," etc., in *Sec. 3.5.1*; and the playful part of those who practice freeganism, when they play around with different practices over time, like the 6 characteristics aforediscussed, in *Sec. 3.5.2*).

And the reverse direction also holds true.

Again, it should be stressed, however, that there are different shades of gray (or different degrees of truth) in the two opposites, and the classification is not necessarily mutually exclusive either. In fact, even when some combinations of the two occur, they only end up sharing the same dialectic relationship, but in a different degree.

Also, there are some other relationships (as a kind of family resemblance) which have something in common with the principle, but they are not exactly the same but only more or less comparable, which varies from case to case.

Good examples of family resemblance in relation to the principle are seriousness-jokingness, seriousness-humorousness, seriousness-wittiness, seriousness-nonseriousness, Apollo-Dionysus, seriousness-restingness, seriousness-relaxedness, seriousness-clownishness, and so forth.

22nd Thesis: The Activeness-Inactiveness Principle

The twenty-second thesis refers to the activeness-inactiveness principle (on the extent of activeness) in the category of "agency" in existential dialectics (which was first worked out in *FPHTRAN*), in that, if there is activeness, there is also inactiveness, especially (though not exclusively) for any agent with some kind of intelligence life, be it about humans, animals, or, later, post-humans. There is no activeness without inactiveness—and vice versa.

Being active can come in all shapes and sizes, just as a person (or an animal) can be inactive because he (or it) is tired, old, injured, sick, sleepy (or hibernating), uninterested, in a state of passive inertia (or disposition), etc.—and an object can be inactive because it is at rest, out of work (or order), etc.

This principle, together with some other principles (especially though not exclusively, like the formalness-informalness principle and seriousness-playfulness principle), have important implications for what constitutes rationality, be it about "rationality of application" at the practical level or "rationality of knowledge" at the meta-theoretical level (as summarized in *Sec. 4.20* of *FPHH*).

For instance, on the one hand, there is activeness (e.g., the alertness of those who are not infected with diseases and remain healthy to actively work, in *Sec. 2.2.1*; the activeness of those "private waste contractors," when they dispose clinical waste in accordance to the regulations under the MWTA and are allowed to stay in business, in *Sec. 2.2.2*; the activeness of those animals which are fortunate enough to live in areas without much litter, so they do not suffer from any sickness or pain due to it, in *Sec. 2.3.1*; the activeness of "the U.S. and Soviet Union in the 1960s and 1970s" when they regularly engaged in "the testing of anti-satellite weapons," in *Sec. 2.3.2*; the activeness of those in "the 1988 Koko case, in which 5 ships transported 8,000 barrels of hazardous waste from Italy to the small town of Koko in Nigeria in exchange for $100 monthly rent which was paid to a Nigerian for the use of his farmland," in *Sec. 2.4.1*; the activeness of those from "the middle and upper class" in the early environmental justice movement, in *Sec. 2.4.2*; the activeness of those ethical consumers who buy "ethical products" which "are favored," in *Sec. 2.5.1*; the activeness of those who work a lot to earn money for what they want in the modern capitalist culture, in *Sec. 2.5.2*; the activeness of those companies which are able to "design for recycling and make products last longer," in compliance with EPR, in *Sec. 3.2.1*; the activeness of those dumpster divers in Italy who are allowed to collect discarded items in dumpsters, because of "a law issued in 2000" which "declared dumpster diving to be legal," in *Sec. 3.2.2*; the activeness of those wastes which can be recycled or converted to energy, in *Sec. 3.3.1*; the activeness of those who can convert certain animal products and animal by-products into foodstuffs and non-foodstufs, in *Sec. 3.3.2*; the activeness of those in "large recycling organizations" to do waste recyle in developed countries nowadays, in *Sec. 3.4.1*; the relative activeness of those new autoclaves nowadays which can compete with waste converters, because they "employ the same method of sterilization as converters, but do so using larger and more complex equipment that also has higher safety… requirement…," so these new versions are increasingly used, in *Sec. 3.4.2*; the activeness, in our time, of those wastes in construction like "windows, doors, and metals, many of which are being saved and resold by reuse" in accordance to Zero Waste, in *Sec. 3.5.1*; and the activeness of those in the corporate capitalist world who are driven to work, work, and work for a good living, in *Sec. 3.5.2*).

On the other hand, there is inactiveness (e.g., the relative inactiveness of those who get sick because of infection with bacteria and therefore have to rest until recovery, in *Sec. 2.2.1*; the relative inactiveness of those "private waste contractors," when they are found "guilty of knowingly endangering the life of another through noncompliance,…face upwards of fifteen years imprisonment," and therefore cannot work in that period because of imprisonment, in *Sec. 2.2.2*; the relative inactiveness of those animals which "get…poisoned with litter in their habitats" and are "in serious discomfort," so they cannot move around freely, in *Sec. 2.3.1*; the relative inactiveness of the U.S. and Russia later on, when, "following the U.S. test in 1985, there was a de-facto moratorium on such tests," in *Sec. 2.3.2*; the relative inactiveness of those who are no longer allowed to engage in hazardous waste dumping under the Basel Convention, so they have to do something else instead, in *Sec. 2.4.1*; the relative inactiveness of those "minorities and working class people" in the early history of the environmental justice movement, because there were excluded, in *Sec. 2.4.2*; the relative inactiveness of those ethical consumers who refuse to buy "products" which violate their moral basis, in *Sec. 2.5.1*; the relative inactiveness of those who are happy with what they need, not what they want, and therefore decide to work much less and earn less in the simple living movement, although they may spend more time on something else more spiritual, in *Sec. 2.5.2*; the relative inactiveness of those companies which cannot comply with EPR, because it is too "expensive if the product contains hazardous materials and does not have a scrap value, such as with CRT televisions"—although they can shift their activities to something else instead, in *Sec. 3.2.1*; the relative inactiveness of those dumpster divers in Germany who are not allowed to collect discarded items because "a dumpster's contents are regarded as the property of the dumpster's owner" and "therefore, taking items from a dumpster is viewed as theft" under the law, although they can do something else, in *Sec. 3.2.2*; the relative inactiveness of those wastes which cannot be reused, recycled, or converted to energy, since "unusable material" is often "dumped in landfill areas" instead, in *Sec. 3.3.1*; the relative inactiveness of those who cannot convert certain animal products and animal by-products into foodstuffs and non-foodstufs, although they can do something else altogether, in *Sec. 3.3.2*; the relative inactiveness of those like "the karung guni, Zabaleen, the rag-and-bone man, waste picker, and junk man" who are "driven out of the recycling and the remanufacturing market" in developed countries nowadays and cannot do waste-recycling much, in *Sec. 3.4.1*; the relative inactiveness of those older autoclaves nowadays which cannot compete with waste converters, because of their technological obsolescence, in *Sec. 3.4.2*; the relative inactiveness, in our time, of those wastes in construction like "wood flooring, brick walls, and structural timbers" which cannot be reused in accordance to Zero Waste, in *Sec. 3.5.1*; and the relative inactiveness of those in the freeganist world who are willing to consume and work less instead, although they spend time on something else, in *Sec. 3.5.2*).

And the reverse direction also holds true.

Again, it should be stressed, however, that there are different shades of gray (or different degrees of truth) in the two opposites, and the classification is not necessarily mutually exclusive either. In fact, even when some combinations of the two occur, they only end up sharing the same dialectic relationship, but in a different degree.

Also, there are some other relationships (as a kind of family resemblance) which have something in common with the principle, but they are not exactly the same but only more or less comparable, which varies from case to case.

Good examples of family resemblance in relation to the principle are alertness-unalertness, activeness-decrepidness, alertness-sleepiness, activeness-tiredness, alertness-unwareness, alertness-uninterestedness, soberness-inactiveness, alertness-inactiveness, soberness-unawareness, awakenness-hibernatedness, and so on.

23rd Thesis: The Selfness-Otherness Principle

The twenty-third thesis refers to the selfness-otherness principle (on the magnitude of self-centeredness) in the category of "agency" in existential dialectics (which is first worked out in *FPHCA*), in that, if there is the tendency to be mostly centered on one's self or group, there is likewise the tendency to be mostly centered on others, especially (though not exclusively) for any agent with some kind of intelligence life, be it about humans, animals, or, later, post-humans. There is no selfness without otherness—and vice versa.

Being self-centered and being other-centered do not have to be mutually exclusive, as the two can come in all shapes and sizes, or in different degrees. And other-centeredness does not entail desirability, any more than self-centeredness implies non-desirability, as both can go either way.

In addition, whether self-centeredness constitutes a moral alternative or, conversely, an alternative to morality is of course debatable (as addressed in my other book titled *Beyond Ethics to Post-Ethics*).

For instance, on the one hand, there is selfness (e.g., the tendency to be mostly centered on one's self or group, as shown in the use of the bubonic plague as a biological weapon by the Imperial Japanese Army Air Service in 1940, in *Sec. 2.2.1*; the tendency to be mostly centered on one's self or group, as shown in those "private waste contractors," when they engage in dumping illegally "to avoid high fees," in *Sec. 2.2.2*; the tendency to be mostly centered on one's self or group, as shown in the littering behavior of some individuals, in *Sec. 2.3.1*; the tendency to be mostly centered on one's self or group, as shown in the numerous tests of "anti-satellite weapons" by the U.S. and Soviet Union in the 1960s and 1970s" for global domination, in *Sec. 2.3.2*; the tendency to be mostly centered on one's self or group, as shown in "the 1988 Koko case in which 5 ships transported 8,000 barrels of hazardous waste from Italy to the small town of Koko in Nigeria in exchange for $100 monthly rent which was paid to a Nigerian for the use of his farmland," which is later condemned as "toxic colonialism" by the critics, in *Sec. 2.4.1*; the tendency to be mostly centered on one's self or group, as shown in the supporters of economic growth without caring for environmental justice in the older days, in *Sec. 2.4.2*; the tendency to be mostly centered on one's self or group, as shown in unethical behavior of those market-oriented capitalists who were not environment-friendly in the older days, in *Sec. 2.5.1*; the tendency to be mostly centered on one's self or group, as shown by those who engage in conspicuous consumption in modern capitalist culture to satisfy what they want, not what they need, as "more is better" for them, in *Sec. 2.5.2*; the tendency to be mostly centered on one's self or group, as shown by those campanies in the older days which produced less sustainable, more toxic, and hardly recyclable electronics, in *Sec. 3.2.1*; the tendency to be mostly centered on one's self or group, as shown by those "people who resort to dumpster-diving" because they "are forced to do so out of economic necessity," in *Sec. 3.2.2*; the tendency to be mostly centered on one's self or group, as shown by those firms which found it profitable to use the older types of incinerators, even when they released pollutants, in *Sec. 3.3.1*; the tendency to be mostly centered on one's self or group, as shown in the sale of animal products and animal by-products by the meat industry, which "allows the meat industry to compete economically with industries selling sources of vegetable protein," in *Sec. 3.3.2*; the tendency to be mostly centered on one's self or group, as shown in those "businesses" which "ignore…[negative externalities] despite the costs imposed on society," in *Sec. 3.4.1*; the tendency to be mostly centered on one's self or group, as shown by different proponents of competing technologies for waste treatment, since each of them advocates his own preferred type, in *Sec. 3.4.2*; the tendency to be mostly centered on one's self or group, as shown by "many commercial or industrial companies" like "Staples, Home Depot, Toyota, General Motors and computer take-back campaigns," which "embrace Zero Waste" but "usually mean no more than a major materials recycling effort, having no bearing on product redesign," in *Sec. 3.5.1*; and the tendency to be mostly centered on one's self or group, as shown in the possession of private property by the capitalists, in *Sec. 3.5.2*).

On the other hand, there is otherness (e.g., the tendency to be mostly centered on others, as shown by the legendary nurse Florence Nightingale to take care of patients, in *Sec. 2.2.1*; the tendency to be mostly centered on others, as shown in those few good individuals who genuinely work under the MWTA to ensure proper clinical waste disposal for the common good, in *Sec. 2.2.2*; the tendency to be mostly centered on others, as shown in the efforts of the government to clean up litter for the common good, in *Sec. 2.3.1*; the tendency to be mostly centered on others, as shown by those working for the Union of Concerned Scientists about the dangers of space debris, for a better human future, in *Sec. 2.3.2*; the tendency to be mostly centered on others, as shown in some environmental NGOs campaigning for the end of toxic colonialism, for the common good of humanity, in *Sec. 2.4.1*; the tendency to be mostly centered on others, as shown in the more mature environmental justice movement nowadays, when they try to include different groups of people in the campaign, to avoid the charge of elitism which had plagued its early history, in *Sec. 2.4.2*; the tendency to be mostly centered on others, as shown in the 2009 finding that "people act more altruistically after…exposure to green than conventional products," although there is the danger of unethical behavior too, in *Sec. 2.5.1*; the tendency to be mostly centered on others, as shown by those in the simple living movement when they work less and spend "the time saved…to help…or volunteer" for those who are homeless and poor, in *Sec. 2.5.2*; the tendency to be mostly centered on others, as shown by the proponents of EPR who ask for a new policy of making products which are "more sustainable, less toxic, and easily recyclable electronics" for the common good of society, in *Sec. 3.2.1*; the tendency to be mostly centered on others, as shown in those academic researchers who point out the issues of health risks and physical hazards in the habit of dumpster diving, for the enhancement of social welfare, in *Sec. 3.2.2*; the tendency to be mostly centered on others, as shown by those in academic research studies nowadays who reveal the polluting effects of the older types of incinerators, for an advancement of the understanding of environmental sustainability, in *Sec. 3.3.1*; the tendency to be mostly centered on others, as shown in the work by academic researchers to show the problems with the use of animal products and their critics, for a better understanding of the complicated issues facing the meat industry, that is, for the advancement of knowledge, in *Sec. 3.3.2*; the tendency to be mostly centered on others, as shown by those economists who include "negative externalities" to society in the "economic analysis of recycling," so as to give a better understanding of the economic measurement of waste recycle, for the enhancement of social welfare, in *Sec. 3.4.1*; the tendency to be mostly centered on others, as shown in the academic works by researchers who try to show the pros and cons of each competing technology for waste treatment, like the use of the waster converter, autoclave, etc., so as to make an advancement in the knowledge of waste treatment, in *Sec. 3.4.2*; the tendency to be mostly centered on others, as shown in those academic works which point out the pros and cons of the debate on Zero Waste for an advancement in the knowledge of reuse, in *Sec. 3.5.1*; and the tendency to be mostly centered on others, as shown in the practice of sharing by the freegans, in *Sec. 3.5.2*).

And the reverse direction also holds true.

Again, it should be stressed, however, that there are different shades of gray (or different degrees of truth) in the two opposites, and the classification is not necessarily mutually exclusive either. In fact, even when some combinations of the two occur, they only end up sharing the same dialectic relationship, but in a different degree.

Also, there are some other relationships (as a kind of family resemblance) which have something in common with the principle, but they are not exactly the same but only more or less comparable, which varies from case to case.

Good examples of family resemblance in relation to the principle are egoisticness-altruisticness, selfishness-unselfishness, selfness-selflessness, egocentricness-othercentricness, and so on.

In Relation to Outcome

The next four theses are in relation to outcome in existential dialectics, as shown below.

24th Thesis: The Regression-Progression Principle

The twenty-fourth thesis is called the regression-progression princple (on the direction of history) in the category of "outcome" in existential dialectics (which was also already worked out in my previous books), in that neither the cyclical nor the linear views are adequate for explaining many phenomena at all levels. There is no regression without progression—and vice versa.

History progresses to more advanced forms, but with a regressive touch. Examples include no freedom without unfreedom, no equality without inequality, and no civilization without barbarity. This is not an inevitable law, but merely a highly likely empirical trend.

For instance, on the one hand, there is regression (e.g., the regression made by the research on the bubonic plague, as shown in the use of the bubonic plague as a biological weapon by the Imperial Japanese Army Air Service in 1940, in *Sec. 2.2.1*; the regression made by the methods of clinical waste disposal and their enforcement, as shown in the problems pointed out by the critics, in *Sec. 2.2.2*; the regression made by different solutions to litter, as shown in the disadvantages of each solution pointed out by the critics, in *Sec. 2.3.1*; the regression made by the proposed solutions to space debris, as shown in the problems pointed out by the critics, in *Sec. 2.3.2*; the regression made by the Basel Convention, as shown in the problems pointed out by the critics, in *Sec. 2.4.1*; the regression made by the environmental justice movement, as shown in the problems pointed out by the critics, in *Sec. 2.4.2*; the regression made by ethical consumerism, as shown in the problems pointed out by the critics, in *Sec. 2.5.1*; the regression made by simple lifestyle, as shown in the problems pointed out by the critics, in *Sec. 2.5.2*; the regression made by EPR, as shown in the problems pointed out by the critics, in *Sec. 3.2.1*; the regression made by dumpster diving, as shown in the problems pointed out by the critics, in *Sec. 3.2.2*; the regression made by WtE methods, as shown in the problems pointed out by the critics, in *Sec. 3.3.1*; the regression made by the use of animal products and animal by-products, as shown in the problems pointed out by the critics, in *Sec. 3.3.2*; the regression made by the cost-benefit analysis of waste recycle, as shown in the problems pointed out by the critics, in *Sec. 3.4.1*; the regression made by waste converters, as shown in the problems pointed out by the critics, in *Sec. 3.4.2*; the regression made by Zero Waste, as shown in the disputes pointed out by the critics, in *Sec. 3.5.1*; and the regression made by freeganism, as shown in the problems pointed out by the critics, in *Sec. 3.5.2*).

On the other hand, there is progress (e.g., the progress made by the research on the bubonic plague, as shown in the use of antibiotics to fight the bubonic plague aforecited, in *Sec. 2.2.1*; the progress made by the methods of clinical waste disposal and their enforcement, as shown in the contribution to the prevention of infectious disease by clinical waste aforecited, in *Sec. 2.2.2*; the progress made by different solutions to litter, as shown in their contribution to anti-litter efforts aforecited, in *Sec. 2.3.1*; the progress made by the proposed solutions to space debris, as shown in their contribution to the reduction of space debris aforecited, in *Sec. 2.3.2*; the progress made by the Basel Convention, as shown in the contribution to a better hazardous waste management aforecited, in *Sec. 2.4.1*; the progress made by the environmental justice movement, as shown in its contribution to environmental justice aforecited, in *Sec. 2.4.2*; the progress made by ethical consumerism, as shown in its contribution to moral buying aforecited, in *Sec. 2.5.1*; the progress made by simple lifestyle, as shown in its contribution to waste reduction and a more sustainable environment aforecited, in *Sec. 2.5.2*; the progress made by EPR, as shown in its contribution to waste recycling aforecited, in *Sec. 3.2.1*; the progress made by dumpster diving, as shown in its contribution to waste reuse and other uses aforecited, in *Sec. 3.2.2*; the progress made by WtE methods, as shown in their contribution to converting wastes into energy aforecited, in *Sec. 3.3.1*; the progress made by the use of animal products and animal by-products, as shown in their contribution to the creation of foodstuffs and non-foodstuffs aforecited, in *Sec. 3.3.2*; the progress made by the cost-benefit analysis of waste recycle, as shown in its contribution to a better understanding of efficiency aforecited, in *Sec. 3.4.1*; the progress made by waste converters, as shown in their contribution to waste disposal aforecited, in *Sec. 3.4.2*; the progress made by Zero Waste, as shown in its contribution to a more "zero waste" community aforecited, in *Sec. 3.5.1*; and the progress made by freeganism, as shown in its contribution to the reclaiming of wastes aforecited, in *Sec. 3.5.2*).

And the reverse direction also holds true.

Again, it should be stressed, however, that there are different shades of gray (or different degrees of truth) in the two opposites, and the classification is not necessarily mutually exclusive either. In fact, even when some combinations of the two occur, they only end up sharing the same dialectic relationship, but in a different degree.

Also, there are some other relationships (as a kind of family resemblance) which have something in common with the principle, but they are not exactly the same but only more or less comparable, which varies from case to case.

Good examples of family resemblance in relation to the principle are undesirability-desirability, risk-opportunity, badness-goodness, cost-benefit, dysfunctionality-functionality, cyclicity-linearity, and so on.

25th Thesis: The Same-Difference Principle

The twenty-fifth thesis refers to the same-difference principle (on the metamorphosis of change) in the category of "outcome" in existential dialectics (which was worked out in *ALD* and other books of mine), in that an entity, as it evolves over time, can be both different from and similar to its opposing alternatives and does not have to be solely more different from them over time. There is no similarity without difference—and vice versa.

Opposites are not absolute in a black-or-white fashion; so, an entity can become relatively more similar to (or more different from) its opposite over time.

This is further constrained by another principle, that is, the symmetry-asymmetry principle about the relationships among existents under the category about agency in existential dialectics, in that if there is symmetry (equality) between two entities, there is likewise asymmetry (inequality) emerging in a different way.

For instance, on the one hand, there is similarity in outcome (e.g., the contribution to the molding and control of beliefs, values, and behaviors, regardless of whether this be done by way of using antibiotics to fight the bubonic plague, or by way of carving "the symbol of the cross onto the front door of a house with the words 'Lord have mercy on us' near it," etc., in *Sec. 2.2.1*; the contribution to the molding and control of beliefs, values, and behaviors, regardless of whether this be done by way of those "private waste contractors to dump clinical waste "to avoid high fees," or by way of the federal government under the MWTA for the enforcement of proper clinical waste disposal, in *Sec. 2.2.2*; the contribution to the molding and control of beliefs, values, and behaviors, regardless of whether this be done by way of anti-litter campaigns, or by way of throwing away waste on the street, in *Sec. 2.3.1*; the contribution to the molding and control of beliefs, values, and behaviors, regardless of whether this be done by way of hunting and gathering in the older days, or by way of space exploration nowadays, in *Sec. 2.3.2*; the contribution to the molding and control of beliefs, values, and behaviors, regardless of whether this be done by way of dumping in "the 1988 Koko case, or by way of the reduction of "the movements of hazardous waste between nations" in the Basel Convention, in *Sec. 2.4.1*; the contribution to the molding and control of beliefs, values, and behaviors, regardless of whether this be done by way of market-oriented economic growth, or by way of environmental justice, in *Sec. 2.4.2*; the contribution to the molding and control of beliefs, values, and behaviors, regardless of whether this be done by way of ethical consumerism, or by way of market-oriented capitalism, in *Sec. 2.5.1*; the contribution to the molding and control of beliefs, values, and behaviors, regardless of whether this be done by way of simple lifestyle in the simple living movement, or by way of conspicuous consumption in modern capitalist mainstream culture, in *Sec. 2.5.2*; the contribution to the molding and control of beliefs, values, and behaviors, regardless of whether this be done by way of EPR, or by way of market-oriented responsibility, in *Sec. 3.2.1*; the contribution to the molding and control of beliefs, values, and behaviors, regardless of whether this be done by way of the habit of dumpster diving by poor people, or by way of the throw-away practice of rich people, in *Sec. 3.2.2*; the contribution to the molding and control of beliefs, values, and behaviors, regardless of whether this be done by way of letting organic wastes to decompose naturally, or by way of using incinerators to convert wastes into energy, in *Sec. 3.3.1*; the contribution to the molding and control of beliefs, values, and behaviors, regardless of whether this be done by way of the vegetarian, kosher, and halaal diets, or by way of mainstream non-human animal diets, in *Sec. 3.3.2*; the contribution to the molding and control of beliefs, values, and behaviors, regardless of whether this be done by way of the fiscal analysis of waste recycle, or by way of the economic analysis of waste recycle, in *Sec. 3.4.1*; the contribution to the molding and control of beliefs, values, and behaviors, regardless of whether this be done by way of waste dumping in overfilled landfills, or by way of the use of waste converters, in *Sec. 3.4.2*; the contribution to the molding and control of beliefs, values, and behaviors, regardless of whether this be done by way of "the literal application of natural processes," or by way of "the creation of industry-specific more efficient reuse modalities," in *Sec. 3.5.1*; and the contribution to the molding and control of beliefs, values, and behaviors, regardless of whether this be done by way of freeganism, or by way of corporate capitalism, in *Sec. 3.5.2*).

On the other hand, there is difference in outcome (e.g., the contribution to the molding and control of beliefs, values, and behaviors by way of using antibiotics to fight the bubonic plague for a relatively more scientific lifeworld—but the contribution to the molding and control of beliefs, values, and behaviors by way of carving "the symbol of the cross onto the front door of a house with the words 'Lord have mercy on us' near it," etc. for a relatively more sacral lifeworld, in *Sec. 2.2.1*; the contribution to the molding and control of beliefs, values, and behaviors by way of those "private waste contractors for a relatively more market-oriented lifeworld—but the contribution to the molding and control of beliefs, values, and behaviors by way of the federal government under the MWTA for a relatively more state-regulated lifeworld, in *Sec. 2.2.2*; the contribution to the molding and control of beliefs, values, and behaviors by way of anti-litter campaigns for a relatively more environment-conscious lifeworld—but the contribution to the molding and control of beliefs, values, and behaviors by way of throwing away waste on the street for a relatively more environment-dominated lifeworld, in *Sec. 2.3.1*; the contribution to the molding and control of beliefs, values, and behaviors by way of hunting and gathering in the older days for a relatively more harmonious co-existence with nature—but the contribution to the molding and control of beliefs, values, and behaviors by way of space exploration nowadays for a relatively more dominative conquest of nature, in *Sec. 2.3.2*; the contribution to the molding and control of beliefs, values, and behaviors by way of dumping for a relatively more market-oriented lifeworld—but the contribution to the molding and control of beliefs, values, and behaviors by way of the reduction of "the movements of hazardous waste between nations" in the Basel Convention for a relatively more legal-regulative lifeworld, in *Sec. 2.4.1*; the contribution to the molding and control of beliefs, values, and behaviors by way of market-oriented economic growth for a relatively more competitive lifeworld—but the contribution to the molding and control of beliefs, values, and behaviors by way of environmental justice for a relatively more egalitarian lifeworld, in *Sec. 2.4.2*; the contribution to the molding and control of beliefs, values, and behaviors by way of ethical consumerism for a relatively more moralistic lifeworld—but the contribution to the molding and control of beliefs, values, and behaviors by way of market-oriented capitalism for a relatively more amoral lifeworld, in *Sec. 2.5.1*; the contribution to the molding and control of beliefs, values, and behaviors by way of simple lifestyle for a relatively more spiritual lifeworld—but the contribution to the molding and control of beliefs, values, and behaviors by way of conspicuous consumption for a relatively more materialistic lifeworld, in *Sec. 2.5.2*; the contribution to the molding and control of beliefs, values, and behaviors by way of EPR for a relatively more environment-friendly worldview—but the contribution to the molding and control of beliefs, values, and behaviors by way of market-oriented responsibility for a relatively more environment-dominated worldview, in *Sec. 3.2.1*; the contribution to the molding and control of beliefs, values, and behaviors by way of the habit of dumpster diving by poor people for a relatively more impoverished lifeworld—but the contribution to the molding and control of beliefs, values, and behaviors by way of the throw-away practice of rich people for a relatively more affluent lifeworld, in *Sec. 3.2.2*; the contribution to the molding and control of beliefs, values, and behaviors by way of letting organic wastes to decompose naturally for a relatively more naturalistic lifeworld—but the contribution to the molding and control of beliefs, values, and behaviors by way of using incinerators to convert wastes into energy for a relatively more technophilic lifeworld, in *Sec. 3.3.1*; the contribution to the molding and control of beliefs, values, and behaviors by way of the vegetarian, kosher, and halaal diets for a relatively more spiritual lifeworld—but the contribution to the molding and control of beliefs, values, and behaviors by way of the mainstream non-human animal diets for a relatively more secular lifeworld, in *Sec. 3.3.2*; the contribution to the molding and control of beliefs, values, and behaviors by way of the fiscal analysis of waste recycle for a relatively more market-oriented economic life—but the contribution to the molding and control of beliefs, values, and behaviors by way of the economic analysis of waste recycle for a relatively more government-regulated economic life, in *Sec. 3.4.1*; the contribution to the molding and control of beliefs, values, and behaviors by way of waste dumping in overfilled landfills for a relatively more individual-materialistic lifeworld—but the contribution to the molding and control of beliefs, values, and behaviors by way of the use of waste converters for a relatively more environment-friendly lifeworld, in *Sec. 3.4.2*; the contribution to the molding and control of beliefs, values, and behaviors by way of "the literal application of natural processes" for a relatively more naturalistic lifeworld—but the contribution to the molding and control of beliefs, values, and behaviors by way of "the creation of industry-specific more efficient reuse modalities" for a relatively more technophilic lifeworld, in *Sec. 3.5.1*; and the contribution to the molding and control of beliefs, values, and behaviors by way of freeganism for a relatively more egalitarian lifeworld—but the contribution to the molding and control of beliefs, values, and behaviors by way of corporate capitalism for a relatively more elitist lifeworldin *Sec. 3.5.2*).

And the reverse direction also holds true.

Again, it should be stressed, however, that there are different shades of gray (or different degrees of truth) in the two opposites, and the classification is not necessarily mutually exclusive either. In fact, even when some combinations of the two occur, they only end up sharing the same dialectic relationship, but in a different degree.

Also, there are some other relationships (as a kind of family resemblance) which have something in common with the principle, but they are not exactly the same but only more or less comparable, which varies from case to case.

Good examples of family resemblance in relation to the principle are homogeneity-heterogeneity, we-they, similarity-contrast, and so forth.

26th Thesis: The Stability-Reaction Principle

The twenty-sixth thesis is called the stability-reaction princple (on the feedback mechanism for further changes) in the category of "outcome" in existential dialectics (which was first worked out in *FPHCHEM*), in that once an outcome becomes stabilized over time (which varies from case to case), it will undergo further changes in its multifaceted interactions with the outside world (e.g., with other entities, energy, space, and so on).

There is no stability without reaction for further changes—and vice versa.

For instance, on the one hand, there is stability in outcome (e.g., the relatively more stable condition, in the older days, of appealing to folkloric and religious methods to fight the bubonic plague, like "walking around with flowers in or around their nose" and carving "the symbol of the cross onto the front door of a house with the words 'Lord have mercy on us' near it," etc., in *Sec. 2.2.1*; the relatively more stable condition of the enforcement of proper clinical waste disposal between 1988 and 1991 under the MWTA, in *Sec. 2.2.2*; the relatively more stable condition of using litter bins to deal with litter in the modern history of anti-litter campaign, in *Sec. 2.3.1*; the relatively more stable condition of space exploration in this space age of ours, when major powers like the USSR/Russia, the U.S.A., and now China send spacecrafts into space, in *Sec. 2.3.2*; the relatively more stable condition of the Basel Convention in its early days, in *Sec. 2.4.1*; the relatively more stable condition of the early environmental movement, when, "during the first half of the 20th century, the focus was wilderness protection and wildlife preservation," in *Sec. 2.4.2*; the relatively more stable condition of market-oriented capitalism in the older days, when moral issues were often not explicitly considered, in *Sec. 2.5.1*; the relatively more stable condition of conspicuous consumption in modern times, when the consumerist mentality in modern capitalist mainstream culture has been domiant, in *Sec. 2.5.2*; the relatively more stable condition of EPR, when it popularly required producer responsibility in recent times, in *Sec. 3.2.1*; the relatively more stable condition of restricting the practice of dumpster diving in Belgium in the older days, as shown in the arrest of "a Belgian dumpster diver and eco-activist nicknamed Ollie" on February 25th, 2009, for "taking food from a dumpster at an AD Delhaize market in Bruges," in *Sec. 3.2.2*; the relatively more stable practice of using the old types of incinerators in the older days, although "they released acidic pollutants and at times "neither recovered energy nor materials," in *Sec. 3.3.1*; the relatively more stable condition of the practice of using animal by-products for pet food in the older days, in *Sec. 3.3.2*; the relatively more stable condition of "the traditional job of recycling," which "is performed by the entrepreneurial poor such as the karung guni, Zabaleen, the rag-and-bone man, waste picker, and junk man" in the older days, in *Sec. 3.4.1*; the relatively more stable condition of the use of waste converters in modern times, in *Sec. 3.4.2*; the relatively more stable condition of waste dumping in the older days, when many dumps were not exceeding carrying capacity, in *Sec. 3.5.1*; and the relatively more stable condition of the practice of corporate capitalism in the older days, when "councils and landlords…would rather keep properties boarded up if they cannot make a sufficient profit from them," and good buildings are left unused, while there were homeless on the streets, in *Sec. 3.5.2*).

On the other hand, there is reaction in outcome (e.g., the old-fashioned folkloric and religious methods to fight the bubonic plague then led to further changes later on, when nowadays, antibiotics are routinely used instead to fight the bubonic plague, in *Sec. 2.2.1*; the enforcement of proper clinical waste disposal between 1988 and 1991 under the MWTA then led to further changes later on, when "the Act was repealed in 1991" and "states were given the responsibility to regulate and pass laws concerning the disposal of medical waste," in *Sec. 2.2.2*; the practice of using litter bins to deal with litter in the modern history of anti-litter campaign then led to further changes later on, when it is discovered that "if the bins are not regularly emptied, then overfilling of bins occurs and can increase litter indirectly" and "some local authorities will only take responsibility for rubbish that is placed in the bins, which means that litter remains a problem," in *Sec. 2.3.1*; the launches of spacecrafts then led to further changes later on, in that, in accordance to the Kessler Syndrome, "instead of the density of debris being a measure of the number of items launched into orbit, it was that number plus any new debris caused when they collided," so "if the new debris did not decay from orbit before impacting another object, the number of debris items would continue to grow even if there were no new launches" and collisions would continue in the cascades, in *Sec. 2.3.2*; the Basel Convention in its early days then led to further changes later on, as shown by the later campaign for the

"Basel Ban Amendment to the Basel Convention," in *Sec. 2.4.1*; the early focus of the environmental movement during the first half of the 20th century then led to further changes later on, when, nowadays, it pays more attention to the issues affecting "minorities and working class people" like the one concerning environmental discrimination, in *Sec. 2.4.2*; market-oriented capitalism in the older days then then to further changes later on, when moral basis is given more attention in capitalism of our time, in *Sec. 2.5.1*; conspicuous consumption in modern times then led to further changes later on, when "a new economics movement has been building since the UN conference on the environment in 1972" for "simple prosperity" instead, in *Sec. 2.5.2*; the implementation of EPR then led to further changes over time, as "producers would be adding recycling costs into the initial price tag," in *Sec. 3.2.1*; the arrest of "a Belgian dumpster diver and eco-activist nicknamed Ollie" in 2009 then led to further changes later on, when his subsequent trial "evoked protests in Belgium against restrictions from taking discarded food items," in *Sec. 3.2.2*; on, when nowadays the industry depends on more environment-friendly incinerators, which use "lime scrubbers and electro-static precipitators on smokestacks," so that "by passing the smoke through the lime scrubbers, any acids that may be in the smoke are neutralized which prevents the acid from reaching the atmosphere and hurting our environment," in *Sec. 3.3.1*; the practice of using animal by-products for pet food in the older days then led to further changes later on, when nowadays, there are "a new category of pet food," which "often emphasizes the use of human-grade meat sources only, with no animal meat by-products," in *Sec. 3.3.2*; "the traditional job of recycling…performed by the entrepreneurial poor" then led to further changes later on, when "large recycling organizations" are doing a lot of waste recycle in developed countries nowadays, in *Sec. 3.4.1*; the use of waste converters in modern times then led to further changes later on, when nowadays, some contemporary autoclaves have been so improved that they "employ the same method of sterilization as converters, but do so using larger and more complex equipment that also has higher safety…requirement…," so these new versions are to compete with waste converters, in *Sec. 3.4.2*; waste dumping in the older days then led to further changes later on, when nowadays, there is an increasing call for reuse like Zero Waste, because "many dumps are currently exceeding carrying capacity," in *Sec. 3.5.1*; and the practice of corporate capitalism in the older days then led to further changes later on, as shown in the rise of freeganism nowadays, since "freegans see this practice [of leaving good buildings unused when there are homeless on the streets] as senseless and a counter-productive use of resources," in *Sec. 3.5.2*).

And the reverse direction also holds true.

Again, it should be stressed, however, that there are different shades of gray (or different degrees of truth) in the two opposites, and the classification is not necessarily mutually exclusive either. In fact, even when some combinations of the two occur, they only end up sharing the same dialectic relationship, but in a different degree.

Also, there are some other relationships (as a kind of family resemblance) which have something in common with the principle, but they are not exactly the same but only more or less comparable, which varies from case to case.

Good examples of family resemblance in relation to the principle are stability-forwardness, normalcy-feedback, balance-forwardness, balance-reaction, consolidation-modification, normalcy-feedback, stability-breakup, rest-motion, and the like.

27th Thesis: The Functionality-Nonfunctionality Principle

The twenty-seventh thesis is called the functionality-nonfunctionality princple (on the presence of function) in the category of "outcome" in existential dialectics (which is first worked out in *FPHCA*), in that, if there is the presence of action for which something or someone is fitted, there is likewise the relative lack of action for which something or someone is fitted. There is no functionality without nonfunctionallity—and vice versa.

A *function* should not be confused with an *alternative function*, which is just another kind of function, just as *nonfunction* should not be confused with *dysfunction* (which has more to do with malfunction and is thus related more to the regression-progression principle).

In addition, being functional (or useful) and being nonfunctional (or useless) do not have to be mutually exclusive, as the two can come in all shapes and sizes, or in different degrees. And what is functional at one time may become nonfunctional at another time, and vice-versa.

For instance, on the one hand, there is functionality in outcome (e.g., the presence of action for which something is fitted, as shown by "walking around with flowers in or around their nose," which served the ethno-medical function in the older days as a way to fight the bubonic plague, in *Sec. 2.2.1*; the presence of action for which something is fitted, as shown in the passing of the MWTA in 1988 to serve the function of enforcing proper clinical waste dispoasal by the federal government, in *Sec. 2.2.2*; the presence of action for which something is fitted, as shown in the habit of some Europeans in the 19th century and before to throw waste on the street, which served the function to deal with waste at the time, in *Sec.*

2.3.1; the presence of action for which something is fitted, as shown in the rivalry between "the U.S. and Soviet Union in the 1960s and 1970s" when they regularly engaged in "the testing of anti-satellite weapons" to serve the political function of global domination, in *Sec. 2.3.2*; the presence of action for which something is fitted, as shown in "the 1988 Koko case, when "5 ships transported 8,000 barrels of hazardous waste from Italy to the small town of Koko in Nigeria in exchange for $100 monthly rent which was paid to a Nigerian for the use of his farmland," which served the function of dumping for those involved at the time, in *Sec. 2.4.1*; the presence of action for which something is fitted, as shown in the focus on "wilderness protection and wildlife preservation" in the early environmental movement during the first half of the 20th century, because it served the function of reflecting the interests of "the middle and upper class" in the organization at the time, in *Sec. 2.4.2*; the presence of action for which something is fitted, as shown in the neglect of environmental concerns in the older days, when it served the function of market-oriented capitalism at the time, in *Sec. 2.5.1*; the presence of action for which something is fitted, as shown in the rise of conspicuous consumption in modern times, when it serves the function to maintain the materialistic capitalist mainstream culture, in *Sec. 2.5.2*; the presence of action for which something is fitted, as shown by those campanies in the older days which produced less sustainable, more toxic, and hardly recyclable electronics, as this served the function of manufacturing at the time, in *Sec. 3.2.1*; the presence of action for which something is fitted, as shown in the practice of throwing away things by those in consumer society in the older days, as it served the function of the "throw-away culture" at the time, in *Sec. 3.2.2*; the presence of action for which something is fitted, as shown in the use of the old types of incinerators in the older days, even though they were polluting the environment, because they served the function of being a technological solution for WtE at the time, in *Sec. 3.3.1*; the presence of action for which something is fitted, as shown in the use of animal products for foodstuffs, since this serves the vital function of food consumption for those involved, in *Sec. 3.3.2*; the presence of action for which something is fitted, as shown in "the traditional job of recycling…performed by the entrepreneurial poor such as the karung guni, Zabaleen, the rag-and-bone man, waste picker, and junk man" in many underdeveloped or poor countries, as they serve the essential function of waste recycle, in *Sec. 3.4.1*; the presence of action for which something is fitted, as shown in the dumping of wastes in overused landfills in the older days, as it served the function of waste disposal at that time, in *Sec. 3.4.2*; the presence of action for which something is fitted, as shown in the practice of waste dumping in the older days, when it served the function of waste treatment at the time, in *Sec. 3.5.1*; and the presence of action for which something is fitted, as shown in the practice of "councils and landlords" in the older days who "would rather keep properties boarded up if they cannot make a sufficient profit from them," because this serves the function of perpetuating corporate capitalism at the time, in *Sec. 3.5.2*).

On the other hand, there is nonfunctionality in outcome (e.g., the relative lack of action for which something is fitted, as shown in the use of antibiotics to fight the bubonic plague nowadays, such that the ethno-medical method of "walking around with flowers in or around their nose" is no longer functional in the battle against the bubonic plague in our time, in *Sec. 2.2.1*; the relative lack of action for which something is fitted, as shown in the repeal of the MWTA in 1991, when the federal government was no longer deemed necessary to serve the function of enforcing proper clinical waste dispoasal, so "states were given the responsibility to regulate and pass laws concerning the disposal of medical waste" instead, in *Sec. 2.2.2*; the relative lack of action for which something is fitted, as shown in the wider use of litter bins, container deposit schemes, litter traps, etc. nowadays, so that the older habit of some Europeans to throw waste on the street is no longer functional in our time, in *Sec. 2.3.1*; the relative lack of action for which something is fitted, as shown in the decision of the U.S. and Russia later on, when, "following the U.S. test in 1985, there was a de-facto moratorium on such tests," so "widespread ASAT testing had ended" and it no longer serves the function of global domination in the post-Cold War era, since the attention is now paid to other things instead, in *Sec. 2.3.2*; the relative lack of action for which something is fitted, as shown in the creation of the Basel Convention, so that the practice of dumping in "the 1988 Koko case" is no longer functional in our time, as it is more appropriate nowadays to reduce "the movements of hazardous waste between nations," in *Sec. 2.4.1*; the relative lack of action for which something is fitted, as shown in the inclusion of the interests of "minorities and working class people" in the environmental movement nowadays, since the focus on wilderness protection and wildlife preservation is no longer functional for the need to address environmental discrimination against minorities at our time, in *Sec. 2.4.2*; the relative lack of action for which something is fitted, as shown in the rise of ethical consumerism nowadays, such that market-oriented capitalism can no longer serve the function of moral purchase and environment-conscious consumption nowadays, in *Sec. 2.5.1*; the relative lack of action for which something is fitted, as shown in the rise of the new economics movement" nowadays, such that conspicuous consumption can no longer satisfy those segments of the population who yearn for simple living for a more spiritual and environmental lifeworld instead, in *Sec. 2.5.2*; the relative lack of action for which something is fitted, as shown in the call for EPR nowadays, so the old-fashioined way of manufacturing can no longer serve the function of manufacturing in our time, which must be "more sustainable, less toxic, and easily recyclable electronics," in *Sec. 3.2.1*; the relative lack of action for which something is fitted, as shown in dumpster diving nowadays, which "becomes an environmentalist endeavor...and is thus practiced by many pro-green communities" as a way to protest "the wastefulness of consumer society," so the practice of throwing away things in the older days is no longer functional in our time for the new mindset for a more sustainable environment, in *Sec. 3.2.2*; the relative lack of action for which something is fitted, as shown in the use of more environment-friendly incinerators nowadays, which no longer pollute as much as the older types, which are no longer functional as a technological solution for WtE at our time, because of higher standards and expectations, in *Sec. 3.3.1*; the relative lack of action for which something is fitted, as shown in the prohibition of using "some non-human animal products" by those for vegetarian, kosher, and halaal diets, so that these non-human animal products can no longer serve the function of food consumption for these other groups, because of their different standards and expectations, mostly for spiritual and religious reasons, in *Sec. 3.3.2*; the relative lack of action for which something is fitted, as shown in "the creation of large recycling organizations that may be profitable, either by law or economies of scale," in such a way that "the poor are...driven out of the recycling and the remanufacturing market," so that these poor people can no longer serve the essential function of waste recycle in many rich, developed countries nowadays, as they could in the older days, in *Sec. 3.4.1*; the relative lack of action for which something is fitted, as shown in the use of waste converters nowadays, so that dumping of wastes in overused landfills can no longer serve the function of waste disposal in our time, due to higher standards for positive environmental impact, in *Sec. 3.4.2*; the relative lack of action for which something is fitted, as shown in the call for Zero Waste instead nowadays, because "many dumps are currently exceeding carrying capacity," so waste dumping is no longer dominant in serving the essential function of waste treatment in our time, in *Sec. 3.5.1*; and the relative lack of action for which something is fitted, as shown in the practice of squatting by the freegans nowadays, such that the practice of "councils and landlords" in the older days for leaving good buildings unused is no longer so functional in our time, because of different standards and expectations, due to "the anti-globalization movement," in *Sec. 3.5.2*).

And the reverse direction also holds true.

Again, it should be stressed, however, that there are different shades of gray (or different degrees of truth) in the two opposites, and the classification is not necessarily mutually exclusive either. In fact, even when some combinations of the two occur, they only end up sharing the same dialectic relationship, but in a different degree.

Also, there are some other relationships (as a kind of family resemblance) which have something in common with the principle, but they are not exactly the same but only more or less comparable, which varies from case to case.

Good examples of family resemblance in relation to the principle are usefulness-nonusefulness, valueness-nonvalueness, usefulness-uselessness, and the like.

28th Thesis: The Intentionality-Nonintentionality Principle

The twenty-eighth thesis is called the intentionality-nonintentionality princple (on the planning for outcomes) in the category of "outcome" in existential dialectics (which is first worked out in *FPHW*), in that, if there is the planning or design of something for a certain outcome, there is likewise the relative absence of planning or design of something for a certain outcome. There is no intentionality without nonintentionallity—and vice versa.

An intention for an outcome does not have to be certain, so the outcome can even be counter-intentional in being the opposite of what was originally planned. In this sense, counter-intentionality is a special case of non-intentionality; in other words, there are unintented consequences which can be just the opposite of what was originally planned.

An intention for an outcome is mostly relevant to intelligent life, but there are non-intelligent life and non-living matter in the universe too, so nonintentionality is also important.

In addition, an intention does not have to be obvious or straightforward, as William Shakspeare once famously said: "I must be cruel only to be kind."

In any event, on the one hand, there is intentionality in outcome (e.g., the planning or design of something for a certain outcome, as shown in some European "armies of the 14th century" which "were recorded catapulting diseased corpses over the walls of towns and villages in order to spread the pestilence" in the history of Black Death, in *Sec. 2.2.1*; the planning or design of something for a certain outcome, as shown in the enactment of governmental regulation for better clinical waste disposal, in *Sec. 2.2.2*; the planning or design of something for a certain outcome, as shown in the effort to prevent littering, in *Sec. 2.3.1*; the planning or design of something for a certain outcome, as shown in the efforts to get rid of space debris, in *Sec. 2.3.2*; the planning or design of something for a certain outcome, as shown in the design of the Basel Convention to handle hazardous waste, in *Sec. 2.4.1*; the planning or design of something for a certain outcome, as shown in the founding of the environment justice movement to protect minorities from environmental discrimination, in *Sec. 2.4.2*; the planning or design of something for a certain outcome, as shown in the creation of ethical consumerism for waste reduction, in *Sec. 2.5.1*; the planning or design of something for a certain outcome, as shown in the practice of simple living for waste reduction, in *Sec. 2.5.2*; the planning or design of something for a certain outcome, as shown in the design of EPR for more producer responsibility, in *Sec. 3.2.1*; the planning or design of something for a certain outcome, as shown in the idea of dumpster diving for waste reuse, in *Sec. 3.2.2*; the planning or design of something for a certain outcome, as shown in the design of WtE for energy recovery, in *Sec. 3.3.1*; the planning or design of something for a certain outcome, as shown in the design of animal products and animal by-products to make better use of animal wastes, in *Sec. 3.3.2*; the planning or design of something for a certain outcome, as shown in the design of waste recycle for a more positive environmental impact, in *Sec. 3.4.1*; the planning or design of something for a certain outcome, as shown in the design of the waste converter for waste recycle in modern time, in *Sec. 3.4.2*; the planning or design of something for a certain outcome, as shown in the design of Zero Waste for more reuse, especially during 1998-2002, in *Sec. 3.5.1*; and the planning or design of something for a certain outcome, as shown in the design of freeganism for a better world based on reclaiming wastes, in *Sec. 3.5.2*).

On the other hand, there is nonintentionality in outcome (e.g., the relative absence of planning or design of something for a certain outcome, such that the outcome can even be counter-intentional in being the opposite of what was originally planned, as shown in the unintended consequence of the Bubonic plague which caused "the Black Death that swept through Europe in the 14th century," in that "the plague killed so many of the working population [that] wages rose and some historians have seen this as a turning point in European economic development," in *Sec. 2.2.1*; the relative absence of planning or design of something for a certain outcome, such that the outcome can even be counter-intentional in being the opposite of what was originally planned, as shown in the unintended consequences of the governmental regulation for clinical waste disposal, in that it can be costly and thus produce incentives to violate it and that the methods of clinical waste disposal have their own limits, in *Sec. 2.2.2*; the relative absence of planning or design of something for a certain outcome, such that the outcome can even be counter-intentional in being the opposite of what was originally planned, as shown in the unintended consequences of the prevention of littering, in that the prevention of littering is too much more complicated in practice than previously thought and that many cases of littering are unintentional and thus it is difficult to control littering behavior, in *Sec. 2.3.1*; the relative absence of planning or design of something for a certain outcome, such that the outcome can even be counter-intentional in being the opposite of what was originally planned, as shown in the unintended consequences of the efforts to get rid of space debris, in that the measurement of space debris is not an accurate business in the first place since the exact nature of the problem is still not very well known, that to minimize space debris by preventing space collisions is easier said than done technologically, and that many of the proposed solutions are too expensive to be cost-effective, in *Sec. 2.3.2*; the relative absence of planning or design of something for a certain outcome, such that the outcome can even be counter-intentional in being the opposite of what was originally planned, as shown in the unintended consequences of the Basel Convention, in that what constitutes hazardous waste is subject to contention, that there are questionable rules about "obligations" in the Basel Convention, and that it does not go far enough to protect less developed countries from the spread of hazardous waste, in *Sec. 2.4.1*; the relative absence of planning or design of something for a certain outcome, such that the outcome can even be counter-intentional in being the opposite of what was originally planned, as shown in the unintended consequences of the environmental justice movement, in that what constitutes fairness or its opposite is in the eyes of the beholder, that the environmental movement has been elitist in producing the opposite effects, that the legal process is too costly, etc., in *Sec. 2.4.2*; the relative absence of planning or design of something for a certain outcome, such that the outcome can even be counter-intentional in being the opposite of what was originally planned, as shown in the unintended consequences of ethical consumerism, in that the actual effect of ethical consumerism is limited, that moral boycott can be based on incorrect or limited information, that it can be counter-productive in encouraging immoral behavior instead, etc., in *Sec. 2.5.1*; the relative absence of planning or design of something for a certain outcome, such that the outcome can even be counter-intentional in being the opposite of what was originally planned, as shown in the unintended consequences of simple living, in that simple living can be decadent instead, that simple living is not possible for everyone, that simple lifestyle has been used by different groups for their ulterior purposes, that it is not so practical as romantically assumed by the advocates, etc., in *Sec. 2.5.2*; the relative absence of planning or design of something for a certain outcome, such that the outcome can even be counter-intentional in being the opposite of what was originally planned, as shown in the consequences that the application of EPR is not effective, that it can be counter-productive, and that it can be misused for something else, in *Sec. 3.2.1*; the relative absence of planning or design of something for a certain outcome, such that the outcome can even be counter-intentional in being the opposite of what was originally planned, as shown in the unintended consequence that it can cost the dumpster diver even more in the end, because it can be illegal, unhealthy, dangerous, etc., in *Sec. 3.2.2*; the relative absence of planning or design of something for a certain outcome, such that the outcome can even be counter-intentional in being the opposite of what was originally planned, as shown in the unintended consequences that the design of WtE can end up destroying valuable resources, that it can actually be polluting, and that its efficiency is essentially limited in practice, in *Sec. 3.3.1*; the relative absence of planning or design of something for a certain outcome, such that the outcome can even be counter-intentional in being the opposite of what was originally planned, as shown in the facts that the use of animal products and animal by-products is actually prohibited by certain groups for different moral and religious reasons and that their use is also not recommended for health reason, for instance, in *Sec. 3.3.2*; the relative absence of planning or design of something for a certain outcome, such that the outcome can even be counter-intentional in being the opposite of what was originally planned, as shown by the work by Steven Landsburg who showed that, for instance, "paper recycling actually reduces tree populations" and is thus bad for the environment instead, in *Sec. 3.4.1*; the relative absence of planning or design of something for a certain outcome, such that the outcome can even be counter-intentional in being the opposite of what was originally planned, as shown by some contemporary autoclaves, which have been so improved that they "employ the same method of sterilization as converters, but do so using larger and more complex equipment that also has higher safety… requirement," as an alternative technology to compete with the waste converter, in *Sec. 3.4.2*; the relative absence of planning or design of something for a certain outcome, such that the outcome can even be counter-intentional in being the opposite of what was originally planned, as shown by "many commercial or industrial companies" like "Staples, Home Depot, Toyota, General Motors and computer take-back campaigns," which "embrace Zero Waste" but "usually mean no more than a major materials recycling effort, having no bearing on product redesign," in *Sec. 3.5.1*; and the relative absence of planning or design of something for a certain outcome, such that the outcome can even be counter-intentional in being the opposite of what was originally planned, as shown in the unintended consequences of freeganism when implemented, in that foragers on the street suffer from disrespect by those in mainstream society, that reclaimed items in dumpsters can be unhealthy and dangerous, that it is not so practical as romantically assumed by the advocates, and that its practice can be criminal, etc., in *Sec. 3.5.2*).

And the reverse direction also holds true.

Again, it should be stressed, however, that there are different shades of gray (or different degrees of truth) in the two opposites, and the classification is not necessarily mutually exclusive either. In fact, even when some combinations of the two occur, they only end up sharing the same dialectic relationship, but in a different degree.

Also, there are some other relationships (as a kind of family resemblance) which have something in common with the principle, but they are not exactly the same but only more or less comparable, which varies from case to case.

Good examples of family resemblance in relation to the principle are purposiveness-nonpurposiveness, intention-accidentalness, design-randomness, etc.

29th Thesis: The Post-Human Transfiguration

And the twenty-ninth thesis is about the role of "post-humans," which I originally proposed in my previous books, starting with the first book titled *The Future of Human Civilization* in 2000 and all others afterwards.

As already pointed out in *Sec. 1.8*, I need to emphasize, as this is something that I used to repeat (*almost verbatim*) from my previous books, two clarifications here about the term "post-human" as a neologism in my works.

Firstly, the word "post-human" here should *not* be confused with another term which looks similar but has a totally different meaning in the literature of postmodernism, namely, "post-humanism"—which constitutes a critique of "humanism" as tradionally understood (especially, though not exclusively, in relation to the idea of progress in science and reason in the Enlightenment project). (WK 2008)

My works reject the project of "postmodernism" and propose the future world of what I orignally called "after-postmodernity" in *FHC* and *FCD*, for instance.

And secondly, the word "post-human" here should also *not* be confused with a similar term which is used to champion the ideology of technology for the future co-existence between humans and cyborgs in "trans-humanism." (WK 2008a)

Instead, my term "post-human" in relation to "posthuman-ism" also rejects "transhumanism" (especially, though not exclusively, in relation to the promises of technology) and refers to something else altogether, that is, the future extinction of humans and its post-human successors in deep space and beyond unto multiverses.

My critique of "transhumanism" was more extensively elabroated in *Sec. 2.4.1* of *BEPE*—just as my critique of different futuristic scenarios of fictional artificial intelligence in science fiction was given in *Sec. 3.5.2* of *FPHCOM.*

With these two clarifications in mind (as summarized in *Table 1.31*)—the post-human transfiguration in the study of waste can be addressed in relation to five main directions, based on my *transfigurative theory of waste*, as analysed in this book (together with my visions as already worked out in my previous books).

Firstly, the post-human transfiguration in the study of waste will learn from my transfigurative theory of waste in relation to the dialectic context of no uselessness without usefulness (and vice versa), in the context of waste.

Secondly, the post-human transfiguration in the study of waste will learn from my transfigurative theory of waste in relation to the four perspectives of the mind, nature, society, and culture.

Thirdly, the post-human transfiguration in the study of waste will learn from my transfigurative theory of waste in relation to the non-privilege of any specific theory over others in the literature and the non-integration of them all (as they are not necessarily compatible with each other).

Fourthly, the post-human transfiguration in the study of waste will have to confront the new challenges as posed by the long-term civilizational development of intelligent life in the distant future, both here on this planet Earth and elsewhere in deep space until multiverses.

As this is something that I regularly pointed out in my previous books for background information (and summarize here, *almost verbatim*, with some updates), I already worked out, in my numerous books, what these new challenges will be and provide different original visions to meet them—especially, in regard to the future evolutions of the mind (e.g., "the supersession computing," "metamorphic humor," "the contrarian personality," "the hyper-martial body," "the hyper-sexual body," "floating consciousness," "hyper-spatial consciousness," "thinking machines," "unfolding unconsciousness," "thinking robots," "genetically altered superior beings," "cyborgs," and others), of nature (e.g., "transdisciplinary architecture," "creational chemistry," "multilateral sound," "resettlement geology," "post-cosmology," "the alteration of space-time," "the creation of new matter-energy," "selective geometry," and the like), of society (e.g., "panoramic transportation," "cyclical-progressive migration," "reflective criminology," "interventive-reshaping geography," "detached gambling," "heterodox education," "multifaceted war and peace," "virtual organizations," different versions of "post-capitalism" and "post-democracy," the movement of "cyclical progression" at both structural and systemic levels, "ambivalent technology," and so forth), and of culture (e.g., "inquisitive culinary art," "interactive semantics," "transdisciplinary performing arts," "comparative-impartial literature," "post-human mind games," "comparative-substitutive religion," "post-ethics," "post-civilization," "transformative aesthetic experience," "contrastive mathematical-logic," and the whatnot) in history (e.g., "multifold history," the age of "after-postmodernity"), in the context of my approaches in relation to methodology (e.g., "sophisticated methodologial holism," "critical-dialectic formal science," "complex data analysis," etc.) and ontology (e.g., "existential dialectics," "contrastive rationality").

Of course, the examples (as listed above) are not exhaustive but illustrative, since my numerous books have worked out many other visions (as already roughly summarized in *Sec.1.8* and *Sec. 1.9*). Many, though not all, of my visions on the mind, nature, society, culture, history, methodology, and ontology (as cited above) are summarized in the tables as shown in Chapter One and Chapter Four (especially in *Table 4.43* about my original theories on numerous topics).

And finally, or fifthly, the post-human transfiguration in the study of waste will learn from my transfigurative theory of waste in regard to the need to go beyond uselessness and usefulness (as further elaborated in the next section).

Towards the Post-Human Transfiguration

This "post-human" transfiguration in the study of waste to transcend both uselessness and usefulness can be understood in terms of nine great future transformations of waste, in relation to the mind, nature, society, and culture—as explained below (and summarized in *Table 1.2*).

Nine Great Future Transformations of Waste

The nine great future transformations of waste (in relation to the mind, nature, society, and culture) can be explained hereafter, with the caveat, however, that the classification in terms of nine is solely aesthetic and utilitarian (illustrative), as there is no objective basis that it must be classified in term of nine (as this is something that I often stressed and repeated in my previous books, albeit in the context of different topics).

The First Great Future Transformation of Waste

The first great future transformation of waste in the post-human era concerns what I originally call *new goals,* in relation to the mind.

This search for "new goals" will give future advanced humans and later their successors, namely, the post-humans, the capacity to be motivated to cope with waste in a way that current humans cannot.

There are already some indications that the future of waste is moving toward this direction.

Consider, for illustration, the different "goals" as already seen in the history of waste, such as the goal of understanding the causes of epidemics (*Sec. 2.2.1*), the quest for the scientific knowledge of infectious disease (*Sec. 2.2.2*), the search for a litter-free cleaner environment (*Sec. 2.3.1*), the research on space debris (*Sec. 2.3.2*), the quest for the end of anti-toxic colonialism (*Sec. 2.4.1*), the fight for environmental justice (*Sec. 2.4.2*), the goal of moral purchasing (*Sec. 2.5.1*), the goal of simple living (*Sec. 2.5.2*), the goals of EPR (*Sec. 3.2.1*), the waste-collecting habit (*Sec. 3.2.2*), the goal of environmental sustainability (*Sec. 3.3.1*), the goal of reusing animal waste (*Sec. 3.3.2*), the goal of waste recycle (*Sec. 3.4.1*), the goal of waste conversion (*Sec. 3.4.2*), the goal of zero waste (*Sec. 3.5.1*), and the goal of reclaiming waste (*Sec. 3.5.2*), for instance.

More importantly, this future quest for new goals should also be understood in relation to my previous works on the future "post-human" supersession of the human mind from the combined perspectives of the mind, nature, society, and culture—with good examples like "the supersession computing," "metamorphic humor," "the contrarian personality," "the hyper-martial body," "the hyper-sexual body," "floating consciousness," "hyper-spatial consciousness," "thinking machines," "unfolding unconsciousness," "thinking robots," "genetically altered superior beings," "cyborgs," and others.

For instance, most advanced forms of future intelligent life in deep space until multiverses (e.g., "floating consciousness") do not need to have the physical bodies that current humans have and therefore do not necessarily produce the kinds of waste that current humans do.

In any event, all these are just some illustrative (relatively primitive) examples in our time, of course, but the important point to remember here is that the search for different new goals will further transform the study of waste at the more advanced level in the post-human future that the human world has never known.

The Second Great Future Transformation of Waste

The second great future transformation of waste in the post-human era concerns what I originally call *new means,* in relation to the mind.

This quest for "new means" will give future advanced humans and later their successors, namely, the post-humans, the capacity to find novel options to achieve the goals of waste treatment in a way that current humans cannot.

There are already some indications that the future of waste is moving toward this direction.

Consider, for illustration, the different "means" as already seen in the history of waste, such as the prevention of epidemics for better social health (*Sec. 2.2.1*), proper clinical waste disposal for disease prevention (*Sec. 2.2.2*), litter traps for cleaner environment (*Sec. 2.3.1*), growth mitigation for space waste clean-up (*Sec. 2.3.2*), the Basel Ban Amendment to do more for less developed countries (*Sec. 2.4.1*), the Civil Rights Movement to help the Environmental Justice movement (*Sec. 2.4.2*), the use of moral basis for environmental sustainability (*Sec. 2.5.1*), working less for a more spiritual lifestyle (*Sec. 2.5.2*), EPR for a waste recycling attitude (*Sec. 3.2.1*), dumpster diving for waste reuse (*Sec. 3.2.2*), source-separated curbside collection for resource recovery (*Sec. 3.3.1*), the use of animal-by-products for animal food (*Sec. 3.3.2*), the use of cost-benefit analysis for waste reccyle (*Sec. 3.4.1*), the use of the waste converter for waste recycle (*Sec. 3.4.2*), the use of governance for zero waste (*Sec. 3.5.1*), and veganism for freeganism (*Sec. 3.5.2*), for instance.

All these are just some illustrative (relatively primitive) examples in our time, of course, but the important point to remember here is that the search for different new means will further transform the study of waste at the more advanced level in the post-human future that the human world has never known.

The Third Great Future Transformation of Waste

The third great future transformation of waste in the post-human era concerns what I originally call *new comparisons*, in relation to nature.

This search for "new comparisons" will give future advanced humans and later their successors, namely, the post-humans, the ability to learn from animals, plants, and the universe, through comparison and contrast, about the study of waste in a way that current humans cannot.

There are already some indications that the future of waste is moving toward this direction.

Consider, for illustration, the different forms of comparison and contrast, such as as already seen in the history of waste, such as the comparison and contrast between humans and rodents in contact with infected fleas (*Sec. 2.2.1*), the effects of litter on both human and natural environments (*Sec. 2.3.1*), the comparison and contrast between used spacecraft fragments and small natural particles in the Earth's orbit (*Sec. 2.3.2*), the comparison between animal rights and human rights (*Sec. 2.5.1*), the comparison between humans and animals in their natural state of living (*Sec. 2.5.2*), the comparison between natural decomposition and biological processing in WtE (*Sec. 3.2.2*), the comparison and contrast between humans in natural evolution and humans now for recycling behavior (*Sec. 3.4.1*), the comparison and contrast between nature and industry for zero waste (*Sec. 3.5.1*), and the comparison and contrast between humans and animals in regard to animals in waste treatment (*Sec. 3.5.2*), for instance.

All these are just some illustrative (relatively primitive) examples in our time, of course, but the important point to remember here is that the search for different new comparisons will further transform the study of waste at the more advanced level in the post-human future that the human world has never known.

The Fourth Great Future Transformation of Waste

The fourth great future transformation of waste in the post-human era concerns what I originally call *new integrations*, in relation to nature.

This search for "new integrations" will give future advanced humans and later their successors, namely, the post-humans, the ability to live and interact with nature for the study of waste in a way that current humans cannot.

There are already some indications that the future of waste is moving toward this direction.

Consider, for illustration, the different integrations for the study of waste as already seen in the history of waste, such as the interconnection between humans and rodents in the bubonic plague (*Sec. 2.2.1*), the effects of litter on natural environment (*Sec. 2.3.1*), the care about animal rights in ethical consumerism (*Sec. 2.5.1*), the care about the environment in simple living (*Sec. 2.5.2*), the design for a more sustainable environment in EPR (*Sec. 3.2.1*), the concern with environmental sustainability in WtE (*Sec. 3.2.2*), the concern with animal cruelty by the critics of animal products (*Sec. 3.3.1*), the inclusion of negative externalities (*Sec. 3.4.1*), the positive environment impact of the waste converter (*Sec. 3.4.2*), the concern with the availability of land for dumping (*Sec. 3.5.1*), and the concern with animal cruelty in veganism (*Sec. 3.5.2*), for instance.

More importantly, this future quest for new integrations should also be understood in relation to my previous works on the future "post-human" adventures into different dimensions of space-time unto multiverses from the combined perspectives of the mind, nature, society, and culture—with good examples like "the alteration of space-time" (*Table 1.3* and *Table 1.4*), "multiverses" (*Table 1.7* and *Table 1.8*), "hyperspace" (*Table 1.9* and *Table 1.25*), "time travel" (*Table 1.10* and *Table 1.11*), and others.

In addition, in my previous book titled *The Future of Post-Human Geology* (2010), I already envisioned the three great future transformations of geology, namely, "solar geology," "galactic geology," and "cosmic geology," in the context of my "resettlement theory of geology," in that humans and later post-humans will explore and migrate to other planets, solar systems, galaxies, and cosmos in the long haul.

Then, in another book of mine titled *The Future of Post-Human Geography* (2011), I linked the two books together, since geography has much to learn from geology (as in "physical geography," which, as already introduced in *Sec. 1.3* of *FPHGEOG*, relies heavily on the earth sciences) and discussed the search for "new explorations" in the context of the three forms of environments outside the planet Earth, namely, what I originally called "solar geography," "galactic geography," and "cosmic geography"—as a further elaboration of the three great future transformations of geology (in *FPHGEOL*).

Later, in still another book of mine titled *The Future of Post-Human Migration* (2012), I envisioned the future integrations in different forms, namely, what I originally called "solar migration," "galactic migration," "clustery migration," "cosmic migration," and "multiversal migration."

And then, in an additional book of mine titled *The Future of Post-Human Transportation* (2012), I envisioned the future integrations in different forms, namely, what I originally called "solar transportation," "galactic transportation," "clustery transportation," "cosmic transportation," and "multiversal transportation."

This extension is consistent with my other theory, namely, "the theory of the cyclical progression of system integration and fragmentation" in *FCD* (2002)—at the social-systemic level (as already clarified in *Sec. 1.8* and shown below):

Local → Regional → Global →Solar →
Galactic → Clustery…→ Multiversal

All these are just some illustrative (relatively primitive) examples in our time, of course, but the important point to remember here is that the search for different new integrations will further transform waste at the more advanced level in the post-human future that the human world has never known.

For instance, most advanced forms of future intelligent life in deep space until multiverses (e.g., "floating consciousness," "hyper-spatial consciousness," "unfolding unconsciousness," etc.) will need to adapt to different environments, with the subsequent impacts on the uselessness and usefulness of their own versions of waste in the distant future.

The Fifth Great Future Transformation of Waste

The fifth great future transformation of waste in the post-human era concerns what I originally call *new praxes*, in relation to society.

This search for "new praxes" will give future advanced humans and later their successors, namely, the post-humans, the ability to engage in practice (like activism, politics, personal experience, etc.) for the study of waste in a way that current humans cannot.

There are already some indications that the future of waste is moving toward this direction.

Consider, for illustration, the different praxes as already seen in history of waste like the practice of using the bubonic plague as a biological weapon by European armies in the 14th century (*Sec. 2.2.1*), the enactment of the Medical Waste Tracking Act (*Sec. 2.2.2*), legislation against litter (*Sec. 2.3.1*), the creation of COPUOS to deal with space debris (*Sec. 2.3.2*), the creation of the Basel Convention (*Sec. 2.4.1*), the creation of the Environmental Justice movement (*Sec. 2.4.2*), the practice of boycott for ethical consumerism (*Sec. 2.5.1*), the grassroots awareness campaign (*Sec. 2.5.2*), the creation of PRO for EPR in practice (*Sec. 3.2.1*), the enactment of the Good Samaritan Food Donation Act (*Sec. 3.2.2*), the Talloires Declaration for resource recovery (*Sec. 3.3.1*), the enactment of Animal By-Products Regulations (*Sec. 3.3.2*), the creation of large recycling organizations (*Sec. 3.4.1*), the applications of the waste converter (*Sec. 3.4.2*), market-based campaigns in action (*Sec. 3.5.1*), and different practices of freeganism in action (*Sec. 3.5.2*), for instance.

All these are just some illustrative (relatively primitive) examples in our time, of course, but the important point to remember here is that the search for different new praxes will further transform the study of waste at the more advanced level in the post-human future that the human world has never known.

The Sixth Great Future Transformation of Waste

The sixth great future transformation of waste in the post-human era concerns what I originally call *new technologies*, in relation to society.

This search for "new technologies" will give future advanced humans and later their successors, namely, the post-humans, the ability to come up with different techniques, materials, and experiments for the study of waste in a way that current humans cannot.

There are already some indications that the future of waste is moving toward this direction.

Consider, for illustration, the use of different technologies as already seen in the history of waste, such as the invention of antibiotics against the bubonic plague (*Sec. 2.2.1*), the technique of "mail-back medical waste disposal" (*Sec. 2.2.2*), the invention of litter bins (*Sec. 2.3.1*), the invention of spacecrafts (*Sec. 2.3.2*), the technique of trading in wastes between the OECD and the U.S. (*Sec. 2.4.1*), the technique of litigation by the Environmental Justice movements (*Sec. 2.4.2*), the technique of strike action for moral boycott (*Sec. 2.5.1*), appropriate technology for simple living (*Sec. 2.5.2*), the technique of buy-back (*Sec. 3.2.1*), the technique of sifting through in dumpster diving (*Sec. 3.2.2*), diferent WtE methods (*Sec. 3.3.1*), the technique of advertising some pet food as containing no animal meat by-products (*Sec. 3.3.2*), the technique of using social pressure for recycling programmes (*Sec. 3.4.1*), the invention of the waste converter (*Sec. 3.4.2*), the technique of construction for zero waste (*Sec. 3.5.1*), and the creation of the ideology of freeganism (*Sec. 3.5.2*), for instance.

All these are just some illustrative (relatively primitive) examples in our time, of course, but the important point to remember here is that the search for new technologies will further transform the study of waste at the more advanced level in the post-human future that the human world has never known.

The Seventh Great Future Transformation of Waste

The seventh great future transformation of waste in the post-human era concerns what I originally call *new stratifications*, in relation to society.

This emergence of "new stratifications" will lead future advanced humans and later their successors, namely, the post-humans, to face new inequalities for the understanding of waste in a way that current humans cannot.

There are already some indications that the future of waste is moving toward this direction.

Consider, for illustration, the different inequalities as already seen in the history of waste, such as the divide between the Imperial Japanese Army and the victims of biological warfare (*Sec. 2.2.1*), the divide between the enforcers and the violators (*Sec. 2.2.2*), the divide between authorized officers and individuals to be given fines for littering (*Sec. 2.3.1*), the divide between those space powers which pollute the Earth's orbit and those small countries which do not have the means to do so (*Sec. 2.3.2*), the divide between the developed countries and the underdeveloped countries in regard to the Basel Convention (*Sec. 2.4.1*), the divide between the minorities and the elites in the Environmental Justice movement (*Sec. 2.4.2*), the divide between ethical consumers and unethical corporations (*Sec. 2.5.1*), the divide between the neo-Luddites and the corporate capitalists (*Sec. 2.5.2*), the divide between the EPR advocates and the corporate sector (*Sec. 3.2.1*), the divide between the law enforcement officers and the dumpster divers to be treated as criminals (*Sec. 3.2.2*), the divide between the advocates of incineration and those of alternative technologies (*Sec. 3.3.1*), the divide between industries selling animal by-products and industries selling sources of vegetable protein (*Sec. 3.3.2*), the divide between those jobs in waste recycle with bad working conditions and those with good working conditions (*Sec. 3.4.1*), the divide between the advocates of waste converters and those of autoclaves (*Sec. 3.4.2*), the divide between those favoring construction and those favoring deconstruction (*Sec. 3.5.1*), and the divide between the freegans and the capitalists (*Sec. 3.5.2*), for instance.

All these are just some illustrative (relatively primitive) examples in our time, of course, but the important point to remember here is that the emergence of new stratifications will further transform the study of waste at the more advanced level in the post-human future that the human world has never known.

The Eighth Great Future Transformation of Waste

The eighth great future transformation of waste in the post-human era concerns what I originally call *new sub-fields*, in relation to culture.

This search for "new sub-fields" will give future advanced humans and later their successors, namely, the post-humans, the ability to explore new specialties for the understanding of waste in a way that current humans cannot.

There are already some indications that the future of waste is moving toward this direction.

Consider, for illustration, the different sub-fields as already seen in the history of waste, such as the field of epidemiology (*Sec. 2.2.1*), the study of waste catalogue codes (*Sec. 2.2.2*), the study of litter (*Sec. 2.3.1*), the study of space debris (*Sec. 2.3.2*), the study of hazardous waste (*Sec. 2.4.1*), the field of environmental justice (*Sec. 2.4.2*), the field of ethical consumerism (*Sec. 2.5.1*), the study of simple living (*Sec. 2.5.2*), the study of EPR (*Sec. 3.2.1*), the study of dumpster diving (*Sec. 3.2.2*), the field of WtE (*Sec. 3.3.1*), the study of animal waste (*Sec. 3.3.2*), the study of recycling behavior (*Sec. 3.4.1*), the study of waste conversion (*Sec. 3.4.2*), the field of zero waste (*Sec. 3.5.1*), and the study of freeganism (*Sec. 3.5.2*), for instance.

All these are just some illustrative (relatively primitive) examples in our time, of course, but the important point to remember here is that the search for new sub-fields will further transform the study of waste at the more advanced level in the post-human future that the human world has never known.

The Ninth Great Future Transformation of Waste

And the ninth great future transformation of waste in the post-human era concerns what I originally call *new interpretations*, in relation to culture.

This search for "new interpretations" will give future advanced humans and later their successors, namely, the post-humans, the ability to come up with different evaluations and views about the study of waste in a way that current humans cannot.

There are already some indications that the future of waste is moving toward this direction.

Consider, for illustration, the different interpretations as already seen in the history of waste, such as the view about the bubonic plague as a biological weapon (*Sec. 2.2.1*), the view of clinical waste disposal as an environmental concern (*Sec. 2.2.2*), the view of imposing fines on littering individuals (*Sec. 2.3.1*), the view of space debris to be cleaned up (*Sec. 2.3.2*), different contentious views about the definition of "hazardous" waste (*Sec. 2.4.1*), the view about elitism in the Environmental Justice movement (*Sec. 2.4.2*), the critique of certain corporations by ethical consumers (*Sec. 2.5.1*), the view of reconsidering technology in simple living (*Sec. 2.5.2*), the producer-centered view of waste recycling (*Sec. 3.2.1*), the view of treating dumpster diving as illegal in some countries (*Sec. 3.2.2*), the view about pollution from the process of waste-to-energy (*Sec. 3.3.1*), the view of those for vegetarian, kosher, and halaal diets against animal products (*Sec. 3.3.2*), the view about the unnaturalness of recycling behaivor (*Sec. 3.4.1*), the view about the positive environmental impact of the waste converter (*Sec. 3.4.2*), the interpretation of zero waste vs. recycling (*Sec. 3.5.1*), and the view of freeganism and its religious roots (*Sec. 3.5.2*), for instance.

All these are just some illustrative (relatively primitive) examples in our time, of course, but the important point to remember here is that the search for new interpretations will further transform the study of waste at the more advanced level in the post-human future that the human world has never known.

Enormous Implications

These nine great future transformations of the study of waste have enormous implications for some soul-searching questions which have challenged some of the best minds in the history of waste.

For illustration, consider the following fundamental questions in the study of waste:

- Ex: to what extent are waste minimization and waste disposal really feasible and desirable (in the context of useflessness)?
- Ex: to what extent are waste reuse, waste recycle, and waste-to-energy really feasible and desirable (in the context of usefulness)?
- Ex: how exactly will the uselessenss and usefulness of waste be transfigured in future history, from the perspectives of the mind, nature, society, and culture?

Questions and disagreements like this will receive new answers in the future history of advanced human and later post-human waste, at least in two fundamental ways.

The first fundamental way is to examine the relationships between uselessness and usefulness—such that their possibility (or impossibility) and desirability (or undesirability) are not to the extent that the respective defenders would like us to believe.

The different views about the study of waste, as already addressed in *Sec. 1.1*, *Sec. 1.5*, and the rest of the book, are just different ways to understand the nature of waste, with each from its own vantage point.

Each views has its own merits and defects and is preferred over others by different proponents with their own different interests, but none of them is possible (or impossible) and desirable (or undesirable) to the extent that its respective advocates would like us to believe.

And the second fundamental way is to inquire about, and then transcend, in the long haul, the debate about uselessness and usefulness, such that it will be shifted to different directions, as shown in the nine great future transformations that I have envisioned and outlined above. To transcend them all is to move into these different directions, in conjunction with the various visions of my previous books (as already summarized in *Sec. 1.8* and also earlier in this chapter, with the further help of the tables at this end of Chapter One and this chapter).

Two examples are worth repeating here. The first is that most advanced forms of future intelligent life in deep space until multiverses (e.g., "floating consciousness," "hyper-spatial consciousness," "unfolding unconsciousness," etc.) do not need to have the physical bodies that current humans have and therefore do not produce the kinds of waste that current humans do. And the second is that these advanced forms of intelligent life will also need to adapt to different alien environments, with the subsequent impacts on the uselessness and usefulness of their own versions of waste in the distant future.

In fact, the current production of space debris (or space waste) for space exploration in the Earth 's orbit and a bit beyond (as described in *Sec. 2.3.2*) is only a small step in the direction of the larger future of waste in radically different environments in outer space and beyond unto multiverses that the Earthlings have never known.

Lest any misunderstanding occurs, one must remember that these nine great future transformations of the study of waste will bring neither utopia nor dystopia, regardless of the current progress about different solutions to space debris (in *Sec. 2.3.2*)—as this is something that I have warned time and again in all of my previous books, almost verbatim.

In any event, as I so often stressed in my previous books, it is not up to us current humans to decide whether or not, or to what extent, the post-human future of waste will be for better and for worse, because it will be judged by the values and beliefs of future post-humans on earth and in deep space unto multiverses—in a way that the entire history of the human world has never known.

Thus is the future of waste—and of its post-human fate.

Table 4.1. Sophisticated Methodological Holism (Part I)

• "My methodological holism implies the partiality-totality principle in the ontology of existential dialectics (see the table on the partiality-totality principle for summary), which is against the varieties of (a) reductionism and (b) reverse-reductionism, in relation to (i) concept, (ii) theory, (iii) methodology, and (iv) ontology." (*FC*)

• "[M]y methodological holism here is not opposed to methodological individualism but includes it (and, for that matter, other methodologies too)…." (*FPHC*) For this reason (and others too, as summarized hereafter), my version of methodological holism is sophisticated—not vulgar as sometimes used by inapt scholars using the same term. (*FC*)

• "[M]y methodological holism does not democratically presume that all levels are equally valid, as all levels are not created equal. In other words, in relation to issue X, level A may be more relevant than level B, but in relation to Y, level B can be more relevant than level A instead." (*FPHC*) One excellent example of this vulgar democratic presumption is what I called in *BNN* "the compromise fallacy." (*FC*)

• My methodological holism does not presume that a lower level of analysis is more important than a higher level, solely because the former serves as the foundation for the latter—and vice versa, for that matter. One excellent example of this reductionistic presumption is what I called in *FPHST* "the foundation fallacy." (*FPHST*, *FC*)

(continued on next page)

Table 4.1. Sophisticated Methodological Holism (Part II)

• "[M]y methodological holism does not make any a-priori postulation that there must be a definite (and, for that matter, indefinite) number of levels" in any analysis. (*FPHC*) Nor does it dogmatically require that there must be a certain combination of levels of analysis in a given inquiry. (*FC*)

• "[M]ethodological holism, in my usage, does not assume that all levels…can necessarily be integrated, since methodological holism is not aimed to search for the holy grail of 'an integral theory…' (as is the case for Wilber). In other words, it allows that sometimes some levels may experience irreducible gaps between them, to be understood, at best, as empirical correlations, not as causal relations…." (*FPHC*)

• "[D]ifferent levels may overlap and even interact with each other in a given context (but sometimes may not), and the fact that I even proposed different ways of re-classifying the levels (whenever needed) in *FDC* reinforces this point….The dual danger here is either forcefully making different levels interact when they are just different (or, metaphorically speaking, apples and oranges) or inappropriately ignoring their interactions when some situations instead require them." (*FPHC*)

• "[T]o understand different levels from their own (unique) perspectives (as required by my methodological holism) is not the same as trying to reduce them to a preferred level in the process of learning from other levels. This second kind of multidisciplinary work is not genuine and does no justice to the unique complexities and merits inherent at each level." (*FPHC*)

(continued on next page)

Table 4.1. Sophisticated Methodological Holism (Part III)

• "[My] methodological holism walks a fine line between the artificial classification (separation) of levels and the simultaneous incorporation of them, if only for the sake of human scholarly endeavor. It should be reminded that nature does not impose upon itself the academic classification of the levels of analysis as humans have. The enterprise of classification is therefore anthropocentric." (*FPHC*)

• "[M]y methodological holism advocates neither epistemic subjectivism nor epistemic non-subjectivism (e.g., realism, idealism, and historicism), neither epistemic relativism (e.g., subjectivism, historicism) nor epistemic absolutism (e.g., realism, positivism), neither epistemic reductionism nor epistemic emergencism, and neither epistemic objectivism (e.g., realism, idealism) nor epistemic historicism….Neither does methodological holism, in my usage, accept the false meta-conceptual dichotomy between nominalism and realism….These false dichotomies…are to be transcended. In other words, methodological holism does not fully accept epistemic realism, positivism (a form of epistemic idealism), historicism, subjectivism, and reductionism in epistemology and philosophy of science but learns from the strengths and weaknesses of all of the opposing approaches without siding with any of them…." (*FPHC*)

• "Sophisticated methodological holism is subject to the constraints as imposed by the syntax of existential dialectics (e.g., the partiality-totality principle and the predictability-unpredictability principle). Even in predictability, outcomes are subject to uncertainty, the degree of which varies from case to case." (*FC*)

(continued on next page)

Table 4.1. Sophisticated Methodological Holism (Part IV)

• "Sophisticated methodological holism—when applied, especially though not exclusively, as illustrated in my numerous works—can enrich the understanding of reality in some distinctive ways. Here are three examples (as revealed in each of my books)." (*FPHCT*; *FPHL*)

—"Firstly, it provides a comprehensive analysis of a subject matter, from which much can be learned about reality," "both in relation to the perspectives of the mind, nature, society, and culture and also in relation to a new classification of the subject matter."

—"Secondly, it suggests some visions of the future in relation to the subject matter in question."

—"And thirdly, it proposes some insights on meta-theory (e.g., methodology and ontology) in general—with the clear understanding, however, of the dilemma of specific vs. general ontology (as shown in the table on the syntax of existential dialectics in the context of the dilemma of ontology). For this very reason, all of these ways are important, without reducing one into the analysis of another."

• "In the end, my meta-theory (both sophisticated methodological holism and existential dialectics) serves as a foundation to unify all domains of knowledge into *an unified theory of everything* (by way of some ontological principles and the comprehensive perspectives of the mind, nature, society, and culture). This is so, without commiting the sins of reductionism and reverse-reductionism (as often seen in many holistic approaches, with the fad of sysems approach—be it about systems theory, chaos theory, complexity theory, or else—as a most recent notorious example, which I debunked in *The Future of Complexity* and also in *The Future of Post-Human Formal Science*). My distinctive approach makes good use of different schools of thought without favoring any of them nor trying to integrate them (as they are not necessarily compatible with each other), so as to adjust for subjectivity, diversity, conflict, and complexity, for example. In this sense, the word "unified" does not have to mean integrative approach, in a narrow sense, nor systems approach, in a broad one, both of which I reject while learning from them." (*FPHFS*; *FPHS*)

Sources: A summary of *Sec.1.2* in *FPHC*—and also from *BNN*, *FPHST*, *ALD*, *FC*, *FPHCT*, and the rest of all other books of mine. See the books for more detail.

Table 4.2. On Reductionism and Reverse-Reductionism (Part I)

• **The Partiality-Totality Principle**
—The partiality-totality principle in the ontology of existential dialectics targets against the varieties of reductionism and reverse-reductionism (as already worked out in my previous books).

• **Against the Varieties of Reductionism**
—*Conceptual Reductionism*
- Some illustrative instances involve myriad dualities like mind vs. body, self vs. world, democracy vs. non-democracy, and the like (as already addressed in *FHC*, *FPHC*, and *BDPD*, for instance).

—*Theoretical Reductionism*
- A fascinating case study concerns what I originally called "the foundation fallacy" in *FPHST*, in any attempt to naively understand space-time from the physical perspective as the foundation and, consequently, to dangerously dismiss other perspectives.
- In *FAE*, I elaborated further these versions of reductionism in the literature on aesthetics (e.g., form vs. content, representation vs. expression, critics vs. artists, and externalism vs. internalism).
- In *FIA*, I revealed other forms of reductionism in the literature on information architecture (e.g., the constructivist argument).
- In *FPHU*, I showed the persistent legacy of reductionism, this time, in the literature on anomalous experience (e.g., the obsession with physics, chemistry, and biology for explaining anomalous experience).
- In *FPHE*, I examined another case of reductionism in action, in the context of engineering (e.g., technical constraints vs. normative constraints).
- In *FPHMM*, I elaborated one more version of reductionism, in relation to the three domains of communication (e.g., the competing views on sending, connecting, and receiving).
- In *FPHCT*, I explored another version of reductionism, in relation to invention and innovation (e.g., the bio-psychological argument vs. the socio-cultural arguments).

(continued on next page)

Table 4.2. On Reductionism and Reverse-Reductionism (Part II)

- **Against the Varieties of Reductionism *(cont'd)***
  —*Theoretical Reductionism (cont'd)*
    - In *FPHG*, I identify another version of reductionism, in relation to infinity, symmetry, and dimensionality (e.g., the Euclidean argument vs. the non-Euclidean arguments).
    - In *FPHUP*, I analyzed another version of reductionism, in relation to density and void (e.g., the engineering argument and the ecology argument).
    - In *FPHL*, I examined another version of reductionism, in relation to structure and context (e.g., the structuralist argument and the contextualist argument).
    - In *PFHO*, I explored another version of reductionism, in relation to communcation, decision-making, and leadership (e.g., the rational-system argument vs. the natural-system argument).
    - In *PFHMA*, I revealed another version of reductionism, in relation to the martial body and spirit (e.g., the spiritual argument vs. the materialist argument vs. the defensive argument).
    - In *PFHS*, I analyzed another version of reductionism, in relation to the sexual body and spirit (e.g., the naturalist argument vs. the constructivist argument).
    - In *PFHLAW*, I examined another version of reductionism, in relation to law (e.g., the necessity argument vs. the contengency argument).
    - In *FPHWP*, I showed another version of reductionism, in relation to war and peace (e.g., the aggressivist argument vs. the pacifist argument).
    - In *BEPE*, I analyzed another version of reductionism, in relation to morality and immorality (e.g., the objectivist argument vs. the non-objectivist argument vs. the skeptical argument).
    - In *BCOS*, I examined another version of reductionism, in relation to the contested beginnings and speculative ends of the universe (e.g., the scientific argument vs. the religious argument vs. the esoteric argument vs. the metaphysical argument).
    - In *FPHP*, I show another version of reductionism, in relation to normality and abnormality (e.g., the natural argument vs. the social argument vs. the cultural argument vs. the mental argument).

(continued on next page)

Table 4.2. On Reductionism and Reverse-Reductionism (Part III)

- **Against the Varieties of Reductionism *(cont'd)***
  —*Theoretical Reductionism (cont'd)*
    - In *FPHGEOL*, I scrutinized another version of reductionism, in relation to statics and dynamics (e.g., the catastrophe argument vs. the uniformity argument vs. the revision argument).
    - In *FPHCHESS*, I showed another version of reductionism, in relation to tactics and strategy (e.g., the natural argument vs. the social argument vs. the cultural argument vs. the mental argument).
    - In *FPHR*, I revealed another version of reductionism, in relation to secularness and sacredness (e.g., the critical argument vs. the skeptical argument vs. the theist argument).
    - In *FPHEDU*, I examined another version of reductionism, in relation to teaching and learning (e.g., the teacher-centered argument vs. the student-centered argument vs. the balanced argument).
    - In *FPHH*, I explored another version of reductionism, in relation to joking and laughing (e.g., the natural argument vs. the social argument vs. the cultural argument vs. the mental argument).
    - In *FPHA*, I proposed another version of reductionism, in relation to sound and silence (e.g., the natural argument vs. the social argument vs. the cultural argument vs. the mental argument).
    - In *FPHGAM*, I proposed another version of reductionism, in relation to risk and caution (e.g., the natural argument vs. the social argument vs. the cultural argument vs. the mental argument).
    - In *FPHCOM*, I worked out another version of reductionism, in relation to hardware and software (e.g., the natural argument vs. the social argument vs. the cultural argument vs. the mental argument).
    - In *FPHLIT*, I proposed another version of reductionism, in relation to fiction and non-fiction (e.g., the natural argument vs. the social argument vs. the cultural argument vs. the mental argument).
    - In *FPHCHEM*, I proposed another version of reductionism, in relation to substances and their changes (e.g., the natural argument vs. the social argument vs. the cultural argument vs. the mental argument).

(continued on next page)

Table 4.2. On Reductionism and Reverse-Reductionism (Part IV)

- **Against the Varieties of Reductionism *(cont'd)***
  —*Theoretical Reductionism (cont'd)*
    - In *FPHDA*, I proposed another version of reductionism, in relation to quantitative and qualitative research (e.g., the natural argument vs. the social argument vs. the cultural argument vs. the mental argument).
    - In *FPHGEOG*, I proposed another version of reductionism, in relation to environments and their interactions (e.g., the natural argument vs. the social argument vs. the cultural argument vs. the mental argument).
    - In *FPHPA*, I proposed another version of reductionism, in relation to the body and its presence (e.g., the natural argument vs. the social argument vs. the cultural argument vs. the mental argument).
    - In *FPHSEM*, I proposed another version of reductionism, in relation to internality and externality (e.g., the natural argument vs. the social argument vs. the cultural argument vs. the mental argument).
    - In *FPHARCH*, I proposed another version of reductionism, in relation to form and function (e.g., the natural argument vs. the social argument vs. the cultural argument vs. the mental argument).
    - In *FPHCRIM*, I proposed another version of reductionism, in relation to heroes and villains (e.g., the natural argument vs. the social argument vs. the cultural argument vs. the mental argument).
    - In *FPHHIST*, I proposed another version of reductionism, in relation to universality and relativity (e.g., the natural argument vs. the social argument vs. the cultural argument vs. the mental argument).
    - In *FPHMIG*, I proposed another version of reductionism, in relation to Sameness, Otherness, and identity (e.g., the natural argument vs. the social argument vs. the cultural argument vs. the mental argument).

(continued on next page)

Table 4.2. On Reductionism and Reverse-Reductionism (Part V)

- **Against the Varieties of Reductionism *(cont'd)***
  —*Theoretical Reductionism (cont'd)*
    - In *FPHTRAN*, I proposed another version of reductionism, in relation to networks and operations (e.g., the natural argument vs. the social argument vs. the cultural argument vs. the mental argument).
    - In *FPHCA*, I proposed another version of reductionism, in relation to ingredients and techniques (e.g., the natural argument vs. the social argument vs. the cultural argument vs. the mental argument).
    - And in *FPHW*, I propose another version of reductionism, in relation to uselessness and usefulness (e.g., the natural argument vs. the social argument vs. the cultural argument vs. the mental argument).

  —*Methodological Reductionism*
    - A good instance concerns the debate between different versions of qualitative and quantitative methods (as already analyzed in *FC* and also *FHC*). In *FPHML*, I examined similar reductionism, this time, in the literature on mathematical logic (e.g., the obsession with consistency, soundness, and completeness). And in *FPHFS*, I also explored the problems of reductionism in the context of formal science (e.g., the analytical argument).

  —*Ontological Reductionism*
    - An excellent example is the debate between emergentism and reductionism in complexity theory and also in psychology (as elaborated in *FPHC*, in the context of Being and Becoming).

- **Against the Varieties of Reverse-Reductionism**
  —*Conceptual Reverse-Reductionism*
    - Any concept of "art" (e.g., fine art, cave art, outsider art, junk art) is deemed acceptable in postmodernism (as already addressed in Ch.4 of *FHC*).

(continued on next page)

Table 4.2. On Reductionism and Reverse-Reductionism (Part VI)

- **Against the Varieties of Reverse-Reductionism** ***(cont'd)***

—*Theoretical Reverse-Reductionism*

- There are numerous art and literary theories co-exist. Take the case of literary studies, as there are now Literary Structuralism, Marxist Literary Criticism, New Criticism, Phenomenology, Hermeneutics, Language-Game Literary Criticism, Feminist Literary Criticism, Reception Theory, Reader Response Criticism, Poststructuralism, Semiotics, Pyschoanalytic Literary Criticism, just to cite some well-known ones, with no one being said to be better than any others (as detailedly analyzed in Ch.4 of *FHC*). (S. Raman 1997) In *BNN*, I even introduced "the compromise fallacy" as another good illustration of theoretical reverse-reductionism, in misleadingly treating both genetic and environmental approaches as equally valid.

—*Methodological Reverse-Reductionism*

- There is the "anything-goes" mentality in postmodernism (e.g., doing art without praxis, doing art with praxis, and doing art by sublation), as analyzed in Ch.4 of *FHC*. And in *FPHFS*, I also exposed the problems of reverse-reductionism in the context of formal science in relation to systems theory.

—*Ontological Reverse-Reductionism*

- There are likewise no privileged ontology, and the door is open for anything in postmodernism (e.g., the equal status of the ontology of Being vs. that of Becoming, as already addressed in Ch.4 of *FHC*—and also in *FPHC).* In *FAE*, I also introduced another version of reverse-reductionism, that is, "the pluralist fallacy," in the context of understanding aesthetic experience, for instance—although this fallacy has been committed not exclusively in relation to the ontological level (but also at the conceptual, theoretical, and methodological ones).

Sources: From my previous books.

Table 4.3. The Concepton of Existential Dialectics (Part I)

- **Sets and Elements**
  - —Sets
    - Ex: the Same
    - Ex: the Others
  - —Elements
    - Ex: whites in 20th century America (in the set of "the Same")
    - Ex: Iraq during the U.S. invasion in 2003 (in the set of "the Others")

- **Relations, Operations, Functions**
  - —Relations (e.g., "belongs," "equals to," "is greater than")
    - Ex: symmetric interactions within the Same (or the Others)
    - Ex: asymmetric interactions between the Same and the Others
  - —Operations (e.g., "and," "or," "not," "if…then")
    - Ex: if the Same oppresses the Others, it will also oppress itself.
    - Ex: the Same is not the Others.
  - —Functions (e.g., goals)
    - Ex: the Same is hegemonic in relation to the Others.

- **Truth Values**
  - —"1" if True (in Symbolic Logic)
    - Ex: the proposition that imperial Japan was hegemonic to China during WWII
  - —"0" if False (in Symbolic Logic)
    - Ex: the proposition that Grenada invaded France in 2003
  - —"1" & "0" if Both True and False (in Dialectic Logic)
    - Ex: the proposition that the rabbit-duck picture refers to a duck
  - —"~1" & "~0" if Neither True Nor False (or N/A, in Dialectic Logic)
    - Ex: the proposition that God really exists

(continued on next page)

Table 4.3. The Conception of Existential Dialectics (Part II)

- **Axioms and Postulates**
  —Axioms
  - Ex: the reflexive axiom—"any quantity is equal to itself"

  —Postulates
  - Ex: the SSS postulate—"if the three sides of a triangle are congruent to their corresponding parts, then the triangles are congruent"

- **Theorems and Principles**
  —Theorems (and Principles) in Existential Dialectics
  - *In Relation to Method*
    –#28: The formalness-informalness principle
    –#27: The absoluteness-relativeness principle
    –#26: The partiality-totality principle
    –#25: The predictability-unpredictability principle
    –#24: The explicability-inexplicability principle
    –#23: The fiction-reality principle
  - *In Relation to Structure*
    –#22: The finiteness-transfiniteness principle
    –#21: The preciseness-vagueness principle
    –#20: The simpleness-complicatedness principle
    –#19: The openness-hiddenness principle
    –#18: The denseness-emptiness principle
  - *In Relation to Process*
    –#17: The change-constancy principle
    –#16: The order-chaos principle
    –#15: The slowness-quickness principle
    –#14: The expansion-contraction principle

(continued on next page)

Table 4.3. The Conception of Existential Dialectics (Part III)

- **Theorems and Principles (*cont'd*)**
  —Theorems (and Principles) in Existential Dialectics
  - *In Relation to Agency*
    –#13: The theory-praxis principle
    –#12: The convention-novelty principle
    –#11: The evolution-transformation principle
    –#10: The symmetry-asymmetry principle
    –#9: The softness-hardness principle
    –#8: The seriousness-playfulness principle
    –#7: The activeness-inactiveness principle
    –#6: The selfness-otherness principle
  - *In Relation to Outcome*
    –#5: The regression-progression principle
    –#4: The same-difference principle
    –#3: The stability-reaction principle
    –#2: The functionality-nonfunctionality principle
    –#1: The intentionality-nonintentionality principle

Notes: The categories and examples in each are solely illustrative (not exhaustive). The comparison is also relative (not absolute), nor are they mutually exclusive. As generalities, they allow exceptions.
Sources: Starting from Ch.6 of *BCPC* and also from other books of mine

Table 4.4. The Syntax of Existential Dialectics I: The Principles (Part I)

- **In Relation to Method**

  —*The Formalness-Informalness Principle*

  (On the Formal Requirements of Logical Systems)

  - The formal requirements of a logical system (e.g., consistency, soundness, and completeness) have both usefulness and non-usefulness, to the extent that, if there are formal systems requiring them, there are alternative ones which do not. Thus, it does *not* exclude classical logics but simply goes beyond both classical and non-classical logics, while learning something from each. There is no formalness without informalness—and vice versa.
  - For instance, existential dialectics can make use of both classical logics under certain conditions (e.g., especially, though not exclusively, when they are clear-cut, etc.) and non-classical logics under alternative conditions (especially, though not exclusively, when they are "unknown," "irrelevant," "ambiguous," "possible," with "different degrees of truth," empirically inconsistent in a desirable way, etc.). (WK 2008u)
  - Family resemblance: e.g., logicalness-nonlogicalness, rationality-nonrationality, etc.
  - Sources: From *FPHML*, *FPHH*, *FPHGAM*, *FPHSEM*, and my later books.

(continued on next page)

Table 4.4. The Syntax of Existential Dialectics I: The Principles (Part II)

- **In Relation to Method** ***(cont'd)***

  —*The Absoluteness-Relativeness Principle*
  (On the Multiplicity of Things)

  - There is the multiplicity of things in reality, be they about entities, qualities (or properties), and relationships, such that what is acceptable from one standpoint may not be so from another. For instance, if there is something absolute, there is likewise something relative. There is no absoluteness without relativeness—and vice versa.
  - Both absoluteness and relativeness here are also relevant to different modalities often cited in the literature on ontology, such as possibility (e.g., something "can" happen) and its opposite (e.g., impossibility), probability (e.g., something "will" happen) and its opposite (e.g., improbability), and necessity (e.g., something "should" happen) and its opposite (e.g., contingency).
  - Family resemblance: e.g., independence-dependence, isolatedness-interconnectiveness, universalness-relativeness, immanence-transcendence, autonomy-interdependency, internalness-externalness, etc.
  - Sources: From *FPHK*. See also my later books.

(continued on next page)

Table 4.4. The Syntax of Existential Dialectics I:
The Principles
(Part III)

- **In Relation to Method *(cont'd)***
  —*The Partiality-Totality Principle*
  (On the Relationships between Whole and Parts)
  - The whole is not the sum of the parts. There is no partiality without totality—and vice versa.
  - Any inquiry about a phenomenon in the work is to guard against the varieties of (a) reductionism and (b) reverse-reductionism.
  - Reductionism and reverse-reductionism can be (i) conceptual, (ii) theoretical, (iii) methodological, and (iv) ontological.
  - Works on emergence (as in *FC*), gestalt psychology (as in *FPHGEOL*), the hermeneutic circle (as in *FPHDA*), and mental language (as in *FPHSEM*), to name just a few of them, have contributed to the understanding of whole and parts.
  - Family resemblance: e.g., individualisticness-holisticness, partness-wholeness, analysis-synthesis, etc.
  - Sources: Especially from *FC* & *FPHGEOL*. See also my later books.

(continued on next page)

Table 4.4. The Syntax of Existential Dialectics I: The Principles (Part IV)

- **In Relation to Method *(cont'd)***
  —*The Predictability-Unpredictability Principle*
  (On the Occurrence of Events)
  - Both predictability and unpredictability have a major role to play in the occurrence of things, so that neither determinism nor indeterminism wins the centuries-old fight. There is no predictability without unpredictability—and vice versa.
  - There are events which are predictable, just as there are those which are not. Or what is regarded as unpredictable at one point in time may turn out to be predictable later, and, conversely, what is deemed as predictable may turn out to not be so predictable. Even in predictability, outcomes are subject to uncertainty, the degree of which varies from case to case.
  - Works on complexity theory (as in *FC*), quantum mechanics (as in *FPHST* and *FPHG*), frequency probability, subjective (or Bayesian) probability, and "special causes" (as in *FPHGAM*), just to name a few of them, have contributed to the understanding of uncertainity.
  - Family resemblance: e.g., sureness-arbitrariness, certainty-uncertainty, deterministicness-indeterministicness, sureness-doubtfulness, likeliness-riskiness, deterministicness-randomness, etc.
  - Sources: Especially from *FC* & *FPHGAM.* See also my later books.

(continued on next page)

Table 4.4. The Syntax of Existential Dialectics I: The Principles (Part V)

- **In Relation to Method** ***(cont'd)***

  —The Explicability-Inexplicability Principle
  (On the Underlying Mechanisms of Things)

  - Both explicability and inexplicability are part of the understanding of things. There is no explicability without inexplicability—and vice versa.
  - This principle tells us the duality of the research dilemma, in that, if reality can be explained in some ways, it also has its other ways which are not quite explainable, at a given point in time.
  - Family resemblance: e.g., underlyingness-regularness, causation-regularness, causation-correlation, explanation-description, explanation-narration, explanation-interpretation, etc.
  - Sources: Especially from *FPHU*—and also *FPHC*. See also my later books.

  —The Fiction-Reality Principle
  (On the different faces of Reality)

  - If there is reality, there is also fiction, especially for any agent with some kind of intelligence life, be it about humans, animals, or, later, post-humans. There is no reality without fiction—and vice versa.
  - The distinction between reality and fiction is not absolute, since there can be a mixture of both. Fiction can have non-fictional elements, just as non-fiction can have fictional ones. So, the distinction is in degree, not in kind.
  - Family resemblance: e.g., fiction-nonfiction, fiction-fact, imagination-fact, subjectivity-objectivity, imagination-nonfiction, etc.
  - Sources: First from *FPHLIT*.

(continued on next page)

Table 4.4. The Syntax of Existential Dialectics I: The Principles (Part VI)

- **In Relation to Structure**
  —*The Finiteness-Transfiniteness Principle*
  (On the Nature of Numbers)
  - If there are finite things, there are likewise transfinite ones. There is no finiteness without transfiniteness—and vice versa.
  - To avoid confusion, my usage of the word "transfinite" here differs radically from the one used by Cantor (and other mathematicians) for "relative" infinity—and is more limited, in light of the problems confronting any attempt to understand the idea of infinity, be it by intuition, imagination, and conception (as detailedly analyzed in *Sec. 2.2.3* of *FPHG*).
  - Instead, by "transfinity," I allow numbers which can be many times larger—or smaller, for that matter—than the finite things that we encounter in daily life, but they do not have to be related to the idea of infinity at all (which may not exist).
  - Of course, there may be some *borderline* cases, in which it is not clear whether the number in question is transfinite (in my usage) or simply a mathematical convenience. A good example of a borderline case is the Planck unit of length for "the smallest space possibly measured in nature," which is "less than billionths of trillionths of trillionths of an inch" (or something like 1.6×10^{-35} meters).
  - That said—my usage of "transfinity" can also resolve (or better, dissolve) an age-old problem in philosophy known as "Zeno's paradoxes" (as already explained in *Sec. 2.2.3* of *FPHG*).
  - Family resemblance: e.g., boundedness-quasiunboundedness, smallness-largeness, boundedness-unboundedness, etc.
  - Source: From *FPHG*. See also my later books.

(continued on next page)

Table 4.4. The Syntax of Existential Dialectics I: The Principles (Part VII)

• **In Relation to Structure** ***(cont'd)***

—*The Preciseness-Vagueness Principle*
(On the Refinement of Things)

- Both preciseness and vagueness are important, not that one is better than the other, but that both are used, in different degrees of preference, in accordance to the contextual application from the perspectives of nature, the mind, culture, and society. Even when both are used in a combination, the dilemma is shifted instead to one of combinational concern. There is no preciseness without vagueness—and vice versa.
- In relation to taxonomy, preciseness has its taxonomic clarity, just as vagueness has its classificatory flexibility, for instance. And in relation to network, vagueness has its explorative liberty, just as preciseness has its conceptual definitiveness, for instance.
- Family resemblance: e.g., clarity-ambiguity, directness-indirectness, quantitativeness-qualitativeness, describability-nondescribability, specificity-obscurity, specificity-generality, thickness-thinness, translatability-untranslatability, clearcutness-transitionalness, concreteness-abstractness, etc.
- Sources: Especially from *FIA*. See also my later books.

(continued on next page)

Table 4.4. The Syntax of Existential Dialectics I: The Principles (Part VIII)

- **In Relation to Structure** ***(cont'd)***
  —*The Simpleness-Complicatedness Principle*
  (On the Interconnection among Things)
  - Both simpleness and complicatedness are vital, without favoring one over the other, in that each is utilized, depending on the basis of the perspectives of nature, the mind, culture, and society. And even when a combination of them is preferred, the dilemma is only shifted to a combinational degree of concern. There is no simpleness without complicatedness—and vice versa.
  - In relation to taxonomy, simpleness has its heuristic usefulness, just as complicatedness has its realistic representation, for instance. And in relation to network, simpleness has its economical attractiveness, just as complicatedness has its practical reliability, for instance.
  - Family resemblance: e.g., inflexibility-flexibility, standardization-specialization, imperfectness-perfectness, superficiality-depth, shallowness-deepness, economicalness-elaboratedness, plainness-circumspection, onesidedness-multisidedness, succinctness-duplicatedness, etc.
  - Sources: Especially from *FIA*. See also my later books.

(continued on next page)

Table 4.4. The Syntax of Existential Dialectics I: The Principles (Part IX)

- **In Relation to Structure** ***(cont'd)***
  —*The Openness-Hiddenness Principle*
  (On the Detection of Things)
  - Reality has its hidden face, just as it is open to outside view in some other ways. There is no openness without hiddenness—and vice versa.
  - For instance, in the context of anomalous experience, certain aspects of reality can be open for examination (e.g., the experiment with SPECT images, the transmission of telepathemic bit, and the use of metonymies and metaphors). Other aspects, however, remain hidden, and examples include the elusive deeper nature of spiritual reality in the intangible realm, the undetectability of different branched universes, the underlying mechanisms of signals traveling between different folds of physical space-time, and the dependence on language and the bias for science and logic).
  - Family resemblance: e.g., overtness-covertness, publicness-privateness, openness-closedness, transparency-secrecy, openness-biasedness, participation-obstruction, exposition-illusion, surface-depth, etc.
  - Sources: Especially from *FPHU*. See also my later books.

(continued on next page)

Table 4.4. The Syntax of Existential Dialectics I: The Principles (Part X)

- **In Relation to Structure** ***(cont'd)***
  —*The Denseness-Emptiness Principle*
  (On the Distribution of Entities in Space)
  - Both density and void are needed, in relation to the mind, nature, culture, and society, albeit in different ways. There is no denseness without emptiness—and vice versa.
  - For clarity, the term "void" is used here only as an approximation of emptiness (depending on the degree of the lack of density), since, in physics, it is well known that "empty" space is not really empty all the way, because it can be full of energy (e.g., random quantum fluctuations at the sub-atomic level, and, for that matter, dark energy in the universe) and matter (e.g., different versions of sub-atomic particles, and, for that matter, dark matter in the universe).
  - Family resemblance: e.g., fullness-voidness, nearness-farness, concentration-dispersion, denseness-nothingness, presence-absence, focus-unfocus, solidness-void, something-nothing, heaviness-lightness, composition-emptiness, emphasis-disemphasis, etc.
  - Sources: First worked out in *FPHUP*. See also my later books.

- **In Relation to Process**
  —*The Change-Constancy Principle*
  (On the Alteration of Things)
  - Change occurs over time, although constancy is also allowed. There is no change without constancy—and vice versa.
  - Asymmetry undergoes changes over time, so does symmetry.
  - Old players fade away, and new ones emerges, with ever new causes and ever new forms.
  - Family resemblance: e.g., dynamicness-staticness, instability-stability, etc.
  - Sources: First named in *BCPC*. Especially from *FHC*, *FCD*, and *FPHC*. See also other books of mine (from that point on).

(continued on next page)

Table 4.4. The Syntax of Existential Dialectics I: The Principles (Part XI)

- **In Relation to Process *(cont'd)***
  —The Order-Chaos Principle
  (On the Pattern of Things)
  - Both order and chaos are vital in the process of change in the world. The preference for order is biased, since it does not give sufficient attention to the vital role of chaos in the transformation of the world (without somehow reducing it for the understanding of order). There is no order without chaos—and vice versa.
  - The scientific search for order in the world is often a hidden bias in its ontological obsession with order, since chaos is often treated as the "bad" guy, with order as the "good" guy (for the end goal of science).
  - Neither order nor chaos is the final end of the world, and one is not to be treated as the means for the other in the transformation of things. Both are fundamental in their recurrent dialectical interactions with each other over time, without reducing one for the other.
  - Family resemblance: e.g., lawfulness-disorder, order-disorder, linearity-nonlinearity, linearity-irreversalness, pattern-chaos, unity-disunity, etc.
  - Sources: Especially from *FC*. See also my later books.

(continued on next page)

Table 4.4. The Syntax of Existential Dialectics I: The Principles (Part XII)

- **In Relation to Process *(cont'd)***
  - *—The Slowness-Quickness Principle (On the Speed of Change)*
    - Both slowness and quickness co-exist, with their own internal tension, to the extent that each fights for its own relevance with the other, in accordance to the perspectives of nature, the mind, culture, and society, without one being the victor and the other being the vanquished in the long haul. Even when both are chosen in other cases, this dilemma is only transferred into something else with a combinational character. There is no slowness without quickness—and vice versa.
    - In relation to taxonomy, quickness has its efficient usability, just as slowness has its aesthetic appeal, for instance. And in relation to network, quickness has its adventurous readiness, just as slowness has its risk-adverse convenience, for instance.
    - Family resemblance: e.g., inconvenience-convenience, passiveness-activenessness, gradualness-abruptness, deceleration-acceleration, difficulty-ease, etc.
    - Sources: Especially from *FIA*. See also my later books.

  - *—The Expansion-Contraction Principle (On the Growth of Things)*
    - Entities in the world can both expand in some ways and contract in other ones, as part of their nature. There is no expansion without contraction—and vice versa.
    - For instance, in the context of anomalous experience, one of its most fascinating expressions is none other than the analysis of different causes of the slow but steady expansion (and, furthermore, if put in a different way, unfolding) of unconsciousness.
    - Family resemblance: e.g., conquest-autarky, rise-fall, spread-shrink, extendingness-shorteningness, widening-narrowing, expansion-destruction, expansion-decline, construction-demolition, creation-destruction, construction-deconstruction, etc.
    - Sources: Especially from *FPHU*. See also my later books.

(continued on next page)

Table 4.4. The Syntax of Existential Dialectics I: The Principles (Part XIII)

- **In Relation to Agency**
  —*The Theory-Praxis Principle*
  (On the Duality of Knowledge))
  - This has to do with the duality of knowledge in existential dialectics, in that, if there is theoretical construction, there is likewise its practical application, both technical and normative. There is no theory without praxis—and vice versa.
  - Family resemblance: e.g., discovery-application, knowledge-action, invention-innovation, knowledge-motivation, theory-intentionality, universality-particularity, discourse-action, knowledge-interest, etc.
  - Source: Especially from *FPHE*. See also my later books.

(continued on next page)

Table 4.4. The Syntax of Existential Dialectics I: The Principles (Part XIV)

- **In Relation to Agency *(cont'd)***
  —*The Convention-Novelty Principle*
  (On the Nature of Creative Thinking)
  - If there is conventional wisdom, there is likewise novel challenge, to the extent that both convergent and divergent thinking are part of life. There is no convention without novelty—and vice versa.
  - In addition, in this context of the convention-novelty principle, there are, in the absence of better words, what I want to call (a) *creative techniques* and (b) *creative traits*, which, when satisfied—in relation to the larger context of the mind, nature, society, and culture—can be used to enhance creative works. The list of creative techniques and traits (as summarized in *Table 4.1* of *FPHCT*) are not exhaustive, of course, but the examples here constitute a great beginning of understanding the structure of creative thinking.
  - Yet, lest the reader gets carried away by the euphoria about creative thinking in our time, it should be reminded that creative thinking has its own possibilities and limits (as shown in *Table 2.1* of *FPHCT* on invention), just as it has its own promises and pitfalls (as shown in *Table 3.1* of *FPHCT* on innovation). In the end, it should also be stressed that creative thinking has its own desirability and dark sides (as shown in *Table 4.2* of *FPHCT*, *BNN*, and *FPHMIG*).
  - This has important daily implications for an average individual, and the word "average" here allows different degrees of variation from one individual to another in a population, at any given point in history.
  - Family resemblance: e.g., normalness-nonnormalness, conformity-nonconformity, convergence-divergence, familiarity-freshness, inertia-newness, etc.
  - Sources: First from *FPHCT*. See also my later books.

(*continued on next page*)

Table 4.4. The Syntax of Existential Dialectics I: The Principles (Part XV)

• **In Relation to Agency** ***(cont'd)***

—*The Evolution-Transformation Principle*
(On the Multiple Kinds of Agency)

- This principle (and the symmetry-asymmetry principle) are both about *the agency of change*. The word "agency," in a formal definition, refers to "a person or thing through which power is exerted or an end is achieved." (MWD 2007b) It therefore does not have to necessarily involve an intelligent lifeform.
- Because of this dual meaning in agency, the evolution-transformation principle is more concerned with *the multiple kinds of agency*, that is, both about the *evolution* in the state of nature (e.g., an object of natural beauty) and the *transformation* in the world of intelligent lifeforms (e.g., a work of art, an air-conditioner, etc.). There is no evolution without transformation—and vice versa.
- And the transformative part of the principle precisely refers to the other dimension in the dual meaning of agency, in giving technology a major role to play in the change of the world, which is something that I extensively analyzed in *FHC* in the context of the technophilic lifeworld, espeically though not exclusively since modern times. The same theme is also elaborated in other books of mine.
- Family resemblance: e.g., nonwillingness-willingness, naturalness-technologicalness, evolution-creation, naturalness-nonnaturalness, inorganicness-volition, naturalness-artificialness, inorganicness-motivation, etc.
- Sources: Especially from *FAE*. See also other books of mine.

(*continued on next page*)

Table 4.4. The Syntax of Existential Dialectics I: The Principles (Part XVI)

- **In Relation to Agency *(cont'd)***
  —*The Symmetry-Asymmetry Principle*
  (On the Relationships among Existents)
  - There is no symmetry without asymmetry—and vice versa.
  - For instance, the Same can be symmetric and asymmetric towards the Others. But in case of asymmetry, oppression and self-oppression can occur. So, when the Same is asymmetric towards the Others, the Same can also be relatively asymmetric towards itself in self-oppression, just as the Others can be likewise towards themselves.
  - The subsequent oppressiveness is dualistic, as much by the Same against the Others and itself, as by the Others against the Same and themselves.
  - Both oppression and self-oppression can be achieved by way of downgrading differences between the Same and the Others and of accentuating them.
  - This is true, even though not all forms of asymmetry have to be about oppression and self-oppression.
  - In addition, from Chapter Three of *FPHG*, symmetry is not perfect, to be understood in an approximate sense under many life circumstances.
  - Family resemblance: e.g., balance-extremity, reflexiveness-unreflexiveness, balance-unbalance, equality-inequality, center-periphery, compatibility-incompatibility, proportion-disproportion, harmoniousness-unharrmoniousness, etc.
  - Sources: From all my books, starting with *FHC*. First named in *BCPC*. See, for instance, *FCD* for more details on the Same and the Others.

(*continued on next page*)

Table 4.4. The Syntax of Existential Dialectics I: The Principles (Part XVII)

- **In Relation to Agency *(cont'd)***
  —*The Softness-Hardness Principle*
  (On the Force of Change)
  - This has to do with *the force of change* under the category about agency in existential dialectics, in that any change by an agent, be it organic (like humans) or non-organic (like natural objects), can occur in a forceful (aggressive) or gentle (pacific) way, which can come in all shapes and sizes, of course. There is no softness without hardness—and vice versa.
  - Family resemblance: e.g., peacefulness-violence, cooperation-competition, acceptance-discrimination, love-hatred, peace-war, pacificity-aggression, dovishness-hawkishness, reward-punishment, kindness-hostility, admiration-belittlement, inclusion-exclusion, friendliness-unfriendliness, etc.
  - Sources: Especially from *ALD*. See also my later books.

(continued on next page)

Table 4.4. The Syntax of Existential Dialectics I: The Principles (Part XVIII)

- **In Relation to Agency *(cont'd)***
  —*The Seriousness-Playfulness Principle*
  (On the Extent of Seriousness)
  - If there is seriousness, there is also playfulness, especially for any agent with some kind of intelligence life, be it about humans, animals, or, later, post-humans. There is no seriousness without playfulness—and vice versa.
  - Lest any misunderstanding occurs, the word "playfulness" here should not be confused with other terms like "laughter," "smile," "joke," "tickleness," and "laugh-like vocalization" (in some animals), for instance.
  - Being playful can come in all shapes and sizes and therefore is not necessarily tied up with any of these terms, alhough it can result from any of them, or a combination of them, or something else altogether.
  - This principle, together with some other principles (especially though not exclusively, like the formalness-informalness principle), have important implications for what constitutes rationality, be it about "rationality of application" at the practical level or "rationality of knowledge" at the meta-theoretical level (as summarized in *Sec. 4.20* of *FPHH*).
  - Family resemblance: e.g., seriousness-jokingness, seriousness-humorousness, seriousness-wittiness, seriousness-nonseriousness, seriousness-relaxedness, seriousness-restingness, Apollo-Dionysus, etc.
  - Sources: First from *FPHH* & *FPHGAM* See also my later books.

(*continued on next page*)

Table 4.4. The Syntax of Existential Dialectics I: The Principles (Part XIX)

- **In Relation to Agency** ***(cont'd)***
  - —*The Activeness-Inactiveness Principle*
    (On the Extent of Activeness)
    - If there is activeness, there is also inactiveness, especially (though not exclusively) for any agent with some kind of intelligence life, be it about humans, animals, or, later, post-humans. There is no activeness without inactiveness—and vice versa.
    - Being active can come in all shapes and sizes, as a person can be inactive because he is tired, old, sick, sleepy, uninterested, etc., just as an object can be inactive because it is at rest, out of work, etc.
    - This principle, together with some other principles (especially though not exclusively, like the formalness-informalness principle and seriousness-playfulness principle), have important implications for what constitutes rationality, be it about "rationality of application" at the practical level or "rationality of knowledge" at the meta-theoretical level (as summarized in *Sec. 4.20* of *FPHH*).
      - Family resemblance: e.g., alertness-unalertness, activeness-decrepidness, alertness-sleepiness, awakenness-hibernatedness, activeness-tiredness, alertness-uninterestedness, alertness-unwareness, soberness-unawareness, soberness-inactiveness, alertness-inactiveness, etc.
    - Sources: First from *FPHTRAN*.

(*continued on next page*)

Table 4.4. The Syntax of Existential Dialectics I: The Principles (Part XX)

- **In Relation to Agency *(cont'd)***
  —*The Selfness-Otherness Principle*
  (On the Magnitude of Self-Centeredness)
  - If there is the tendency to be mostly centered on one's self or group, there is likewise the tendency to be mostly centered on others, especially (though not exclusively) for any agent with some kind of intelligence life, be it about humans, animals, or, later, post-humans. There is no selfness without otherness—and vice versa.
  - Being self-centered and being other-centered do not have to be mutually exclusive, as the two can come in all shapes and sizes, or in different degrees. And other-centeredness does not entail desirability, any more than self-centeredness implies non-desirability, as both can go either way.
  - In addition, whether self-centeredness constitutes a moral alternative or, conversely, an alternative to morality is of course debatable (as addressed in my other book titled *Beyond Ethics to Post-Ethics*).
  - Good examples of family resemblance in relation to the principle are egoisticness-altruisticness, selfishness-unselfishness, selfness-selflessness, egocentricness-othercentricness, and so on.
  - Source: First from *FPHCA*

(*continued on next page*)

Table 4.4. The Syntax of Existential Dialectics I: The Principles (Part XXI)

- **In Relation to Outcome**

  —*The Regression-Progression Principle*
  (On the Direction of History)

  - Neither the cyclical nor the linear views are adequate for explaining many phenomena at all levels. There is regression without progression—and vice versa.
  - History progresses to more advanced forms, but with a regressive touch. Examples include no freedom without unfreedom, no equality without inequality, and no civilization without barbarity.
  - This is not an inevitable law, but merely a highly likely empirical trend.
  - Family resemblance: e.g., cost-benefit, undesirability-desirability, badness-goodness, risk-opportunity, dysfunctionality-functionality, cyclicity-linearity, etc.
  - Sources: From all my books, starting with *FHC*. First named in *BCPC*.

  —*The Same-Difference Principle*
  (On the Metamorphosis of Change)

  - An entity, as it evolves over time, can be both different from and similar to its opposing alternatives and does not have to be solely more different from them over time. There is no similarity without difference—and vice versa.
  - Opposites are not absolute in a black-or-white fashion; so, an entity can become relatively more similar to (or more different from) its opposite over time.
  - Family resemblance: e.g., homogeneity-heterogeneity, we-they, similarity-contrast, etc.
  - Sources: Especially from ALD. See also my later books

(*continued on next page*)

Table 4.4. The Syntax of Existential Dialectics I:
The Principles
(Part XXII)

- **In Relation to Outcome** ***(cont'd)***
  —*The Stability-Reaction Principle*
  (On the Feedback Mechanism for Further Changes)
  - Once an outcome becomes stabilized over time (which varies from case to case), it will undergo further changes in its multifaceted interactions with the outside world (e.g., with other entities, energy, space, and so on).
  - There is no stability without reaction for further changes—and vice versa.
  - Family resemblance: e.g., stability-forwardness, balance-reaction, balance-forwardness, rest-motion, normalcy-feedback, stability-breakup, consolidation-modification, etc.
  - Sources: First from *FPHCHEM*.

(*continued on next page*)

Table 4.4. The Syntax of Existential Dialectics I: The Principles (Part XXIII)

- **In Relation to Outcome** ***(cont'd)***
  - —*The Functionality-Nonfunctionality Principle (On the Presence of Function)*
    - If there is the presence of action for which something or someone is fitted, there is likewise the relative lack of action for which something or someone is fitted. There is no functionality without nonfunctionallity—and vice versa.
    - A *function* should not be confused with an *alternative function*, which is just another kind of function; just as *nonfunction* should not be confused with *dysfunction* (which has more to do with malfunction and is thus related more to the regression-progression principle).
    - In addition, being functional (or useful) and being nonfunctional (or useless) do not have to be mutually exclusive, as the two can come in all shapes and sizes, or in different degrees. And what is functional at one time may become nonfunctional at another time, and vice-versa.
    - Good examples of family resemblance in relation to the principle are usefulness-nonusefulness, usefulness-uselessness, valueness-nonvalueness, and the like.
    - Source: First from *FPHCA*

(*continued on next page*)

Table 4.4. The Syntax of Existential Dialectics I: The Principles (Part XXIV)

- **In Relation to Outcome *(cont'd)***
  - —*The Intentionality-Nonintentionality Principle (On the Planning for Outcomes)*
    - If there is the planning or design of something for a certain outcome, there is likewise the relative absence of planning or design of something for a certain outcome. There is no intentionality without nonintentionallity—and vice versa.
    - An intention for an outcome does not have to be certain, so the outcome can even be counter-intentional in being the opposite of what was originally planned. In this sense, counter-intentionality is a special case of non-intentionality; in other words, there are unintented consequences which can be just the opposite of what was originally planned.
    - An intention for an outcome is mostly relevant to intelligent life, but there are non-intelligent life and non-living matter in the universe too, so nonintentionality is also important.
    - In addition, an intention does not have to be obvious or straightforward, as William Shakspeare once famously said: "I must be cruel only to be kind."
    - Good examples of family resemblance in relation to the principle are purposiveness-nonpurposiveness, intention-accidentalness, design-randomness, etc.
    - Source: First from *FPHW*

Notes: The features in each principle are solely illustrative (not exhaustive). The comparison is also relative (not absolute), nor are they mutually exclusive. As generalities, they allow exceptions.

Table 4.5. The Syntax of Existential Dialectics II: The Principles as Short Cuts (Part I)

• The principles should be treated with caution, lest misunderstanding occurs, since they do not constitute rigid dualities (or dichotomies).

• The reason is that each pair in an ontological principle consists of two opposites, which are, however, merely short cuts both for multiple variations and degrees, as well as for different interactions with multiple other entities unlike them. In this light, each pair can end up having hundred (or even thousand, if not more) different versions, which interact with hundred (or even thousand, if not more) other entities. There are two clarifications here.

• Firstly, this conception of shortcuts is not mutually exclusive nor absolute, in that the opposites can come in all shapes and sizes, with different degrees. For instance, by analogy, just as there are different degrees of the two colors "white" and "black"—there are likewise different degrees of the opposites in each ontological pair, to the extent that there can be multiple entities (not only two) interacting with each other in each pair.

• And secondly, this conception of shortcuts do not ignore other possible entities in interacting with the two opposites (with their different versions) in each pair. For instance, by analogy, there are not only the two colors "white" and "black" as opposites, since there are other colors too besides them like "yellow," "green," "purple," or else. The same logic can be applied to each ontological pair, in that they also interact with other entities, not with only two of them (with their different versions).

• Therefore, with these two clarifications in mind—each pair in an ontological principle serves only as an abbreviation for something more complicated and, therefore, although it contains two opposites, it should not be confused as a duality (dualism).

(continued on next page)

Table 4.5. The Syntax of Existential Dialectics II: The Principles as Short Cuts (Part II)

• Existential dialectics rejects any dualism (or dichotomy) as too rigid and instead allows the multiplicity of entities, to the extent that between the two opposites in each pair exist many other alternatives to choose from. They are named in that short form for aesthetic elegance, instead of listing all possible entities between the two opposites in the title.

• To be dialectic is to go beyond any rigid dichotomy and transcend into something different altogether in the long haul. One may be tempted to call the dialectic logic here with a different name like existential "multilectics" (instead of "dialectics"), but this naming is incorrect (or even misleading), for the two reasons aforestated.

• After all, each ontological pair come in all shapes and sizes (with different degrees) and do not exist by themselves but also interact with other entities unlike them. The virtue of revealing an ontological pair is to show how they relate within themselves (in multiple versions) and also interact with others unlike them (also in multiple versions).

• This will be clear in the table on "the dialectic constraints imposed by the principles."

Source: From Ch.1 of *FPHK*

Table 4.6. The Syntax of Existential Dialectics III: The Principles as Family Resemblances

• Each ontological principle is generic, with some other comparable ontological pairs to be put in the same family (like a *family resemblance*). Ludwig Wittgenstein (1953) once suggested the idea of "family resemblance" in explaining different games classified under the same family called *games*.

• Why should, for instance, playing football and chess as playing "games," when it is well understood that football is not the same as chess? The answer is that, although each game is different and has different rules, many of them (though not all) share, more or less, some commonalites (e.g., scoring as necessary for winning).

• And this is so, even though some games share more than some others in any given selection of critieria, and no two games are exactly identical. So, his point here is that there is no essential core which is common to all games, and the best that one can look for is some characteristics which are common to many (but not all) games. (A. Biletzki 2006)

• By the same logic—in the previous section on selection criteria, the flexibility-inflexibility pair can be put in the family resemblance of the simpleness-complicatedness principle, although the two pairs are not exactly identical. Likewise, the directness-indirectness pair can be put in the family resemblance of the preciseness-vagueness principle, although, again, the two pairs are not exactly identical.

• In this sense, which specific pair in a family should be used to designate the name of the family can be at times a bit arbitrary, but with good reason.

• For illustration, in the context of method, the partiality-totality principle can take the different form like individualisticness-holisticness, just as the explicability-inexplicability principle can take the different form like underlyingness-regularness—although each two pairs are not exactly identical and have slightly different meanings and usages. A more comprehensive listing of this family resemblance for all other principles is shown in the table on the syntax of existential dialectics.

Source: From Ch.1 of *FPHK*

Table 4.7. The Syntax of Existential Dialectics IV: The Dialectic Constraints Imposed by the Principles (Part I)

• **Co-Existent and Asymmetric**

—The principles, as they constitute the syntax of existential dialectics, are dialectic in character, such that, when they are applied, they impose dialectic constraints on how reality is to be understood. Consider, say, the symmetry-asymmetry principle as an illustration here, in order to summarize two main characters of the dialectic constraints in question.

—Firstly, to be dialectic here is to go beyond the narrow dichotomies (and, for that matter, any rigid multi-dimensional classificatory scheme), be they about "self" vs. "world," "freedom" vs. "unfreedom," "barbarity" vs. "civilization," "individuality" vs. "communality," and so on.

—One way to do so (to go beyond) is to consider them all in terms of co-existence (without favoring one over the rest). For instance, my theory of "post-civilization" (to be summarized later in the section on the pragmatics of existential dialectics) is to go beyond barbarity and civilization in terms of understanding barbarity and civilization as being co-existent. And the same logic can be said in relation to my theories of "post-democracy," "post-capitalsm," and others (also to be introduced later in the section on the pragmatics of existential dialectics), in regard to freedom vs. unfreedom, equality vs. inequality, communality vs. individuality, spirituality vs. materiality, and so on.

—But to consider them all (in the dichotomies—and, for that matter, in any rigid multi-dimensional classifactory schme) as co-existent is not the same as to imply that the opposites in any classificatory scheme are all equal, since, in accordance to the symmetry-asymmetry principle (as an illustration here), if they are equal in terms of being considered as co-existent, they are asymmetric in terms of being unequal in dialectic interaction (e.g., X can be more dominant than Y in case A, or Y is more dominant than X in case B).

(continued on next page)

Table 4.7. The Syntax of Existential Dialectics IV: The Dialectic Constraints Imposed by the Principles (Part II)

—For this reason, there are different versions of "post-democracy" and "post-capitalism" in my theories. As an illustration, in version I of the theory of post-democracy, freedom is more dominant than equality, whereas in version II of the theory of post-democracy, equality is more so than freedom.

—But this "X more than Y" has to be understood in the context of dialectic logic (not in conventional logic), in that both "X" and "Y" are important in post-democracy (in the context of dialectic logic), but in an asymmetry way. By contrast, in conventional logic, it often favors one over the other—be it in regard to privileging freedom over equality in Fascism, favoring freedom relatively more than equality in Liberal Democracy, or favoring equality relatively more than freedom in Socialist Democracy. In the latter two cases (about Liberal Democracy and Socialist Democracy), the difference between dialectic logic and conventional logic can be one in degree, not in kind—in this sense, albeit not in other senses.

—The same logic can be said about the relatonships between individuality and communality, between spirituality and materiality, and between formal legalism and informal legalism in the different versions of my theory of post-capitalism.

• **Transcendent**

—Secondly, to be dialectic is to go beyond the narrow dichotomies (and, for that matter, any rigid multi-dimensional classificatory scheme) in another way, this time, in a transcendent way, that is, in exploring other possibilities or even other issues not considered within the narrow confines of narrow dichotomies (and, for that matter, any rigid multi-dimensional classificatory scheme).

(continued on next page)

Table 4.7. The Syntax of Existential Dialectics IV: The Dialectic Constraints Imposed by the Principles (Part III)

—As an analogy, to go beyond the narrow color dichotomy of "black" and "white" is not just to choose both "black" and "white" (as in the first meaning) but also to explore other color options (e.g., "green," "purple," "blue," etc.—and, alternatively, "shade," "line," "curve," etc.). By the same logic, to go beyond "democracy" is to transcend democracy (as in version III of the theory of "post-democracy") and to explore other possibilties of lifeforms (e.g., floating consciousness, hyper-spatial consciousness, etc., to live beyond the narrow obsession with freedom and equality).

—This dialectic character of the principles in existential dialectics has important implications for the pragmatics of existential dialectics (as will be clear shortly, in the section on the pragmatics of existential dialectics).

Notes: The examples here are solely illustrative (not exhaustive). The comparison is also relative (not absolute), nor are they mutually exclusive. As generalities, they allow exceptions.
Sources: First explicitly stated in *ALD*. Also from all other books of mine.

Table 4.8. The Syntax of Existential Dialectics V: Further Clarifications (Part I)

• **Pioneering**

—Firstly, the total number of ontological principles is unknown, to be discovered later, as our knowledge of the world becomes more advanced.

—Consequently, the principles as introduced in my books are not exhaustive, with new ones being added, whenever more of them are discovered in later research.

—At least, future generations can pick up where I leave off and continue the discovery.

—In this sense, my work should be treated as a pioneering effort for the development of a systematic, comprehensive analysis of a new general ontology for the future of knowledge.

• **Flexible**

—Secondly, the principles are not rigidly classified, as they can be reclassified in a different way.

—For instance, the preciseness-vagueness principle is classified under the category of "structure" but can be reclassified under the category of "method," although in so doing, it has a different meaning in the context of method.

—The same logic applies to the same-difference principle under the category of "outcome," which can be reclassified under the category of "structure," although in so doing, once more, it has a different meaning in the context of structure.

• **Mutually Constraining**

—Thirdly, the principles are to be understood together, not that each prinicple is to be analyzed independently of others. After all, the principles are mutually constraining, in that they work together as a whole.

(continued on next page)

Table 4.8. The Syntax of Existential Dialectics V: Further Clarifications (Part II)

- **Selectively Useful**
  —Fourthly, the principles are relevant to all subject matters, but some principles are more useful to some subject matters than others—as implied in the symmetry-asymmetry principle. This is true, even if different studies of the same kind can yield different views about the degree of relevance for each ontological principle, depending on the specific nature of a research in question, needless to say.

- **Anti-Reductionistic**
  —And fifthly, the principles constitute only two levels of analysis, this time, at the ontological and methodological levels—while other levels of analysis (from the perspectives of the mind, nature, society, and culture) are also needed, in order to understand reality in its totality (as explained in the section on "sophisticated methodological holism," so as to avoid the dual dangers of reductionism and reverse-reductionism).

Notes: The examples here are solely illustrative (not exhaustive). The comparison is also relative (not absolute), nor are they mutually exclusive. As generalities, they allow exceptions.
Sources: From my previous books like *FHC*, *FCD*, *FPHC*, *BDPD*, *BCPC*, *BNN*, *FC*, *FAE*, and so on.

Table 4.9. The Syntax of Existential Dialectics VI: The Dilemma of Specific vs. General Ontology (Part I)

• Ontology is often more complicated and imposes some seemingly insurmountable difficulties concerning what constitutes a kind of ontology which can be valid enough to be accepted by the wider intellectual community in question. An excellent example concerns what I want to call *the dilemma of ontology* in relation to the relationship between "specific ontology" and "general ontology." (M. Bunge 1999; R. Corazzon 2007)

• On the one hand, "general ontology," as Mario Bunge (1999) put it, "studies all existents." But, on the other hand, "special ontology studies one genus of thing or process—physical, chemical, biological, social, etc." (M. Bunge 1999; R. Corazzon 2007)

• Consequently, the relationship between the former and the latter is that, as an illustration, "whereas general ontology studies the concepts of space, time, and event, the ontology of the social investigates such general sociological concepts as those of social system, social structure, and social change." (M. Bunge 1999; R. Corazzon 2007)

• With this formal definition in mind, the relationship between the two forms of ontology only poses *the dilemma of ontology*, in that general ontology is possible to the extent that its general contructs must be broad enough to serve as the "lowest common denominator" for all disciplines, but this character of the lowest common denominator is not only hard to find but also its usefulness is limited. (WK 2008b)

• But specific ontology fares no better either, in that its specific constructs primarily serve a specific discipline in question, although other disciplines can appropriate them for their own purposes, but in a different context (especially, though not exclusively, when used as a metaphor from one field to another). For instance, the Kantian category of understanding under the heading of "quantity" constitutes a type of specific ontology, which is specific to the discipline of mathematics, although other disciplines can make use of the categories in their own context.

(continued on next page)

Table 4.9. The Syntax of Existential Dialectics VI: The Dilemma of Specific vs. General Ontology (Part II)

• A solution lies in my proposal of "sophisticated methodological holism," which requires all levels of analysis in any subject matter. See the tables on sophisticated methodological holism, and on reductionism and reverse-reductionism, for more details.

Source: From *Sec. 1.4* of *FPHK*

Table 4.10. The Syntax of Existential Dialectics VII: Types of Inappropriate Family Resemblances (Part I)

- **Inadequate Family Resemblances**
  —Ex: consistency-inconsistency: the formalness-informalenss principle
  —Ex: cognition-noncognition: the formalness-informalness principle

- **Compound Family Resemblances**
  —Ex: reality-nonreality: the formalness-informalness principles, the absoluteness-relativeness principle
  —Ex: normality-abnormality: the symmetry-asymmetry principle, the density-emptiness principle, the convention-novelty principle, the same-difference principle
  —Ex: structure-context: the simpleness-complicatedness principle, the theory-praxis principle
  —Ex: freedom-unfreedom: the expansion-contraction principle
  —Ex: morality-immorality: the denseness-emptiness principle, the symmetry-asymmetry, the same-difference principle
  —Ex: particularness-universalness: the preciseness-vagueness principle, the simpleness-complicatedness principle
  —Ex: pureness-mixedness: the absoluteness-relativeness principle, the simpleness-complicatedness principle, the same-difference principle
  —Ex: shortsightedness-foresightedness: the partiality-totality principle, the simpleness-complicatedness principle
  —Ex: shorttermness-longtermness: the partiality-totality principle, the simpleness-complicatedness principle, the slowness-quickness principle

- **One-Sided Family Resemblances**
  —Ex: sound-silence: asymmetry (in the symmetry-asymmetry principle)
  —Ex: vision-blindness: asymmetry (in the symmetry-asymmetry principle)

(continued on next page)

Table 4.10. The Syntax of Existential Dialectics VII: Types of Inappropriate Family Resemblances (Part II)

- **Vague Family Resemblances**
  —Ex: shortness-longness, narrowness-broadness, smallness-largeness, smallness-greatness, microness-macroness, and littleness-muchness: three different possibilities, namely, (a) finiteness (in the finiteness-transfiniteness principle), if both opposites are defined on a finite scale, (b) transfiniteness (in the finiteness-transfiniteness principle) if both opposites are defined on a transfinite scale, and (c) the finiteness-transfiniteness principle if one opposite is defined on a finite scale and the other opposite is defined on a transfinite scale.

- **One-Sided and Compound Family Resemblances**
  —Ex: extroversion-introversion: one-sided, in being part of the side (category) of "normality" in the normality-abnormality pair. And compounded, because the normality-abnormality pair is in turn related to the symmetry-asymmetry principle, the density-emptiness principle, the convention-novelty principle, the same-difference principle
  —Ex: organizedness-spontaneousness: one-sided, in being part of the side (category) of "normality" in the normality-abnormality pair. And compounded, because the normality-abnormality pair is in turn related to the symmetry-asymmetry principle, the density-emptiness principle, the convention-novelty principle, the same-difference principle
  —Ex: risk-caution: one-sided, in being part of the side (category) of "normality" in the normality-abnormality pair. And compounded, because the normality-abnormality pair is in turn related to the symmetry-asymmetry principle, the density-emptiness principle, the convention-novelty principle, the same-difference principle

Notes: The examples here are solely illustrative (not exhaustive). The comparison is also relative (not absolute), nor are they mutually exclusive. As generalities, they allow exceptions.
Source: First summarized in *BEPE* (and updated in later books)

Table 4.11. The Semantics of Existential Dialectics (Part I)

• **Abstract Structure vs. Specific Meanings**

—The syntax of existential dialectics so understood in terms of ontological principles only gives us the structure of ontology in the world, in an abstract (general) sense. These principles by themselves do not tell us the specific meanings in a given context.

—In order to grasp the specific meanings of the principles in a given context, it is necessary to study the semantics of existential dialectics. And the analysis of the ontological principles as family resemblances in the section on syntax is only a starting point (and thus overlaps a bit with the semantics of existential dialectics here).

—With this caveat in mind, the reason that I have often gone in great lengths in my previous books on different subjects is to explain the specific meanings of the principles when applied in different contexts.

• **Specific Meanings in Specific Fields**

—For instance, in *FPHST*, I used the first three principles (i.e., the change-constancy principle, the regression-progression principle, and the symmetry-asymmetry principle) to propose "the perspectival theory of space-time," for a better way to understand space and time—especially, though not exclusively, in relation to future post-human history (as summarized in *Table 3.6*, *Table 3.7*, *Table 3.8*, and *Table 3.9*). In so doing, I had to introduce concepts and theories specific to the field of physics and other related fields (e.g., "absolute space" and "absolute time" in "classical mechanics" and "relative space-time" in "the theory of relativity").

—In *BNN*, I also exploited the three principles to propose the "transcendent" approach to the study of genes and memes as a new way to understand the interaction between nature and nurture. In so doing, I had to explore concepts and theories in the world of evolutionary theory (e.g., "mutation," "variation," "adaptation," "selection," and "inheritance" in Darwinian evolutionary theory) and neural biology (e.g., "chromosome," "gene," "DNA," "RNA," "protein," "neuron," "neural network," and "behavior").

(continued on next page)

Table 4.11. The Semantics of Existential Dialectics (Part II)

- **Two Types of Translation (Interpretation)**
  —In the technical jargons of semantics, these exercises in my books perform two essential functions of translation (or interpretation), namely, (a) "intensive" interpretation (about the properties or relations of the principles in existential dialectics), and (b) "extensive" interpretation (about the classes or cases to which the properties and relations of the principles in existential dialectics are applied). (WK 2011oo)

- **Limits of Translation (Interpretation)**
  —But there are limits of translation (or interpretation), and an excellent example is that "there is a problem with the 'objectivity' of translation, as it is relative to different cultural (or sub-cultural) contexts. For instance, one of the still unresolved problems in linguistics is: 'Is there an objective gauge for the quality of translation?'" (*FPHCOM* & WK 2011pp)
  —This is one serious problem about the logic of translation and interpretation that I already addressed in *Sec. 3.5.1* of *FPHCOM (*and also, for that matter, in *FHC*, in the context of hermeneutics).

Notes: The categories and examples in each are solely illustrative (not exhaustive). The comparison is also relative (not absolute), nor are they mutually exclusive. As generalities, they allow exceptions.
Sources: First explicitly stated in *ALD* (and later revised in *FPHCOM*). Also from all other books of mine.

Table 4.12. The Pragmatics of Existential Dialectics (Part I)

• **The Two-Way Street Connecting Theory and Meta-Theory**

—The pragmatics of existential dialectics is not a one-way street (that is, using the ontological principles for theoretical insights in praxis) but a two-way one, that is, (a) from meta-theory to theory, and (b) from theory to meta-theory.

—(a) On one side of the street, the ontological principles can inspire some theoretical insights in praxis, that is, in relation to some specific fields.

—(b) On the other (opposing) side of the street, however, the study of a subject matter in the specific fields in turn reveals some more hitherto unknown ontological principles to be discovered and identified. For this reason, three new principles were added in *FC*, one in *FAE*, and two in *ALD*, on top of the original three in *BCPC*—after some research on the specific subject matters.

• **Direct and Indirect Applications**

—*Direct*

- The logic of existential dialectics can shed some theoretical insights on diverse phenomena in the world, and good instances are the pertinent use of the principles of existential dialectics for the theoretical insights on the freedom/unfreedom dialectics, the equality/inequality dialectics, and the wealth/poverty dialectics in my previous works.
- My latest books like *FPHST* and *BNN* also use the principles to reveal some theoretical insights on the perspectives of space and time (as in *FPHST*) and of nature and nurture (as in *BNN*).

—*Indirect*

- The theoretical insights can further be used to reveal other phenomena directly from them (viz., the theoretical insights) and therefore indirectly from the principles themselves. A good instance is the use of the theoretical insights on the freedom/unfreedom and equality/inequality dialectics for the understanding of the civilization/barbarity dialectics.

(continued on next page)

Table 4.12. The Pragmatics of Existential Dialectics (Part II)

- **Direct and Indirect Applications (*cont'd*)**
  - Even in indirect applications, however, a phenomenon under study can still be directly related back to the principles themselves. In the example as cited above, the civilization/barbarity dialectics can be directly related to the principles of existential dialectics without the intermediate role of the freedom/unfreedom and equality/inequality dialectics.

- **Multiple Levels of Application**

  —The theoretical insights can be applied to different levels of analysis, even though in a given example, it may refer to one level only. For instance, in the example concerning the freedom/unfreedom dialectics, it can be used at the structural level (e.g., in relation to the theory of cyclical progression of hegemony), but it can be exploited as well for other levels (e.g., the theory of post-capitalism at the institutional level).

Notes: The categories and examples in each are solely illustrative (not exhaustive). The comparison is also relative (not absolute), nor are they mutually exclusive. As generalities, they allow exceptions.
Sources: From Ch.6 of *BCPC*. See also other books of mine.

Table 4.13. The Freedom-Unfreedom Dialectics (Part I)

- **On Having**
  - —*In Relation to the Technological*
    - (1) if freer from submission to Nature, then less free from ecological degradation (Deep and Social Ecology), even if in a hi-tech form
    - (2) if freer from technological inconvenience / backwardness, then less free from technological control and the loss of privacy
    - (3) if freer from technological (material) backwardness, then less free from the abusive (barbaric) maltreatment of the primitive Others
  - —*In Relation to the Everyday*
    - (1) if freer from abject poverty, then less free from artificial needs/discontents (Frankfurt School)
    - (2) if freer from sensual suppression, then less free from violent sublimation (Freud)
    - (3) if freer from the snobbishness of high culture, then less free from the shabbiness (leveling-off effect) of mass culture (Tocqueville)
    - (4) if freer from the inefficiency of traditional "compassionate economy," then less free from the bondage of a "ruthless [competitive] economy" (Keynes)
    - (5) if freer from anarchy in the state of nature (system fragmentation), then less free from government regulations and controls in system integration

---

(continued on next page)

Table 4.13. The Freedom-Unfreedom Dialectics (Part II)

• **On Belonging**

—*In Relation to the Good and the Just*

(1) if freer from disciplinary society, then less free from society of control (Foucault)

(2) if freer from the tyranny of one or a few, then less free from the tyranny of the majority (or sometimes, minority veto)

(3) if freer from elitist decision making, then less free from political gridlock/cleavage

(4) if freer from arbitrary (discretionary) administration, then less free from bureaucratic irrationality (Weber) and legal trickery (loopholes)

• **On Being**

—*In Relation to the True*

(1) if freer from unscientific dogmas, then less free from instrumental abyss (nihilism). Or conversely, if freer from meaninglessness, then less free from dogmas.

(2) if freer from the bondage of partiality/partisanship (e.g., prejudice, discrimination), then less free from the danger of impartiality and neutrality (e.g., opportunism, unrealisticness, lack of compassion, inaction)

(3) if freer from making generalizations, then less free from being unable to understand much of anything

—*In Relation to the Holy*

(1) if freer from collective conscience, then less free from social loneliness

(2) if freer from religious absoluteness, then less free from spiritual emptiness

—*In Relation to the Beautiful/Sublime*

(1) if freer from artistic non-autonomy, then less free from aesthetic disillusion (deconstruction)

(continued on next page)

Table 4.13. The Freedom-Unfreedom Dialectics (Part III)

Notes: The examples in each category are solely illustrative (not exhaustive), and the comparison is relative (not absolute), nor are they necessarily mutually exclusive. And some can be easily re-classified elsewhere. As generalities, they allow exceptions.
Sources: A reconstruction from Ch.10 of *FCD*, based on *FHC*

Table 4.14. The Equality-Inequality Dialectics (Part I)

• **On Having**

—*In Relation to the Technological*

(1) if more equal in treating Nature with spiritual unity, then less equal in suppressing the dominant drive to transcend it altogether

—*In Relation to the Everyday*

(1) if more equal in building social plurality, then less equal in leveling-off effects (e.g., the subsequent relative intolerance of high/intellectual ethos in mass culture industry)

(2) if more equal in socioeconomic distribution beyond a certain point, then less equal in efficiency (e.g. resentment, the erosion of work ethics)

(3) if more equal in urging an affirmative action program, then less equal in creating victim mentality (in oneself), stigma (from others), reverse discrimination (against the once privileged), and mediocracy (against the more able)

• **On Belonging**

—*In Relation to the Good and the Just*

(1) if more equal in banning monarchic/oligarchic exclusion, then less equal in producing "the tyranny of the majority" or of "minority veto"

(2) if more equal in encouraging participatory decision making, then less equal in inducing political divisiveness (gridlock/cleavage in power blocs) and organizational oligarchy

(3) if more equal in institutionalizing a decentralized bureaucracy, then less equal in falling into more territorial/turf politics (intrigues)

(continued on next page)

Table 4.14. The Equality-Inequality Dialectics (Part II)

- **On Being**
  - —In Relation to the Beautiful / Sublime
    - (1) if more equal in accepting diverse styles ("anything goes" mentality), then less equal in artistic good quality (in leveling-off effects against the best)
  - —In Relation to the True
    - (1) if more equal in tolerating multiple viewpoints (no matter how extreme), then less equal in epistemic standards
  - —In Relation to the Holy
    - (1) if more equal in celebrating any cults and sects (no matter how questionable), then less equal in spiritual depth and authenticity

---

Notes: The examples in each category are solely illustrative (not exhaustive), and the comparison is relative (not absolute), nor are they mutually exclusive. And some can be easily reclassified elsewhere. As generalities, they allow exceptions.

Sources: A reconstruction from Ch.10 of *FCD*, based on *FHC*

Table 4.15. The Duality of Oppression in Existential Dialectics: Oppression and Self-Oppression (Part I)

- **From the Same to the Others and Itself**
  - —The Oppression by the Same against the Others
    - *By way of downgrading differences*
      - –Ex: on judiciary caprice for corporate crimes (*Sec.2.2.1.2.1*)
      - –Ex: on the deceptive politics of liberation (*Sec.3.5*)
      - –Ex: on the humanitarian mystique (*Sec.4.4*)
      - –Ex: on the fad of emotional intelligence (*Sec.5.3*)
    - *By way of accentuating differences*
      - –Ex: on the legal sophistry of self-defense (*Sec.2.3*)
      - –Ex: on the legal semantics of proportionality (*Sec.2.4*)
      - –Ex: on the tricky politics of external threat (*Sec.3.4*)
      - –Ex: on the appeal of the Far Right for democracy (*Sec.5.4*)
      - –Ex: on the democratic axis of evil (*Sec.5.5*)
      - –Ex: on the democratic way of brutality and revenge (*Sec.5.6*)
      - –Ex: on democratic autocracy (*Sec.6.4*)
  - —The Oppression by the Same against Itself
    - *By way of downgrading differences*
      - –Ex: on the politics of fear (*Sec.2.2*)
      - –Ex: on the trickery of compassionate conservatism (*Sec.3.2*)
      - –Ex: on the deceptive politics of patriotism (*Sec.3.3*)
    - *By way of accentuating differences*
      - –Ex: on the caprice of due process on domestic suspects (*Sec.2.2*)
      - –Ex: on the false security/freedom dilemma (*Sec.6.5.2*)

(continued on next page)

Table 4.15. The Duality of Oppression in Existential Dialectics: Oppression and Self-Oppression (Part II)

- **From the Others to the Same and Themselves**
  - —The Oppression by the Others against the Same
    - *By way of downgrading differences*
      - –Ex: on judiciary caprice in the reverse direction (*Sec.2.2.1.2.2*)
      - –Ex: on equal pay (*Sec.6.2.1.1*)
      - –Ex: on equal representation (*Sec.6.2.1.2*)
      - –Ex: on affirmative action program (*Sec.6.3.1.1*)
      - –Ex: on same-sex marriage (*Sec.6.3.1.2*)
    - *By way of accentuating differences*
      - –Ex: on sexual harassment (*Sec.6.2.2.1*)
      - –Ex: on physical violence (*Sec.6.2.2.2*)
      - –Ex: on sexual exploitation (*Sec.6.2.2.3*)
  - —The Oppression by the Others against Themselves
    - *By way of downgrading differences*
      - –Ex: on the reverse-class mystique (*Sec.4.2*)
      - –Ex: on the reverse-black mystique (*Sec.4.3*)
      - –Ex: on self-discrimination by downgrading (*Sec.6.3.2.2*)
    - *By way of accentuating differences*
      - –Ex: on self-discrimination by accentuating (*Sec.6.3.2.1*)

Notes: The examples are solely illustrative (not exhaustive), nor are they mutually exclusive. As generalities, they allow exceptions. Also, both forms of oppression co-exist in all of the examples, so the listing of them are only meant in a relative, not absolute, sense.
Source: A summary of the sections (as cited) in Chs.2-6 of *BDPD*. See text for more info and references.

Table 4.16. The Structure of Existential Dialectics I: The Freedom/Unfreedom and Equality/Inequality Dialectics

• Each freedom and equality produces its own unfreedom and inequality, regardless of whether the pair occurs in political society (with the nation-state), in civil society (with some autonomy from the state), or elsewhere (e.g., in the private sphere of individual homes)—and regardless of whether freedom and equality are understood as "negative" or "positive."

• Oppression is dualistic, as much by the Same against the Others and itself, as by the Others against the Same and themselves.

• Both forms of oppression and self-oppression can be achieved by way of downgrading differences (between the Same and the Others) and of accentuating them.

• The relationships are relatively asymmetric between the Same and the Others and relatively symmetric within them. This is true, even when the Same can be relatively asymmetric towards itself in self-oppression, just as the Others can be likewise towards themselves.

• Symmetry and asymmetry change over time, with ever new players, new causes, and new forms.

Notes: The examples in each category are solely illustrative (not exhaustive) nor necessarily mutually exclusive, and the comparison is relative (not absolute). As generalities, they allow exceptions. "Negative" freedom is freedom "from" (e.g., freedom from poverty), whereas "positive" freedom is freedom "to" (e.g., freedom to the state of enlightenment). "Negative" equality is "procedural" equality (e.g., equality of opportunity), while "positive" equality is "substantive" equality (e.g., equality of outcome). Existential dialectics impose constraints on freedom and equality in democracy, non-democracy, and post-democracy. There is no utopia, in the end; even should there be one, dystopia would exist within it.
Sources: From *Table 1.5* of *BDPD*—and also from *FHC*, *FCD*, and *FPHC*

Table 4.17. The Structure of Existential Dialectics II: The Wealth/Poverty Dialectics

• There is no wealth without poverty, just as there is no poverty without wealth.

• The wealth/poverty dialectics occurs in the realms of having, belonging, and being, in relation to the material, relational, and spiritual.

• The wealth/poverty dialectics also expresses itself at the multiple levels of analysis in accordance to methodological holism, be they about the micro-physical, the chemical, the biological, the psychological, the organizational, the institutional, the structural, the systemic, the cultural, and the cosmological.

• The wealth/poverty dialectics is a different manifestation of existential dialectics in general, subject to the principles in its logic of ontology—just as the freedom/unfreedom and equality/inequality dialectics are likewise.

• There is no economic utopia, in the end; even should there be one, dystopia would exist within it.

Notes: The main points here are solely illustrative (not exhaustive) nor necessarily mutually exclusive, and the comparison is relative (not absolute). As generalities, they allow exceptions.
Sources: From *BCPC*. See also *FCD* and *FHC*.

Table 4.18. The Structure of Existential Dialectics III: The Civilization/Barbarity Dialectics

• There is no civilization without barbarity.

• The civilization/barbarity dialectics applies in the four civilizing processes (e.g., the rationalizing process, the pacifying process, the stewardizing process, and the subliming process).

• The civilization/barbarity dialectics is another (different) manifestation of existential dialectics in general, subject to the principles in its logic of ontology—just as the freedom/unfreedom and equality/inequality dialectics and the wealth/poverty dialectics are likewise.

• There is no utopia, in the end; even should there be one, dystopia would exist within it.

Notes: The main points here are solely illustrative (not exhaustive) nor necessarily mutually exclusive, and the comparison is relative (not absolute). As generalities, they allow exceptions.
Sources: From *BCIV*. See also *FCD*, *FHC*, and *BDPD*.

Table 4.19. The Double Sides of Virtual Organizations (Part I)

- **Psychological**
  —Ex: virtual psychosis
  —Ex: impersonality and loneliness in quaternary social relations

- **Organizational**
  —Ex: the race for power and interests
  —Ex: the world of unequal successes
  —Ex: the bureaucratic life of its own
  —Ex: lesser accountability and transparency

- **Economic**
  —Ex: the economic divides
  —Ex: the erosive impact of commercialization

- **Political**
  —Ex: the anti-authoritarian myth
  —Ex: different power struggles among groups (e.g., Conservative, Reformist, Radical)

- **Structural**
  —Ex: different localities (e.g., climates, scenery)
  —Ex: divided domains (e.g., core, peripheral, sub-peripheral)
  —Ex: substitution effect of social capital

- **Systemic**
  —Ex: uneven advances in transportation and communications, and geographical migration

- **Cultural**
  —Ex: conflicting civilizational fabrics (e.g., Confucian, Islamic, Western)
  —Ex: the bias of the Liberal Democratic agenda
  —Ex: the addiction to moral fanaticism

(continued on next page)

Table 4.19. The Double Sides of Virtual Organizations (Part II)

Notes: These categories and examples are solely illustrative (not exhaustive), and some of the items can be reclassified somewhere else. Nor are they mutually exclusive. Since they are generalities, exceptions are expected.
Source: From Ch.7 of *FCD*

Table 4.20. Beyond the World of Titans, and the Remaking of the World Order

- **Hyper-Empires**
  —Ex: The Chinese Union
  —Ex: The Indian Union

- **Meso-Empires**
  —Ex: The European Union
  —Ex: The North American Union

- **Micro-Empires**
  —Ex: The Latin American Union
  —Ex: The Middle Eastern Union

- **The Rest of the World**
  —Odd Powers
    - Ex: Japan
    - Ex: Russia

  —The Poor Club
    - Ex: The African Union

  —Ambivalent Regions
    - Ex: Southeast Asia
    - Ex: Oceania
    - Ex: South Asia
    - Ex: Central Asia
    - Ex: Southern/Eastern Europe
    - Ex: North Africa
    - Ex: Central America
    - Ex: Others (e.g., the Korean peninsula)

Source: A summary of Chs.2-5 (of *BWT*)

Table 4.21. The Origins of Authoritarian Liberal Democracy

- **The Geopower of Nature (Ch.4 of *ALD*)**
  —Ex: Power Character and Geographical Strategy
  —Ex: Living Space and Territorial Expansion
  —Ex: Strategic Heartland and Containment

- **The Biopsychology of the Mind (Ch.5 of *ALD*)**
  —Ex: The Bell Curve and Mass Intelligence
  —Ex: Group Analysis and Mass Knowledge
  —Ex: Groupthink and Elite (Mis)calculation

- **The Disciplinary Control of Society (Ch.6 of *ALD*)**
  —Ex: Social Organizations of Ruthlessness
  —Ex: Social Institutions of Greed
  —Ex: Social Structure of Exclusion
  —Ex: Social Systems of Violence

- **The Molding Force of Culture (Ch.7 of *ALD*)**
  —Ex: The Tradition of Conquest
  —Ex: The Rationalization of Unreason

Notes: The examples in each category are solely illustrative (not exhaustive), and they are also relative (not absolute), nor are they mutually exclusive. As generalities, they allow exceptions.
Source: A summary of Chs.4-7 of *ALD*

Table 4.22. The Theory of Post-Democracy I: The Priority of Freedom over Equality (Part I)

- **Differences**
  - —*For the aggressive Lions (the strong Elitists)*
    - •Setting up rank distinctions among unequals (e.g., between inferior humans and superior post-humans, or later among inferior post-humans and superior ones, relatively speaking)
    - •Yearning for being not only distinguished from unequals, but also the first among equals (the best of the very best)
    - •Soul-searching for a high spiritual culture (not the trashy one for the masses). Mass culture is a dirty joke for them.
  - —*For the manipulative Foxes (the weak Counter-Elitists)*
    - •Seeking a gentle hegemony by way of more communitarian concerns (for inferior humans and, later, inferior post-humans)
    - •Being more sympathetic to less formal-legalistic institutions and values

- **Similarities**
  - —*For both Lions and Foxes*
    - •Exploring different spheres of non-human consciousness in the cosmos (something vastly superior than the human one)
    - •Recognizing the democratic illusions (e.g., no freedom without unfreedom, no equality without inequality, or simply no justice without injustice, and vice versa)

(continued on next page)

Table 4.22. The Theory of Post-Democracy I: The Priority of Freedom over Equality (Part II)

Notes: The two callings and examples in each category are solely illustrative (not exhaustive), since there will be many different post-human value ideals in the distant future of post-human civilization. The comparison is also relative (not absolute) towards post-democracy, so this is not just a version of free-market democracy (nor Fascism/Nazism, as shown in the table later on democracy, non-democracy, and post-democracy). Nor are they mutually exclusive. As generalities, they allow exceptions. And the specific forms of post-human post-democratic ideals need to be further developed in future after-postmodern history, as they will be different from the ones we now know. The point here is to solely give an extremely rough picture of a small part of the world to come that we have never known.
Source: From Ch.10 of *FCD*. Refer to text for more info and references.

Table 4.23. The Theory of Post-Democracy II: The Priority of Equality over Freedom

- **Hybrid Versions of**
  —Ex: the Trans-Feminine Calling
  —Ex: the Trans-Sinitic Calling
  —Ex: the Trans-Islamic Calling
  —Ex: the Trans-Outerspace Calling

- **Qualifications**
  —These four versions of post-capitalist value ideals need not automatically be post-democratic, just as capitalism does not necessarily mean democracy. They are two different entities—though closely related.
  —But up to a certain threshold of elevating equality at the farther expense of freedom, the democratic ideals will be overcome and cease to exist.
  —The overcome will not be socialist or communist, but post-democratic with no freedom without unfreedom and no equality without inequality, subject to the constraints of existential dialectics.

Notes: The callings are solely illustrative (not exhaustive), since there will be many different post-human value ideals in the distant future of post-human lifeforms. The comparison is also relative (not absolute), nor are they mutually exclusive. As generalities, they allow exceptions. And the specific forms of post-human post-democratic ideals need to be further developed in future after-postmodern history, as they will be different from the ones we now know. The point here is to solely give an extremely rough picture of a small part of the world to come that we have never known.
Source: From Ch.10 of *FCD*. Refer to text for more info and references.

Table 4.24. The Theory of Post-Democracy III: The Transcendence of Freedom and Equality (Part I)

- **Transcending Freedom in Floating Existence**
  —*Freedom*: seeking an ultimate elimination of the body. Being without the body. The aim is to transcend freedom in the end into a metaphysical state (i.e., beyond the physique).
  —*Unfreedom*: yet facing difficult trade-offs. The sacrifice of bodily existence and its joyfulness. An eternal boredom in floating existence in dark deep space, though with alternative pleasures. There is no free lunch even in the state of transcending freedom.

- **Transcending Equality in the Rivalry of Cosmic Hegemony**
  —*Inequality*: competing to outlast other lifeforms in floating existence, or just marginalizing them for one's hegemonic expansiveness in the rest of the cosmos (and even beyond). Universalism is only for the mediocre.
  —*Equality*: accepting only those of one's rank as equal partners in the vast spacetime for cosmic supremacy. Even here, the aim is to transcend equality into a metaphysical state.

(continued on next page)

Table 4.24. The Theory of Post-Democracy III: The Transcendence of Freedom and Equality (Part II)

Notes: Do not confuse this transcendence of freedom and equality (as one version of post-democracy) with the naïve temptation to transcend the freedom/unfreedom and equality/inequality dialectics. Existential dialectics hold true for freedom and equality in all cultures and societies—past, present, or future (i.e., democracy, non-democracy, and post-democracy), regardless of whether freedom and equality are conventionally understood as "negative" or "positive."
Also, the two features and examples in each are solely illustrative (not exhaustive), since there will be many different post-human value ideals in the distant future of post-human lifeforms. The comparison is also relative (not absolute), nor are they mutually exclusive. As generalities, they allow exceptions. And the specific forms of post-human ideals even for these radically alien floating lifeforms (and others unknown to us) need to be further developed in future after-postmodern history, as they will likely be different from the ones herein illustrated. The point here is to solely give a very rough picture of a small part of the extremely alien world to come that we have never known.
Source: From Ch.10 of *FCD*. Refer to text for more info and references.

Table 4.25. Democracy, Non-Democracy, and Post-Democracy (Part I)

• **Democracy**

—*Theoretical Constructs*

- •The pursuit of freedom and equality (in various degrees), regardless of whether freedom and equality can be understood as "negative" or "positive"
  - (1) more equality than freedom: The relative priority of the good over the right
  - (2) more freedom than equality: The relative priority of the right over the good

—*Types*

- •Only (1): Different versions of communitarian moral universalism
- •Only (2): Different versions of liberal moral universalism
- •(1) or (2): Different versions of anarchic (non-nation-state) moral universalism
- •(1) or (2): Different versions of postmodern moral localism

• **Non-Democracy**

—*Theoretical Constructs*

- •The focus on (1') equality or (2') freedom, but not both, regardless of whether freedom and equality can be understood as "negative" or "positive"

—*Types*

- •Only (1'): Different versions on the Far Left (e.g., Stalinism, Robespierrianism)
- •Only (2'): Different versions on the Far Right (e.g., Nazism, absolute monarchism)

(continued on next page)

Table 4.25. Democracy, Non-Democracy, and Post-Democracy (Part II)

- **Post-Democracy**
  - —*Theoretical Constructs*
    - •The priority of (1'') equality over freedom, or (2'') freedom over equality, or (3'') the transcendence of freedom and equality, regardless of whether freedom and equality are "negative" or "positive." In degree, (1'') or (2'') is less than (1') or (2') but more than (1) or (2)—respectively.
    - •Like democracy and non-democracy, post-democracy is also subject to the freedom/unfreedom and equality/inequality dialectics (or existential dialectics in general). Unlike them, post-democracy acknowledges the constraints of existential dialectics and no longer value freedom and equality as sacred virtues. There is no utopia, in the end; even were there one, dystopia would exist within it.
  - —*Types*
    - •(1''): Different versions of trans-Sinitic value ideals
    - •(1''): Different versions of trans-feminine value ideals
    - •(1''): Different versions of trans-Islamic value ideals
    - •(1''): Different versions of trans-outerspace value ideals
    - •(2''): Different versions of post-human elitist value ideals
    - •(3''): Different versions of the value ideals of floating consciousness (etc.)

(continued on next page)

Table 4.25. Democracy, Non-Democracy, and Post-Democracy (Part III)

Notes: The examples are solely illustrative (not exhaustive), nor are they mutually exclusive. As generalities, they allow exceptions. "Negative" freedom is freedom "from" (e.g., freedom from poverty), whereas "positive" freedom is freedom "to" (e.g., freedom to the state of enlightenment). "Negative" equality is "procedural" equality (e.g., equality of opportunity), while "positive" equality is "substantive" equality (e.g., equality of outcome). Existential dialectics impose constraints on freedom and equality in democracy, non-democracy, and post-democracy, regardless of whether freedom and equality can be understood as "negative" or "positive" in conventional discourse. Therefore, do not confuse the transcendence of freedom and equality in (3'') with the naïve temptation to transcend existential dialectics. There is no utopia, in the end; even should there be one, it would not exist without dystopia embedded within it.
Sources: A summary, based on my previous works, especially Ch.5 of *FHC*, Chs.5-10 of *FCD*, Chs.2-4 of *FPHC*, and Chs.1 & 7 of *BDPD*. The reader should consult the books for more analysis, as this is only a summary here.

Table 4.26. Multiple Causes of the Emergence of Post-Democracy (Part I)

- **At the Micro-Physical Level**
  —Ex: intelligent life without the human physical-chemical system
  —Sources: Ch.7 of *FHC*; Chs.9-10 of *FCD*; Ch.1 of *FPHC*

- **At the Chemical Level**
  —Ex: space radiation and toxins
  —Sources: Ch.7 of *FHC*; Chs.9-10 of *FCD*

- **At the Bio-Psychological Level**
  —Ex: exo-biological evolution in deep space
  —Ex: genetic engineering of new beings
  —Ex: limits of cognitive partiality
  —Ex: illusions of emotional neutrality
  —Ex: human biological inequality
  —Ex: the rise of unfolding unconsciousness
  —Sources: Ch.2 & Chs.9-10 of *FCD*; Ch.7 of *FHC*; Ch.4 of *BCPC*; *FPHU*

- **At the Institutional Level**
  —Ex: the flawed logic of equality
  —Ex: the conflicting nature of governance
  —Sources: Ch. 5 of *FHC*; Chs. 6 & 10 of *FCD*; Ch. 3 of *FPHC*; Chs.2-5 of *BDPD*

- **At the Organizational Level**
  —Ex: e-civic alienation
  —Ex: the dark sides of formal-legalistic routines
  —Sources: Ch.3 of *FHC*; Ch.7 of *FCD*; Ch.3 of *FPHC*

- **At the Structural Level**
  —Ex: ever new forms of inequities, at home and abroad
  —Ex: the emergence of China, women, and Islam as major actors
  —Sources: Chs.5-6 of *FHC*; Chs.7, 9 & 10 of *FCD*; Chs.4-5 of *BDPD*

(continued on next page)

Table 4.26. Multiple Causes of the Emergence of Post-Democracy (Part II)

• **At the Cultural Level**
—Ex: freedom/unfreedom dialectics
—Ex: equality/inequality dialectics
—Ex: system fragmentation and integration
—Sources: Ch.5 of *FHC*; Chs. 3, 9 & 10 of *FCD*; Ch.4 of *FPHC*; Ch.1 of *BDPD*; Ch.4 of *BCPC*

• **At the Systemic Level**
—Ex: space habitats (in zero-gravity) and colonization
—Ex: ultra advanced future info systems
—Ex: qualitative demography
—Sources: Ch.7 of *FHC*; Chs.9 &10 of *FCD*

• **At the Cosmological Level**
—Ex: the colonization of multiverses
—Ex: the alteration of space-time and the creation of new matter-energy
—Ex: the expansion of floating consciousness
—Ex: the spread of hyper-spatial consciousness
—Sources: Ch.7 of *FHC*; Chs.9 &10 of *FCD*; Ch.4 of *FPHC*; *FPHST*

Notes: The examples in each category are solely illustrative (not exhaustive), and some of the items can be reclassified somewhere else. Nor are they always mutually exclusive. Since they are generalities, exceptions are expected.
Sources: Especially from *FHC*, *FCD*, *FPHC*, *BCPC*, *BDPD*, *FPHST*, and *FPHU*. See also other books and my perspectives on civilizational holism.

Table 4.27. Some Clarifications about Post-Capitalism and Post-Democracy (Part I)

• The prefix "trans-" in the first category of post-capitalism (with its four versions) refers to something "going beyond" (not "uniting" or "combining"). Ex: *Sec.10.3.3* of *FCD*; *Sec.2.4* & *Sec.4.4* of *FPHC*; *Sec.7.2* of *BCPC*

• Such terms like "post-democracy," "post-capitalism," "post-human elitist," "trans-feminine calling," and the like as used in my works are more for our current intellectual convenience than to the liking of future humans and post-humans, who will surely invent more tasteful neologisms to call their own eras, entities, and everything else, for that matter. But the didactic point here is to use the terms to foretell what the future might be like, not that its eras and entities must be called so exactly and permanently.
Ex: *Sec.11.1* of *FCD*; *Sec.7.2* of *BCPC*

• The four versions in the first category of post-capitalist value ideals need not automatically be post-democratic, just as capitalism does not necessarily mean democracy. They are two different entities—though closely related. But up to a certain threshold of elevating equality at the farther expense of freedom, the democratic ideals will be overcome and cease to exist. The same is true for the post-human elitist calling in the second category of post-capitalism in relation to post-democracy, depending on the extent to which freedom is elevated at the expense of equality.
Ex: *Sec.10.4.3.3* of *FCD*; *Table 3.9* of *FPHC*; *Table 7.6* of *BDPD*

• The comparison in each of the three realms of existence in all forms of post-capitalism is not absolute, but relative. Examples include "communal" vs. "individualistic," and the like.
Ex: Notes in *Table 10.8*, *Table 10.9*, *Table 10.10*, & *Table 10.11* of *FCD*; Chs.2-4 of *FPHC*; *Sec.7.2* of *BCPC*

(continued on next page)

Table 4.27. Some Clarifications about Post-Capitalism and Post-Democracy (Part II)

• The emergence of post-capitalism (and post-democracy, for that matter) has multiple causes (to not be reduced to one or only a few).
Ex: Ch.10 of *FCD*, Chs.2-4 of *FPHC*; *Sec.1.3* & *Sec.7.2* of *BCPC* (or *Table 1.8* & *Table 7.11*)

• The specific forms of post-capitalism (and post-democracy, for that matter) need to be further developed in future after-postmodern history, as they will be different from the ones we now know. The point here is to solely give an extremely rough sketch of a world to come that we have never known.
Ex: *Sec.10.3.3* & *Sec.10.4.3.3* of *FCD*; *Table 10.14* & *Table 10.15* of *FCD*; *Sec.7.2* of *BCPC*

• All forms of post-capitalism are not part of a "teleological law," but of "historical trends" only. The same is also true for all forms of post-democracy.
Ex: *Sec.7.1* of *FHC*; *Sec.9.5.3.2* & *Sec.10.3.4.2* of *FCD*; *Sec.7.2* of *BCPC*

• Post-capitalism is not better than capitalism in an "absolute" sense but only fits in better, on the basis of the historical contingency of culture, society, nature, and the mind in some future eras. The same is true for post-democracy in relation to democracy. The term "better" is historically relative.
Ex: *Sec.10.3.3* of *FCD*; *Sec.1.7* of *BDPD*; *Sec.1.5* of *BCPC*

(continued on next page)

Table 4.27. Some Clarifications about Post-Capitalism and Post-Democracy (Part III)

• All forms of post-capitalism and post-democracy are subject to the constraints of existential dialectics. In the process, the dialectic direction is to go beyond the conventional "either-or" dichotomies (e.g., freedom vs. unfreedom, equality vs. inequality, freedom vs. equality, individuality vs. communality, spirituality vs. materiality, formal legalism vs. informal legalism, etc.). As is true in post-civilization, to go beyond the dichotomies is to acknowledge the co-existence of both in each dichotomy, although the degree of scaling one over the over varies from case to case (e.g., the theory of post-capitalism I, the theory of post-capitalism II, the theory of post-democracy I, the theory of post-democracy II, etc.)—but is not to be extreme in largely favoring one over the other, *on average* (all things considered). There is no utopia to be had in the end; even should there be one, dystopia would exist within it.
Ex: Ch.5 of *FHC*; *Sec.10.4.4.2* of *FCD*; *Sec.1.5* of *BDPD*; *Sec.1.3* of *BCPC*; *BCIV*

• All forms of post-capitalism, however different from each other though they are, share one common feature, in that they all inspire for a higher spiritual culture. The same is also true for post-democracy.
Ex: *Sec.10.3*, *Sec.10.4* & *Sec.10.5* of *FCD*; *Chs.2-4* of *FPHC*; *Sec.7.2* of *BCPC*

• All forms of post-capitalism try to avoid the excess in capitalist consumerism by favoring more basic than artificial needs in having, but the quality and quantity of these "basic" needs will be measured by future standards, not by our current ones. Standards are historically relative.
Ex: *Sec.10.3*, *Sec.10.4* & *Sec.10.5* of *FCD*; Ch.2 of *FPHC*; *Sec.7.2* of *BCPC*

(continued on next page)

Table 4.27. Some Clarifications about Post-Capitalism and Post-Democracy (Part IV)

• All forms of post-capitalism make use of a different degree of political authority with advanced info systems in future history and strives for higher spiritual cultures (especially in the post-human age), while acknowledging the constraints of existential dialectics and no longer valuing free market (as in capitalism) and economic control (as in non-capitalism) as sacred virtues.
Ex: *Sec.10.3.4.2*, *Sec.10.3*, *Sec.10.4* & *Sec.10.5* of *FCD*; Chs.2-4 of *FPHC*; *Sec.1.5* of *BDPD*; *Sec.7.2* of *BCPC*

Notes:: The main points here are solely illustrative (not exhaustive) nor necessarily mutually exclusive, and the comparison is relative (not absolute). As generalities, they allow exceptions. The sections as cited are only illustrative (not exhaustive).
Sources: From *FHC*, *FCD*, *FPHC*, and *BDPD*

Table 4.28. The Theory of Post-Capitalism I.1: By Group— Ex: Spiritual/Communal in the Trans-Feminine Calling

- **More Communal Than Individual**
  —*Sharing*: learning from others, as different ideas mutually enrich
  —*Cooperative*: encouraging a sense of shared leadership and teamwork

- **More Informal-Legalistic Than Formal-Legalistic**
  —*Specific*: listening more from the heart than from the head, to know a person as a concrete, not as an abstract, unit
  —*Affective*: thinking and acting with others on a more affective tone. Business can mix with an emotional touch.
  —*Ascriptive*: hiring (or firing) can be done on the basis of merit (or lack of it), but deep solidarity (sisterhood) is important too.
  —*Particularistic*: making decisions on the basis of cost-benefit analysis, but a given group relationship is vital

- **More Spiritual Than Secular**
  —*Long-Term Looking*: sharing for a long-term relationship (e.g., love, friendship), not just for a short-term gain
  —*Loving/Caring*: showing compassion for the sufferings of others, without quickly blaming and pre-judging
  —*Respectful*: showing acceptance about others' feelings (and thoughts)

Notes: The categories and examples are solely illustrative, since there can be different versions, and the comparison is relative (not absolute), nor are they mutually exclusive. As generalities, they allow exceptions. The specific forms of the trans-feminine version need to be further developed in future after-postmodern history, as they will be different from the ones we now know, since the prefix "trans-" here means going beyond or deconstructing the feminine values, while using them as the inspirational point at the beginning.
Source: From Ch.10 of *FCD*. Refer to text for more info and references.

Table 4.29. The Theory of Post-Capitalism I.2: By Nation-State— Ex: Spiritual/Communal in the Trans-Sinitic Calling

- **More Communal Than Individualistic**
  —*Centralized*: being more top-down in management
  —*Collective*: encouraging more group cooperation
  —*Social*: investing in trust and connection

- **More Informal-Legalistic Than Formal-Legalistic**
  —*Specific*: knowing more of those related or connected
  —*Affective*: behaving in a paternalistic, hierarchical way
  —*Ascriptive*: favoring family members and those related
  —*Particularistic*: building connection (guanxi) as imperative

- **More Spiritual Than Secular**
  —*Expansionist*: diffusing civilizational values (e.g., the superiority complex of civilizationalism)
  —*Holistic*: synthesizing things into a panoramic horizon
  —*Historical*: learning from the lessons of the ancient past
  —*Respectful*: deferential to elders and superiors

Notes: The categories and examples are solely illustrative, since there can be different versions, and the comparison is relative (not absolute), nor are they mutually exclusive. As generalities, they allow exceptions. The specific forms of the trans-Sinitic version need to be further developed in future after-postmodern history, as they will be different from the ones we now know, since the prefix "trans-" here means going beyond or deconstructing the Sinitic values, while using them as the inspirational point at the beginning.
Source: From Ch.10 of *FCD*. Refer to text for more info and references.

Table 4.30. The Theory of Post-Capitalism I.3: By Region— Ex: Spiritual/Communal in the Trans-Islamic Calling

- **More Communal Than Individualistic**
  —*Collective*: building the webs of relationships to bind individuals
  —*Sharing*: cultivating the established "wisdom" through common experience
  —*Cooperative*: stressing harmony, solidarity, and commonality

- **More Informal-Legalistic Than Formal-Legalistic**
  —*Specific*: making efforts to know well the participants (family and larger community) in matters of common concern
  —*Affective*: mixing work with language and ritual on explicit religious (Islamic) ideals, texts, stories, and examples
  —*Ascriptive*: privileging local history and custom on relationships among kinship groups
  —*Particularistic*: preferring an unbiased insider with ongoing connections to all parties

- **More Spiritual Than Secular**
  —*Historical*: learning from the lessons of the past as a source of stability and guidance
  —*Deferential*: showing respect for age, experience, status, and leadership in communal affairs
  —*Honorable*: emphasizing face, dignity, prestige, and fairness
  —*Compassionate*: giving mercy and charity ("Zahah") to others

Notes: The categories and examples are solely illustrative (not exhaustive), and the comparison is relative (not absolute), nor are they mutually exclusive. As generalities, they allow exceptions. The specific forms of the trans-Islamic version need to be further developed in future after-postmodern history, as they will be different from the ones we now know, since the prefix "trans-" here means going beyond or deconstructing the Islamic values, while using them as the inspirational point at the beginning.
Sources: From Ch.10 of *FCD*. Refer to text for more info and references, especially from the works by George Irani (2000) and C. Murphy (September 19, 2001).

**Table 4.31. The Theory of Post-Capitalism I.4:
By Universe—
Ex: Spiritual/Communal in the Trans-Outerspace Calling**

- **More Communal Than Individual**
  —*Cooperative*: requiring teamwork in small space habitats
  —*Sharing*: learning from, and enjoying being with, each other in a small group in outer space

- **More Informal-Legalistic Than Formal-Legalistic**
  —*Specific*: knowing more about each other to facilitate living and working together in space, both as fellow astronauts and space-mates
  —*Affective*: being friendly and social to each other as vital to working and living in small space quarters
  —*Ascriptive*: nurturing comaraderie among fellow astronauts as if they are family members over time
  —*Particularistic*: building work relationship with enduring memory in a space mission

- **More Spiritual Than Secular**
  —*Long-Term*: looking beyond selfish materialistic concerns in a precarious space environment with potential life or death
  —*Loving/Caring*: cultivating deep bondage for the success of a long term space mission
  —*Transcendent*: searching for life meaning in outer space

Notes: The calling and examples in each category are solely illustrative (not exhaustive), since there will be many different outer-space value ideals in the distant future of space colonization. The comparison is also relative (not absolute), nor are they mutually exclusive. As generalities, they allow exceptions. And the specific forms of trans-outer-space calling need to be further developed in future after-postmodern history, as they will be different from the ones we now know, since the prefix "trans-" here means going beyond or deconstructing the current outer-space values, while using them as the inspirational point at the beginning. The point here is to solely give an extremely rough picture of a small part of the world to come that we still do not know much about.
Source: From Ch.10 of *FCD*. Refer to text for more info and references.

Table 4.32. The Theory of Post-Capitalism II: Spiritual/Individualistic in the Post-Human Elitist Calling (Part I)

- **More Individualistic Than Communal**
  —Setting up rank distinctions among unequals (e.g., between inferior humans and superior post-humans, or later among inferior post-humans and superior ones, relatively speaking)
  —Yearning for being not only distinguished from unequals, but also the first among equals (the best of the very best)
  —Recognizing the constraints of equality/inequality dialectics (or existential dialectics in general)

- **More Spiritual Than Secular**
  —Soul-searching for a high spiritual culture (not the trashy one for the masses). Mass culture is a dirty joke for them.
  —Exploring different spheres of non-human consciousness in the cosmos (something vastly superior than the human one)
  —Recognizing the constraints of freedom/unfreedom dialectics (or existential dialectics in general)

- **Qualifications**
  —Although post-human elitist post-democracy is comparable to post-human elitist post-capitalism in some respects, the former does not necessarily imply the latter (post-human elitist post-capitalism), just as democracy does not have to entail capitalism. They are two different (though related) entities.
  —But up to a certain threshold of incorporating government intervention with advanced info systems in future civilizations for higher spiritual concerns at the expense of the free market and materialist pursuit, the capitalist ideal will be overcome.
  —The overcome will not be Fascist or feudalistic, but post-capitalist, subject to the constraints of existential dialectics.

(continued on next page)

Table 4.32. The Theory of Post-Capitalism II: Spiritual/Individualistic in the Post-Human Elitist Calling (Part II)

Notes: The calling and examples in each category are solely illustrative (not exhaustive). The comparison is also relative (not absolute), nor are they mutually exclusive. As generalities, they allow exceptions. And the specific forms of post-human elitist post-capitalism need to be further developed in future after-postmodern history, as they will be different from the ones we now know, while using them as the inspirational point at the beginning. The point here is to solely give an extremely rough picture of a small part of the world to come that we still do not know much about.
Sources: From Ch.10 of *FCD* (and also *FPHC*, *BDPD*, and *BCPC*). Refer to the text for more info and references.

Table 4.33. Capitalism, Non-Capitalism, and Post-Capitalism (Part I)

- **Capitalism**
  - —*Theoretical Constructs*
    - •Allocation of scarce resources among alternative wants largely by free market for competition (whose characteristics in its ideal form include, for instance, no barrier to entry or exit, homogeneity, perfect information, a large number of buyers/sellers, and perfect factor mobility)
    - •More formal-legalistic than informal-legalistic, more individualistic than communal, and more material (secular) than spiritual
    - •Either (1) minimal government or (2) relatively active government
  - —*Types*
    - •Only (1): Different versions of market capitalism (e.g., the U.S.)
    - •Only (2): Different versions of welfare capitalism (e.g., Sweden)

- **Non-Capitalism**
  - —*Theoretical Constructs*
    - •Allocation of scarce resources among alternative wants mainly by political authority for policies (which can be regulative, redistributive, symbolic, and participatory)
    - •More informal-legalistic than formal-legalistic
    - •Either (1') more individualistic (for the elites), often (though not always) for material (secular) concerns, or (2') more communal (for the masses), often (though not always) for spiritual concerns
  - —*Types*
    - •Only (1'): Different versions on the Right (e.g., Fascist corporate-state economy for the glory of the new Rome, medieval lord-vassal-serf economy for the power of the feudalistic order)
    - •Only (2'): Different versions on the Left (e.g., Soviet command economy for the creation of the New Socialist Man)

(continued on next page)

Table 4.33. Capitalism, Non-Capitalism, and Post-Capitalism (Part II)

• **Post-Capitalism**

—*Theoretical Constructs*

- •Allocation of scarce resources among alternative wants largely by political authority with advanced info systems in future civilizations, subject to existential dialectics. In degree of allocating by authority, post-capitalism is more than capitalism but less than non-capitalism.
- •More spiritual than secular (material)
- •Either (1'') more individualistic or (2") more communal
- •Like capitalism and non-capitalism, post-capitalism is also subject to the freedom/unfreedom and equality/inequality dialectics (or existential dialectics in general). There is no utopia, in the end; even were there one, dystopia would exist within it.
- •Unlike capitalism and non-capitalism, post-capitalism makes use of a different degree of political authority with advanced info systems in future civilizations and strives for higher-spiritual cultures (especially in the post-human age), while acknowledging the constraints of existential dialectics and no longer valuing free market (as in capitalism) and economic control (as in non-capitalism) as sacred virtues.

—*Types*

- •Only (1"): Different versions of post-human elitist value ideals
- •Only (2''): Different versions of trans-Sinitic value ideals
- •Only (2''): Different versions of trans-feminine value ideals
- •Only (2''): Different versions of trans-Islamic value ideals
- •Only (2''): Different versions of trans-outerspace value ideals

---

(continued on next page)

Table 4.33. Capitalism, Non-Capitalism, and Post-Capitalism (Part III)

Notes: The calling and examples in each category are solely illustrative (not exhaustive). The comparison is also relative (not absolute), nor are they mutually exclusive. As generalities, they allow exceptions. And the specific forms of each calling need to be further developed in future after-postmodern history, as they will be different from the ones we now know, while using them as the inspirational point at the beginning. The point here is to solely give an extremely rough picture of a small part of the world to come that we still do not know much about.
Source: From Ch.10 of *FCD*. Refer to the text for more info and references.

Table 4.34. Multiple Causes of the Emergence of Post-Capitalism (Part I)

- **At the Micro-Physical Level**
  —Ex: intelligent life without the human physical-chemical system
  —Ex: mastering of quantum mechanics, electromagnetism, and other fields for the understanding of a broad range of anomalous experiences and the application for artificial intelligence for spiritual quest
  —Sources: Ch.7 of *FHC*; Chs.9-10 of *FCD*; Ch.1 of *FPHC*

- **At the Chemical Level**
  —Ex: space radiation and toxins
  —Sources: Ch.7 of *FHC*; Chs.9-10 of *FCD*

- **At the Bio-Psychological Level**
  —Ex: exo-biological evolution in deep space
  —Ex: genetic engineering of new beings
  —Ex: limits of human cognition
  —Ex: the rise of unfolding unconsciousness
  —Sources: Ch.2 & Chs.9-10 of *FCD*; Ch.7 of *FHC*; *FPHU*

- **At the Institutional Level**
  —Ex: the flawed logic of the free market
  —Ex: the need of a post-autistic economics
  —Sources: Ch.10 of *FCD*

- **At the Organizational Level**
  —Ex: the dark sides of formal-legalistic routines
  —Sources: Ch.3 of *FHC*; Ch.7 of *FCD*; Ch.3 of *FPHC*

- **At the Structural Level**
  —Ex: ever new forms of inequities, at home and abroad
  —Ex: the emergence of China, women, and Islam as major actors
  —Sources: Chs.5-6 of *FHC*; Chs.7, 9 & 10 of *FCD*; Chs.4-5 of *BDPD*

(continued on next page)

Table 4.34. Multiple Causes of the Emergence of Post-Capitalism (Part II)

- **At the Cultural Level**
  —Ex: freedom/unfreedom dialectics
  —Ex: equality/inequality dialectics
  —Sources: Ch.5 of *FHC*; Chs.3 & 10 of *FCD*; Ch.4 of *FPHC*; Ch.1 of *BDPD*

- **At the Systemic Level**
  —Ex: space habitats (in zero-gravity) and colonization
  —Ex: ultra advanced future info systems
  —Ex: qualitative demography
  —Sources: Ch.7 of *FHC*; Chs. 9 & 10 of *FCD*

- **At the Cosmological Level**
  —Ex: the colonization of multiverses
  —Ex: the alteration of space-time and the creation of new matter-energy
  —Ex: the expansion of floating consciousness
  —Ex: the spread of hyper-spatial consciousness
  —Sources: Ch.7 of *FHC*; Chs. 9 & 10 of *FCD*; Ch.4 of *FPHC*; *FPHST*

Notes: The examples in each category are solely illustrative (not exhaustive), and some of the items can be reclassified somewhere else. Nor are they always mutually exclusive. Since they are generalities, exceptions are expected.
Sources: Especially from *FHC*, *FCD*, *FPHC*, *BCPC*, *BDPD*, *FPHST*, and *FPHU*. See also other books and my perspectives on civilizational holism.

Table 4.35. The Theoretical Debate on Civilization

- **The Progressive Theory of Civilization**
  —*Thesis*: The "civilizing" process is "good," as opposed to the "barbarizing" process as something "bad," relatively speaking.
  —*Discourse*: Especially, though not exclusively, in the Enlightenment era and a bit before. Example: Thomas Hobbes—in that the tribes in primitive societies were "savages."

- **The Romantic Theory of Barbarity**
  —*Thesis*: The "civilizing" process is "bad," as opposed to the "barbarizing" process as something "good," relatively speaking.
  —*Discourse*: Especially, though not exclusively, in the Counter-Enlightenment circle. Example: Jean-Jacques Rousseau—in that civilization "corrupts" men, and the "savages" are in fact "noble."

- **The Moderate Theory of Civilization**
  —*Thesis*: The "civilizing" process is "good," but there is a price to pay, especially in systematic (compulsive) self-control.
  —*Discourse*: Especially, though not exclusively, in some late modern and postmodern circles. Example: Norbert Elias—in that social manners become more refined in the civilizing process, but self-control also becomes more systematic.

- **The Theory of Post-Civilization**
  —*Thesis*: The civilizing process is as evil and good as barbarity, and each cannot exist without the other, to be eventually superseded by post-civilization unto the post-human age.
  —*Discourse*: Proposed by Peter Baofu. See the rest of *BCIV* for more analysis.

Source: From *BCIV* on the theoretical debate

Table 4.36. No Freedom Without Unfreedom in the Civilizing Processes (Part I)

- **The Rationalizing Process (at the Level of Culture)**
  —if freer from the dominance of unreason (as in barbarism) in the civilizing process, then less free from the rationalizing process (be it in the form of the principle of either transcendence or immanence)
  —if freer from the principle of immanence in the rationalizing process, then less free from the inclination to commit terror in the name of reason and the relative underdevelopment of non-reason (e.g., in relation to yoga and meditation)
  —if freer from the principle of transcendence in the rationalizing process, then less free from the relative underdevelopment of reason (e.g., in relation to systematic methodology) and the occurrence of oppression in the name of non-reason

- **The Pacifying Process (at the Level of Society)**
  —if freer from the dominance of pillage (as in savagery) in the civilizing process, then less free from the pacifying process (be it in the form of external control or self-control)
  —if freer from self-control in the pacifying process, then less free from the temptation of expansionist oppression and rebellious mindset in external control
  —if freer from external control in the pacifying process, then less free from the gruesome psychological self-torture and conformism in self-control

- **The Stewardizing Process (at the Level of Nature)**
  —if freer from the dominance of nature (as in the state of nature) in the civilizing process, then less free from the stewardizing process (be it in the form of the stewardship of creation or the covenant with nature)
  —if freer from the stewardship of creation in the stewardizing process, then less free from material underdevelopment, relatively speaking, and spiritual exclusion in the covenant with nature
  —if freer from the covenant with nature in the stewardizing process, then less free from ecological degradation and spiritual disconnection from nature in the stewardship of creation

(continued on next page)

Table 4.36. No Freedom Without Unfreedom in the Civilizing Processes (Part II)

- **The Subliming Process (at the Level of the Mind)**
  —if freer from the dominance of spontaneity (as in the wild state of the mind) in the civilizing process, then less free from the subliming process, be it in the form of (cyclical-centric) self-refinement or (linear-centric) self-discipline
  —if freer from (cyclical-centric) self-refinement in the subliming process, then less free from the (linear-centric) self-regimen (as a form of neurosis)
  —if freer from (linear-centric) self-discipline in the subliming process, then less free from the (cyclical-centric) self-torture (equally as a form of neurosis)

Notes: The examples in each category are solely illustrative (not exhaustive), and the comparison is relative (not absolute), nor are they necessarily mutually exclusive. And some can be easily re-classified elsewhere. As generalities, they allow exceptions.
Sources: From *BCIV*. See also *FHC*, *FCD*, *FPHC*, *BDPD*, and *BCPC*.

Table 4.37. No Equality Without Inequality in the Civilizing Processes (Part I)

• **The Rationalizing Process (at the Level of Culture)**
—if more equal for the role of rationalization in the rationalizing process (of civilizational making), then less equal for that of mythicization (as in barbarism)
—if more equal for the principle of transcendence in (linear-centric) rationalizing process, then less equal for the principle of immanence
—if more equal for the principle of immanence in (cyclical-centric) rationalizing process, then less equal for the principle of transcendence

• **The Pacifying Process (at the Level of Society)**
—if more equal for pacification in civilizational making, then less equal for the institution of pillaging and others (as in savagery)
—if more equal for external control, relatively speaking, in pacifying process, then less equal for self-control
—if more equal for self-control, relatively speaking, in pacifying process, then less equal for external-control

• **The Stewardizing Process (at the Level of Nature)**
—if more equal for stewardship in the stewardizing process (of civilizational making), then less equal for reverent (submissive) existence (as in barbarism)
—if more equal for the stewardship of creation in (linear- centric) stewardizing process, then less equal for the (cyclical-centric) covenant with nature for harmonious co-existence
—if more equal for the (cyclical-centric) covenant with nature in the stewardizing process, then less equal for the (linear-centric) stewardship of nature for domination

---

(continued on next page)

Table 4.37. No Equality Without Inequality in the Civilizing Processes (Part II)

- **The Subliming Process (at the Level of the Mind)**
  —if more equal for the role of reason in the subliming process, then less equal for that of unreason (as in the natural state of wildness)
  —if more equal for the primacy of reason in (linear-centric) subliming process, then less equal for other faculties (e.g., intuition, existential feelings, and analogous thinking) in cyclical-centric one
  —if more equal for the exercise of other faculties (e.g., intuition, existential feelings, and analogous thinking) in cyclical-centric subliming process, then less equal for the role of reason in linear-centric counterpart

Notes: The examples in each category are solely illustrative (not exhaustive), and the comparison is relative (not absolute), nor are they mutually exclusive. And some can be easily reclassified else-where. As generalities, they allow exceptions.
Sources: From *BCIV*. See also *FHC*, *FCD*, *FPHC*, *BDPD*, and *BCPC*.

Table 4.38. Five Theses on Post-Civilization

• Post-civilization no longer treats civilization as good and barbarity as evil (relatively speaking), nor does it nostalgically regard barbarity as good and civilization as evil (relatively speaking again). Civilization is as evil and good as barbarity.

• Post-civilization also no longer accepts the dichotomy between civilization and barbarity. Civilization cannot exist without barbarity. It is no longer necessary to preserve civilization, any more than it is imperative to destroy barbarity. To go beyond civilization and barbarity is to acknowledge the co-existence of both, although the degree of scaling one over the over varies from case to case—but is not to be extreme in largely favoring one over the other, *on average* (subject to the constraints of existential dialectics).

• Post-civilization is thus subject to the constraints of existential dialectics. There is no freedom without unfreedom, and no equality without inequality, for instance. There will be no utopia; even should there be one, there would be dystopia embedded within it.

• Post-civilization will eventually replace civilization (as a form of life settlement), to be dominated by post-capitalist and post-democratic lifeforms here on earth and in deep space (besides other alien lifeforms that we have never known), unto the post-human age in multiverses. Those few post-humans who keep civilization will live in a "post-human civilization," while the rest (the majority), who choose post-civilization, will evolve towards the state of "post-human post-civilization." One therefore should not confuse "post-human civilization" with "post-human post-civilization," as the two are not the same.

• Post-civilization will confront psychosis as a primary problem in the culture of virtuality unto the post-human age, just as civilization has neurosis as a primary one of its own (although both neurosis and psychosis are major problems in both).

Notes: The comparison in each category is relative (not absolute), nor are they necessarily mutually exclusive. And some can be easily re-classified elsewhere. As generalities, they allow exceptions.
Sources: From *BCIV*. See also *FHC*, *FCD*, *FPHC*, *BDPD*, and *BCPC*.

Table 4.39. Barbarity, Civilization, and Post-Civilization

- **The Rationalizing Process (at the Level of Culture)**
  —*Barbarity*
    - More mythicizing than rationalizing, relatively speaking

  —*Civilization*
    - More rationalizing than mythicizing, relatively speaking

  —*Post-Civilization*
    - Beyond the dichotomy, subject to existential dialectics

- **The Pacifying Process (at the Level of Society)**
  —*Barbarity*
    - More pillaging than pacifying, relatively speaking

  —*Civilization*
    - More pacifying than pillaging, relatively speaking

  —*Post-Civilization*
    - Beyond the dichotomy, subject to existential dialectics

- **The Stewardizing Process (at the Level of Nature)**
  —*Barbarity*
    - More revering than stewardizing, relatively speaking

  —*Civilization*
    - More stewardizing than revering, relatively speaking

  —*Post-Civilization*
    - Beyond the dichotomy, subject to existential dialectics

- **The Subliming Process (at the Level of the Mind)**
  —*Barbarity*
    - More impulsing than subliming, relatively speaking

  —*Civilization*
    - More subliming than impulsing, relatively speaking

  —*Post-Civilization*
    - Beyond the dichotomy, subject to existential dialectics

Notes: The comparison in each category is relative (not absolute), nor are they necessarily mutually exclusive. And some can be easily re-classified elsewhere. As generalities, they allow exceptions.
Sources: From *BCIV*. See also *FHC*, *FCD*, *FPHC*, *BDPD*, and *BCPC*.

Table 4.40. Types of Super Civilization in the Cosmos (Part I)

- **Type I**
  —a civilization which gains control of and uses the total energy output "falling on its planet from its sun for interstellar communication" (or, in general, space colonization). For N. Kardashev, who proposed the first three types, human civilization is currently Type Zero (Type O), which is below even Type I, since its present energy consumption for all purposes, let alone for interstellar communication, is still 10,000 times less.

- **Type II**
  —a civilization which gains control of and uses directly the total energy output of its sun for interstellar communication (or, in general, space colonization).

- **Type III**
  —a civilization which gains control of and uses the total energy output of its galaxy for interstellar communication (or, in general, space colonization).

- **Type IV**
  —a civilization which gains control of and uses the total energy output of its cluster of galaxies for interstellar communication (or, in general, space colonization).

- **Type V**
  —a civilization which gains control of and uses the total energy output of its supercluster of galaxies for interstellar communica-tion (or, in general, space colonization).

(continued on next page)

Table 4.40. Types of Super Civilization in the Cosmos (Part II)

- **Type…n**
  —So continues the series in what I call the cyclical progression of hegemony in the cosmos and beyond.

Notes: The Russian astrophysicist Nikolai Kardashev proposed the first three types of super civilization in terms of total energy out-put for interstellar communication. (CSM 1979) I extend his argument further to propose Type IV, Type V, Type VI, and Type…n, in the context of my claim about the cyclical progression of he-gemony in the cosmos and beyond.
Sources: From *Table 9.4* of *FCD.* See *FHC*, *FCD*, and *FPHC* for more info.

Table 4.41. The Civilizational Project from Pre-Modernity to After-Postmodernity (Part I)

	Pre-Modern	*Modern*	*Postmodern*	*After-Postmodern*
Main narratives	•Sacralness •Courtliness •Vitalism •Animism	•Freedom •Equality •Fraternity	•Multiplicity •Hybridization	•Naked contingency •Cyclical progres-sion of hegemony
Main institutions	•Monarchy •Aristocracy •Feudalism •Holy order •Primitivism	•Capitalism •Liberalism •Socialism •Nazism •Fascism	•Capitalism •Liberalism •Postmodern politics of difference	•Post-Capitalism •Post-Democracy •Others
Main techno-logical and economic revolutions	•Agricultural	•Service •Industrial	•Informational	•Biological •Material •Energy •Space •Others

(continued on next page)

Table 4.41. The Civilizational Project from Pre-Modernity to After-Postmodernity (Part II)

	Pre-Modern	*Modern*	*Postmodern*	*After-Postmodern*
Main agents	•Males •Upper strata •Mini-states	•Males •Upper strata •Whites •Empires	•Males •Upper strata •Whites •Others •Supra-states •IO's	•Post-humans •Humans •Others
Main impacts	•Local	•Inter-national	•Global	•Outer-space •Multiverse
Main outcomes	•Towards moderntiy •Rise of linear- & cyclical-centric civiliza-tions	•Towards post-modernity •Dominance of linear-centric civilization	•Towards after-post-moderntiy •Linear-centric civiliza-tion in crisis	•Towards human (& maybe post-human) extinction •Rise of post-civiliza-tion, especially in post-human forms of space-time

Notes: The examples in each category are solely illustrative (not exhaustive) nor necessarily mutually exclusive, and the comparison is relative (not absolute). As generalities, they allow exceptions.
Sources: From *Table 10.16* of *FCD*—and also from *BCIV* on post-civilization (and *FPHST*)

Table 4.42. Civilizational Holism (Part I)

- **At the Micro-Physical Theoretical Level**
  —Ex: Mastering of quantum mechanics, electromagnetism, and other fields for the understanding of a broad range of anomalous experiences and theapplication for artificial intelligence (*Sec.1.4.1* of *FPHC*)

- **At the Chemical Theoretical Level**
  —Ex: Unprecedented expansion of (and violence to) the mind through ever new forms of drugs (and virtual technologies, for that matter) (Ch.9 of *FCD*)

- **At the Biological Theoretical Level**
  —Ex: Humans are not biologically equal, on the basis of race, gender, ethnicity, age, and whatnot. (*Sec.2.6* & Ch.10 of *FCD*; *BNN*) And post-humans will experience the same fate, in an even more amazing way.

- **At the Psychological Theoretical Level**
  —Ex: Human cognitive impartiality and emotional neutrality are quite limited. (*Secs.2.4-2.5* of *FCD*)
  —Ex: Rise of Floating Consciousness (Ch.10 of *FCD*; Chs.1 & 4 of *FPHC*) and Unfolding Unconsciousness (*FPHU*)

- **At the Organizational Theoretical Level**
  —Ex: Administrative colonization of deep space, with less legal-formalism in some corners. (Chs.9-10 of *FCD*)

- **At the Institutional Theoretical Level**
  —Ex: Both capitalism and democracy will not last, to be superseded by different versions of post-capitalism and post-democracy in after-postmodernity. (Ch.10 of *FCD*)

(continued on next page)

Table 4.42. Civilizational Holism (Part II)

- **At the Structural Theoretical Level**
  —Ex: Social stratification reappears in ever new forms, also with new causes and new players in the cyclical progression of hegemony. (Chs.8-10 of *FCD*)
  —Ex: The world of hyper-empires, and the union of the unions (*BWT*)

- **At the Systemic Theoretical Level**
  —Ex: Outerspace expansion: local → regional → global → solar → galactic → clustery → multiversal (Ch.9 of *FCD*)
  —Ex: Demographic transition: human extinction, and the rise of post-humans (e.g., cyborgs, thinking machines, thinking robots, genetically altered superior beings, floating consciousness, hyper-spatial consciousness) (Ch.4 of *FPHC*; Ch.10 of *FCD*; & Ch.7 of *FHC*)
  —Ex: New technological forces in material sciences, electronic and communication sciences, energy sciences, biosciences, manufacturing and engineering sciences, and space sciences (Ch.10 of *FCD* & Ch.7 of *FHC*)
  —Ex: Systematic dominance towards nature for space colonization (Chs.9-10 of *FCD*; Chs.2 & 7 of *FHC*)

- **At the Cultural Theoretical Level**
  —Ex: The post-human transcendence of freedom and equality (Ch.10 of *FCD*)
  —Ex: Methodological Holism (Ch.1 of *FCD*; Ch.1 of *FPHC*; *Sec.2.1* & *Sec.2.5* of *BCPC*)
  —Ex: The Evolution from Barbarity to Post-Civilization (*BCIV*)

- **At the Cosmological Theoretical Level**
  —Ex: Mastering of dark matter and dark energy, and the exploration of multiverses (Ch.4 of *FPHC*; Ch.10 of *FCD*; & Ch.7 of *FHC*)
  —Ex: Alternation of space-time (*FPHST*)
  —Ex: The emergence of hyper-spatial consciousness (*FPHC*)

(continued on next page)

Table 4.42. Civilizational Holism (Part III)

- **At Other Levels**
  —Ex: Historical: pre-modernity → modernity → postmodernity → after-postmodernity (human distinction, and the rise of post-humans, including floating consciousness, hyper-spatial consciousness, and unfolding unconsciousness) (Ch.7 of *FHC*; Ch.10 of *FCD*; *FPHC*; *FPHU*)

Notes: These examples are solely illustrative (not exhaustive), and some of the items can be reclassified somewhere else. Nor are they always mutually exclusive. Since they are generalities, exceptions are expected. And the comparison is relative, not absolute.
Sources: From *Table 5.1* of *FPHC*—with details from *FHC*, *FCD*, and the rest of my books. Not every aspect in each category as discussed in all my books are presented here, since there are too many issues. For more info, also consult the table on theories on civilizational holism and, of course, the books themselves.

Table 4.43. Theories on Civilizational Holism (Part I)

I. Theories in Relation to Nature

—*At the Chemical Theoretical Level*

• 62. Creational Theory of Chemistry (Peter Baofu) (*FPHCHEM*)

—*At the Macro-Physical (Cosmological) Theoretical Level*

• 61. Resettlement Theory of Geology (Peter Baofu) (*FPHGEOL*)

• 60. Theory of Post-Cosmology (Peter Baofu) (*BCOS*)

• 59. Theory of Hyper-Spatial Consciousness (Peter Baofu) (Ch.4 of *FPHC*; *FPHG*)

(• 57). Panoramic Theory of Transportation (Peter Baofu) (*FPHTRAN*)

(• 55). Multilateral Theory of Acoustics (Peter Baofu) (*FPHA*)

(• 54). Selective Theory of Geometry (Peter Baofu) (*FPHG*)

(• 53). Perspectival Theory of Space-Time (Peter Baofu) (*FPHST*)

(• 52). Dialectic Theory of Complexity (Peter Baofu) (*FC*)

(• 30). Theory of Floating Consciousness (Peter Baofu) (Ch.10 of *FCD*; Chs.1 & 4 of *FPHC*)

• 58. Theory of the Geopower of Nature (Ch.4 of *ALD*)—refined as the Interventive-Reshaping Theory of Geography (Peter Baofu) (*FPHGEOG*)

(continued on next page)

Table 4.43. Theories on Civilizational Holism (Part II)

I. Theories in Relation to Nature (*cont'd*)

—At the Micro-Physical Theoretical Level

- 57. Panoramic Theory of Transportation (Peter Baofu) (*FPHTRAN*)
- 56. Transcendent Theory of Architecture (Peter Baofu) (*FPHARCH*)
- 55. Multilateral Theory of Acoustics (Peter Baofu) (*FPHA*)
- 54. Selective Theory of Geometry (Peter Baofu) (*FPHG*)
- 53. Perspectival Theory of Space-Time (Peter Baofu) (*FPHST*)
- 52. Dialectic Theory of Complexity (Peter Baofu) (*FC*)

II. Theories in Relation to Culture

—At the Cultural Theoretical Level

- 51. Inquisitive Theory of Culinary Art (Peter Baofu) (*FPHCA*)
- 50. Interactive Theory of Semantics (Peter Baofu) (*FPHSEM*)
- 49. Transdicsiplinary Theory of Performing Arts (Peter Baofu) (*FPHPA*)
- 48. Comparative-Impartial Theory of Literature (Peter Baofu) (*FPHLIT*)

(continued on next page)

Table 4.43. Theories on Civilizational Holism (Part III)

II. Theories in Relation to Culture (*cont'd*)

—At the Cultural Theoretical Level (*cont'd*)

- • 47. Mediative-Variative Theory of Chess (Peter Baofu) (*FPHCHESS*)
- • 46. Theory of Post-Ethics (Peter Baofu) (*BEPE*)
- • 45. Dualistic Theory of Mass Culture (Peter Baofu) (Ch.2 of *FHC*)
- • 44. Comparative Theory of Religion—also known as the Comparative-Substitutive Theory of Religion (Peter Baofu) (Ch.3 of *FHC*; Ch.9 of *FCD*; Ch.1 of *FPHK*; *FPHR*)
- • 43. Theory of Post-Civilization (Peter Baofu) (*BCIV*)
- • 42. Theory of the Trinity of Modernity to Its After-Postmodern Counterpart (Peter Baofu) (*FHC*; Ch.10 of *FCD*)
- • 41. Transformative Theory of Aesthetic Experience (Peter Baofu) (*FAE*)
- (• 20). Theory of Post-Capitalism (Peter Baofu) (Ch.10 of *FCD*; Chs.2 & 4 of *FPHC*; *BCPC*)
- (• 19). Theory of Post-Democracy (Peter Baofu) (Ch.10 of *FCD*; Chs.3 & 4 of *FPHC*; *BDPD*)
- (• 5). Theory of Methodological Holism (Peter Baofu) (Ch.1 of *FCD*; Ch.1of *FPHC*; *Sec.2.1* & *Sec.2.5* of *BCPC*; *FC*; *FPHK*; etc.)

(continued on next page)

Table 4.43. Theories on Civilizational Holism (Part IV)

II. Theories in Relation to Culture (*cont'd*)
—At the Cultural Theoretical Level (*cont'd*)
(• 4). Theory of Existential Dialectics, or the Holistic Theory of Knowledge (Peter Baofu) (*FHC*; *FCD*; *FPHC*; *BDPD*; *FC*; *FAE*; *ALD; FIA; FPHK*; etc,)
(• 3). Contrastive Theory of Rationality (Peter Baofu) (*FPHML*)

III. Theories in Relation to the Mind
—At the Biological Theoretical Level
• 40. Theory of Contrastive Advantages (Peter Baofu) (*Sec.2.6* & Ch.10 of *FCD; BNN*)
(• 30).Theory of Floating Consciousness (Peter Baofu) (Ch.10 of *FCD*; Chs.1 & 4 of *FPHC*)
—At the Psychological Theoretical Level
• 39. Complex Theory of Data Analysis (Peter Baofu) (*FPHDA*)
• 38. Supersession Theory of Computing (Peter Baofu) (*FPHCOM*)
• 37. Metamorphic Theory of Humor (Peter Baofu) (*FPHH*)
• 36. Contrarian Theory of Personality (Peter Baofu) (*FPHP*)
• 35. Theory of Virtual Sexuality (Peter Baofu) (*FPHS*)
• 34. Expansive-Contractive Theory of Martial Arts (Peter Baofu) (*FPHMA*)
• 33. Multilogical Theory of Learning (Peter Baofu) (*FPHL*)
• 32. Comprehensive Theory of Creative Thinking (Peter Baofu) (*FPHCT*)

(continued on next page)

Table 4.43. Theories on Civilizational Holism (Part V)

III. Theories in Relation to the Mind (*cont'd*)

—*At the Psychological Theoretical Level* (*cont'd*)

- 31. Theory of Unfolding Unconsciousness—also known as the Unfolding Theory of Anomalous Experience (Peter Baofu) (*FPHU*)
- 30. Theory of Floating Consciousness (Peter Baofu) (Ch.10 of *FCD*; Chs.1 & 4 of *FPHC*)
- 29. Theory of Cognitive Partiality (Peter Baofu) (*Sec.2.4* of *FCD*; *Sec.4.5.1.1* of *BCPC*)
- 28. Theory of Emotional Non-Neutrality (Peter Baofu) (*Sec.2.5* of *FCD*; *Sec.4.5.2* of of *BCPC*)
- 27. Theory of Behavioral Alteration (Peter Baofu) (*Sec.4.5.3* of *BCPC*)

IV. Theories in Relation to Society

—*At the Organizational Theoretical Level*

- 26. Theory of E-Civic Alienation (Peter Baofu) (Ch.7 of *FCD*)
- 25. Combinational Theory of Organization (Peter Baofu) (*FPHO*; Ch.6 of *ALD*)

—*At the Institutional Theoretical Level*

- 24. Reflective Theory of Criminality (Peter Baofu) (*FPHCRIM*)
- 23. Detached Theory of Gambling (Peter Baofu) (*FPHGAM*)
- 22. Heterodox Theory of Education (Peter Baofu) (*FPHEDU*)
- 21. Reconstruction Theory of Law (Peter Baofu) (*FPHLAW*)
- 20. Theory of Post-Capitalism (Peter Baofu) (Ch.10 of *FCD*; Chs.2 & 4 of *FPHC*; *BCPC*)
- 19. Theory of Post-Democracy (Peter Baofu) (Ch.10 of *FCD*; Chs.3 & 4 of *FPHC*; *BDPD*)

(continued on next page)

Table 4.43. Theories on Civilizational Holism (Part VI)

IV. Theories in Relation to Society (*cont'd*)

—*At the Institutional Theoretical Level (cont'd)*

• 18. Dynamic Theory of Comparative Political Systems (Peter Baofu) (*ALD*)

(• 8.) Theory of the Cyclical Progression of Empire-Building (Peter Baofu) (*BWT*)

—*At the Systemic Theoretical Level*

• 17. Transfigurative Theory of Waste (Peter Baofu) (*FPHW*)

• 16. Theory of the Cyclical Progression of Migration (Peter Baofu) (*FPHMIG*)

(• 61). Resettlement Theory of Geology (Peter Baofu) (*FPHGEOL*)

• 15. Contingent Theory of Urban Planning (Peter Baofu) (*FPHUP*)

• 14. Totalistic Theory of Communication (Peter Baofu) (*FPHMM*; *FCD*; *FHC*)

• 13. Ambivalent Theory of Technology (Peter Baofu) (*FPHE*; *FCD*; *FHC*)

• 12. Multifaceted Theory of War and Peace (Peter Baofu) (Ch.9 of *FCD*; Ch.1 of *FPHK*)

• 11. Theory of Post-Humanity (Peter Baofu) (Ch.7 of *FHC*; Chs.3, & 10 of *FCD*; Chs.1, 3 & 4 of *FPHC*; and other books of mine)

• 10. Theory of the Cyclical Progression of System Integration and Fragmentation (Peter Baofu) (Chs.9-10 of *FCD*)

• 9. Synthetic Theory of Information Architecture (Peter Baofu) (*FIA*)

—*At the Structural Theoretical Level*

• 8. Theory of the Cyclical Progression of Hegemony (Peter Baofu) (Chs.9-10 of *FCD*; Chs.1, 3 & 4 of *FPHC*; *BDPD*)

• 7. Theory of the Cyclical Progression of Empire-Building (Peter Baofu) (*BWT*)

(continued on next page)

Table 4.43. Theories on Civilizational Holism (Part VII)

V. Meta-Theories (in Relation to Theories)

—*At the Methodological Meta-Theoretical Level*

(• 50).Interactive Theory of Semantics (Peter Baofu) (*FPHSEM*)

(• 39). Complex Theory of Data Analysis (Peter Baofu) (*FPHDA*)

• 6. Critical-Dialectic Theory of Formal Science (Peter Baofu) (*FPHFS*)

• 5. Theory of Methodological Holism (Peter Baofu) (Ch.1 of *FCD*; Ch.1of *FPHC*; *Sec.2.1* & *Sec.2.5* of *BCPC*; *FC*; *FPHK*; etc.)

—*At the Ontological Meta-Theoretical Level*

• 4. Theory of Existential Dialectics, or the Holistic Theory of Knowledge (Peter Baofu) (*FHC*; *FCD*; *FPHC*; *BDPD*; *FC*; *FAE*; *ALD; FIA; FPHK*; etc,)

• 3. Contrastive Theory of Rationality (Peter Baofu) (*FPHML*)

VI. Theories in Relation to the Rest

—*At the Meta-Historical Level*

• 2. Multifold Theory of History (Peter Baofu) (*FPHHIST*)

—*At the Empirical-Historical Level*

• 1. Theory of the Evolution from Pre-Modernity to After-Postmodernity (Peter Baofu) (*FHC*; Ch.9-10 of *FCD*; *FPHC*)

Notes: All these theories are my constructions, as some of the main contributions of my grant project on civilization and its future. These examples are solely illustrative (not exhaustive), and some of the items can be reclassified somewhere else. Nor are they always mutually exclusive. Since they are generalities, exceptions are expected.

Sources: From my previous books.

Table 4.44. Three Great Future Transformations of Mind Games

- **Virtual Games**
  —Ex: online chess
  —Ex: virtual experience

- **Novel Games**
  —Ex: new chess variants
  —Ex: new chess engines

- **Post-Human Mind Games**
  —Ex: the quest for broader/deeper mental benefits of chess playing
  —Ex: games designed for the evolution of the mind into different body-less forms (e.g., "floating consciousness," "hyper-spatial consciousness," "unfolding unconsciousness")

Notes: The examples in the categories are solely illustrative (not exhaustive), and the comparison is relative (not absolute), nor are they necessarily mutually exclusive. And some can be easily re-classified elsewhere. As generalities, they allow exceptions.
Sources: From *Sec. 4.16* of *FPHCHESS*. See text for more info. Also, consult *FCD* for strategy and tactics in warfare and *FPHO* for strategy and tactics in organization.

BIBLIOGRAPHY

ABC News (ABC). 2007. "Debris Wreaks Havoc on the Road - ABC News" (May 15). <http://abcnews.go.com/GMA/story?id=3175688#.T-dGRZHSTId>.

AllCountries.org (ALL). 2012. "Plague, Overview" (June 23). Health Topics A to Z. <http://allcountries.org/health/plague.html>.

Allison, Cyndeth. 2007. "Dumpster Diving" (May 08). *North Denver News.* <http://northdenvernews.com/content/view/766/2/>.

Anselmo, L.; & C. Pardini. 2000. "Collision Risk Mitigation in Geostationary Orbit." *Space Debris*, 2 (2): 67-82. <http://www.springerlink.com/content/r2642668j47q71k4/?MUD=MP>.

Argento, Mike. 2010. "Illegal Dump Survey Finds a Lot of Trash—and a Buffalo Head" (August 17). <http://www.ydr.com/ci_15429915>.

Auchmutey, Jim. 2009. "Trying on Carbon-Neutral Trend" (January 26). *Atlanta Journal-Constitution.* <http://www.ajc.com/services/content/printedition/2009/01/26/carbon0126b.html>.

Baird, Colin. 1995. *Environmental Chemistry*. NY: W. H. Freeman.

Baofu, Peter. 2012. *The Future of Post-Human Culinary Art: Towards a New Theory of Ingredients and Techniquess*. Cambridge, England: Cambridge Scholars Publishing.

_____.2012a. *The Future of Post-Human Transportation: A Preface to a New Theory of Networks and Operations*. Cambridge, England: Cambridge Scholars Publishing.

_____.2012b. *The Future of Post-Human Migration: A Preface to a New Theory of Sameness, Otherness, and Identity*. Cambridge, England: Cambridge Scholars Publishing.

_____.2012c. *The Future of Post-Human History: A Preface to a New Theory of Universality and Relativity*. Cambridge, England: Cambridge Scholars Publishing.

_____.2012d. *The Future of Post-Human Criminality: A Preface to a New Theory of Heroes and Villains*. Cambridge, England: Cambridge International Science Publishing, Ltd.

_____.2012e. *The Future of Post-Human Architecture: A Preface to a New Theory of Form and Function*. Cambridge, England: Cambridge International Science

Publishing, Ltd.

_____.2012f. *The Future of Post-Human Semantics: A Preface to a New Theory of Internality and Externality*. Cambridge, England: Cambridge Scholars Publishing.

_____.2012g. *The Future of Post-Human Performing Arts: A Preface to a New Theory of the Body and its Presence*. Cambridge, England: Cambridge Scholars Publishing.

_____.2011. *The Future of Post-Human Geography: A Preface to a New Theory of Environments and their Interactions*. Cambridge, England: Cambridge International Science Publishing, Ltd.

_____.2011a. *The Future of Post-Human Data Analysis: A Preface to a New Theory of Quantiative and Qualitative Research*. Cambridge, England: Cambridge International Science Publishing, Ltd.

_____.2011b. *The Future of Post-Human Chemistry: A Preface to a New Theory of Substances and their Changes*. Cambridge, England: Cambridge Scholars Publishing.

_____.2011c. *The Future of Post-Human Literature: A Preface to a New Theory of Fiction and Non-Fiction*. Cambridge, England: Cambridge International Science Publishing, Ltd.

_____.2011d. *The Future of Post-Human Computing: A Preface to a New Theory of Hardware, Software, and the Mind*. Cambridge, England: Cambridge International Science Publishing, Ltd.

_____.2011e. *The Future of Post-Human Gambling: A Preface to a New Theory of Risk and Caution*. Cambridge, England: Cambridge International Science Publishing, Ltd.

_____.2011f. *The Future of Post-Human Acoustics: A Preface to a New Theory of Sound and Silence*. Cambridge, England: Cambridge International Science Publishing, Ltd.

_____.2011g. *The Future of Post-Human Humor: A Preface to a New Theory of Joking and Laughing*. Cambridge, England: Cambridge International Science Publishing, Ltd.

_____.2011h. *Beyond Ethics to Post-Ethics: A Preface to a New Theory of Morality and Immorality*. Charlotte, NC: Information Age Publishing.

_____.2011i. *The Future of Post-Human Education: A Preface to a New Theory of Teaching and Learning*. Cambridge, England: Cambridge International Science Publishing, Ltd.

_____.2010. *The Future of Post-Human Religion: A Preface to a New Theory of Spirituality*. Cambridge, England: Cambridge International Science Publishing, Ltd.

_____.2010a. *The Future of Post-Human Chess: A Preface to a New Theory of*

Tactics and Strategy. Cambridge, England: Cambridge International Science Publishing, Ltd.

_____.2010b. *The Future of Post-Human Geology: A Preface to a New Theory of Statics and Dynamics*. Cambridge, England: Cambridge International Science Publishing, Ltd.

_____.2010c. *Beyond Cosmology to Post-Cosmology: A Preface to a New Theory of Different Worlds*. Cambridge, England: Cambridge International Science Publishing, Ltd.

_____.2010d. *The Future of Post-Human Personality: A Preface to a New Theory of Normality and Abnormality*. Cambridge, England: Cambridge International Science Publishing, Ltd.

_____.2010e. *The Future of Post-Human War and Peace: A Preface to a New Theory of Aggression and Pacificity*. Cambridge, England: Cambridge Scholars Publishing.

_____.2010f. *The Future of Post-Human Formal Science: A Preface to a New Theory of Abstraction and Application*. Cambridge, England: Cambridge Scholars Publishing.

_____.2010g. *The Future of Post-Human Law: A Preface to a New Theory of Necessity, Contingency, and Justice*. Cambridge, England: Cambridge Scholars Publishing.

_____.2010h. *The Future of Post-Human Sexuality: A Preface to a New Theory of the Body and Spirit of Love Makers*. Cambridge, England: Cambridge Scholars Publishing.

_____,2009. *The Future of Post-Human Martial Arts: A Preface to a New Theory of the Body and Spirit of Warriors*. Cambridge, England: Cambridge Scholars Publishing.

_____.2009a. *The Future of Post-Human Organization: A Preface to a New Theory of Communication, Decision-Making, and Leadership*. Cambridge, England: Cambridge Scholars Publishing.

_____.2009b. *The Future of Post-Human Language: A Preface to a New Theory of Structure, Context, and Learning*. Cambridge, England: Cambridge Scholars Publishing.

_____.2009c. *The Future of Post-Human Urban Planning: A Preface to a New Theory of Density, Void, and Sustainability*. Cambridge, England: Cambridge Scholars Publishing.

_____.2009d. *The Future of Post-Human Geometry: A Preface to a New Theory of Infinity, Symmetry, and Dimensionality*. Cambridge, England: Cambridge Scholars Publishing.

_____.2009e. *The Future of Post-Human Creative Thinking: A Preface to a New Theory of Invention and Innovation*. Cambridge, England: Cambridge Scholars Publishing.

_____.2009f. *The Future of Post-Human Mass Media: A Preface to a New Theory of Communication*. Cambridge, England: Cambridge Scholars Publishing.

_____.2009g. *The Future of Post-Human Engineering: A Preface to a New Theory of Technology*. Cambridge, England: Cambridge Scholars Publishing.

_____.2008. *The Future of Post-Human Mathematical Logic: A Preface to a New Theory of Rationality*. Cambridge, England: Cambridge Scholars Publishing.

_____.2008a. *The Future of Post-Human Knowledge: A Preface to a New Theory of Methodology and Ontology*. Oxford, United Kingdom: Chandos Publishing (Oxford) Ltd.

_____.2008b. *The Future of Post-Human Unconsciousness: A Preface to a New Theory of Anomalous Experience*. Cambridge, England: Cambridge Scholars Publishing.

_____.2008c. *The Future of Information Architecture: Conceiving a New Way to Understand Taxonomy, Network, and Intelligence*. Oxford, United Kingdom: Chandos Publishing (Oxford) Ltd.

_____.2007. *The Rise of Authoritarian Liberal Democracy: A Preface to a New Theory of Comparative Political Systems*. Cambridge, England: Cambridge Scholars Publishing.

_____.2007a. *The Future of Aesthetic Experience: Conceiving a Better Way to Beauty, Ugliness, and the Rest*. Cambridge, England: Cambridge Scholars Publishing.

_____.2007b. *The Future of Complexity: Conceiving a Better Way to Understand Order and Chaos*. London, United Kingdom: World Scientific Publishing Co.

_____.2007c. *Beyond the World of Titans, and the Remaking of World Order: A Preface to a New Logic of Empire-Building*. Cambridge, England: Cambridge Scholars Publishing.

_____.2006. *Beyond Nature and Nurture: Conceiving a Better Way to Understand Genes and Memes*. Cambridge, England: Cambridge Scholars Publishing.

_____.2006a. *The Future of Post-Human Space-Time: Conceiving a Better Way to Understand Space and Time*. New York: Peter Lang Publishing, Inc.

_____.2006b. *Beyond Civilization to Post-Civilization: Conceiving a Better Model of Life Settlement to Supersede Civilization*. New York: Peter Lang Publishing, Inc.

_____.2005. *Beyond Capitalism to Post-Capitalism: Conceiving a Better Model of Wealth Acquisition to Supersede Capitalism*. New York: The Edwin Mellen Press.

_____.2004. *The Future of Post-Human Consciousness*. New York: The Edwin Mellen Press.

_____.2004a. *Beyond Democracy to Post-Democracy: Conceiving a Better Model of Governance to Supersede Democracy*. Volume 1. New York: The Edwin Mellen Press.

_____.2004b. *Beyond Democracy to Post-Democracy: Conceiving a Better Model of Governance to Supersede Democracy*. Volume 2. New York: The Edwin Mellen Press.

_____.2002. *The Future of Capitalism and Democracy*. Maryland: The University Press of America.

_____.2000. *The Future of Human Civilization*. Volume 1. New York: The Edwin Mellen Press.

_____.2000a. *The Future of Human Civilization*. Volume 2. New York: The Edwin Mellen Press.

Barenblatt, Daniel. 2004. *A Plague upon Humanity*. New York: HarperCollins.

Barrow, John D. 1998. *Impossibility: The Limits of Science and the Science of Limits*. Oxford: Oxford University Press.

BBC News (BBC). 2011. "International Space Station in Debris Scare" (June 28). <http://www.bbc.co.uk/news/science-environment-13949956>.

_____.2011a. "Less is More: Simple Living in Small Spaces" (December 28). <http://www.bbc.co.uk/news/magazine-16348594>.

_____.2009. "Junk Alert for Space Station Crew" (March 12). <http://news.bbc.co.uk/2/hi/science/nature/7940431.stm>.

Biletzki, Anat; & Anat Matar. 2006. "Ludwig Wittgenstein" (Nov. 17). *Stanford Encyclopedia of Philosophy*. <http://plato.stanford.edu/entries/wittgenstein/>.

bionomicfuel.com (BIO). 2012. "Cost Effective Waste to Energy Technologies" (Friday, October 15). <http://www.bionomicfuel.com/cost-effective-waste-to-energy-technologies/>.

Blue Ridge Environmental Defense League (DREDL). 2009. "Waste Gasification: Impacts on the Environment and Public Health" (February). <http://www.bredl.org/pdf/wastegasification.pdf>.

BottleBill.org (BOT). 2010. "New York's 'Bottle Bill' Brings in $120M" (November 03). <http://www.bottlebill.org/news/articles/2010/NY-11-3-NYsBBBringsIn. htm>.

Bowsky, William. 1971. *The Black Death: A Turning Point in History?*. NY: Holt, Rinehart and Winston.

Brace, Alison. 2007. "Freeloading" (March 2). *Times Educational Supplement*: 28–29.

brainyquote.com (BRA). 2012. "Norman Cousins" (July 08). <http://www.brainyquote.com/quotes/keywords/waste_8.html>.

Bridbury, A. R. 1983. *Economic Growth: England in the Later Middle Ages*. Westport, CT: Greenwood Press.

Brown, M. T.; & V. Buranakarn. 2003. "Emergy Indices and Ratios for

Sustainable Material Cycles and Recycle Options." *Resources, Conservation and Recycling*, 38 (1): 1–22. <http://www.thaiscience.info/Article%20for%20ThaiScience/Article/4/Ts-4%20emergy%20indices%20and%20ratios%20for%20sustainable%20material%20cycles%20and%20recycle%20options.pdf>.

Bullard, Robert D. 1992. "The Quest for Environmental Equity: Mobilizing the African-American Community for Social Change." *American Environmentalism: the U.S. environmental movement, 1970-1990*. New York: Taylor & Francis, Inc.

Bunge, Mario. 1999. *Dictionary of philosophy*. Amherst, MA: Prometheus Books.

Burn, Shawn. 2006. "Social Psychology and the Stimulation of Recycling Behaviors: The Block Leader Approach." *Journal of Applied Social Psychology*, 21.8 (2006): 611–629.

Calrecycle.ca.gov (CAL). 2010. "Waste Prevention Terms and Definitions" (August 24). <http://www.calrecycle.ca.gov/ReduceWaste/Define.htm#WastePrev>.

capacity.org.uk (CAP). 2005. "Environmental Justice: A Snapshot" (January). <http://www.capacity.org.uk/downloads/snapshot_for_ej.pdf>.

Carlson, Erika, et al. 1990. "Final Design of a Space Debris Removal System." NASA/CR-189976.

Carlson, Tucker. 2006. "'Freegans' Choose to Eat Garbage" (February 03). MSNBC. <http://www.msnbc.msn.com/id/11154276/>.

Carpenter, Susan. 2010. "Television Review: 'Trash Inc.'" (September 29). *Los Angeles Times*. <http://articles.latimes.com/2010/sep/29/entertainment/la-et-secret-garbage-20100929>.

Carroll, Joseph. 2002. "Space Transport Development Using Orbital Debris" (December 02). NASA Institute for Advanced Concepts. <http://www.niac.usra.edu/files/studies/final_report/800Carroll.pdf>.

Cerin, P.; & L. Karlson. 2002. Business Incentives for Sustainability: A Property Rights Approach." *Ecological Economics*, 40: 13-22.

Chase, Scott. 1993. "Tachyons" (March). *Physics FAQ*. <http://math.ucr.edu/home/baez/physics/ParticleAndNuclear/tachyons.html>.

Chun, Clayton. 1999. *Shooting Down a Star: America's Thor Program 437, Nuclear ASAT, and Copycat Killers*. Maxwell AFB Base, AL: Air University Press.

CNW Newswire (CNW). 2011. "Intelsat Picks MacDonald, Dettwiler and Associates Ltd. for Satellite Servicing" (March 15). <http://www.newswire.ca/en/story/730363/intelsat-picks-macdonald-dettwiler-and-associates-ltd-for-satellite-servicing>.

Cohn, Samuel K. 2002. "The Black Death: End of a Paradigm." *American Historical Review*, 107 (3): 703-737.

Colon, Dalia. 2008. "One Man's Trash is Another Man's...Lunch" (April 11). *St. Petersburg Times.* <http://www.tampabay.com/features/consumer/article 451134.ece>.

Columbia University (COL). 2008. "Waste Council Attracts Experts Worldwide" (October 31). <http://engineering.columbia.edu/waste-council-attracts-experts-worldwide>.

Corazzon, Raul. 2007. "Definitions of Ontology: From Nicolai Hartmann to the Present Time." <http://www.formalontology.it/section_4_two.htm>.

Corbett, Dan. 2008. "Introduction to String Theory" (November 23). *ThinkQuest.* <http://library.thinkquest.org/27930/stringtheory1.htm>.

Corr, Anders. 1999. *No Trespassing: Squatting, Rent Strikes, and Land Struggles Worldwide*. Cambridge, MA: South End Press.

Cutter, Susan L. 1995. "Race, Class and Environmental justice." *Progress in Human Geography*, 19.111: 111-22. <http://geography.ssc.uwo.ca/faculty/ baxter/readings/Cutter_environmental_justice_PIHG_1995.pdf>.

D'Andrea. Niki. 2009. "Dumpster Dining: For Freegans, Eating Garbage Is Getting Downright Trendy" (August 20). <http://www.phoenixnewtimes.com/2009-08-20/news/dumpster-dining-for-freegans-eating-garbage-is-getting-down right-trendy/>.

David, Leonard. 2007. "China's Anti-Satellite Test: Worrisome Debris Cloud Circles Earth" (February 02). *Space.com.* <http://www.space.com/3415-china-anti-satellite-test-worrisome-debris-cloud-circles-earth.html>.

Davoudi, S. Evans. 2005. "The Challenge of Governance in Regional Waste Planning." *Environment and Planning C: Government and Policy*, 23: 493–517.

Dornfeld, Ann. 2008. "Recycling Suddenly Gets Expensive" (December 23). <http://www.npr.org/templates/story/story.php?storyId=98638760>.

Dufel, Susan, et al. 2011. "Plague" (June 03). <http://emedicine.medscape.com/ article/829233-overview>.

DW. 2009. "Dumpster Diving Stays on Trend in Germany" (November 11). <http://www.dw.de/dw/article/0,,4881647,00.html>.

Echlin, Helena. 2006. *Yoga Journal* (December). <http://books.google.co.uk/ books?id=vekDAAAAMBAJ&pg=PA92&lpg=#v=onepage&q&f=false>.

Economist (ECO). 2007. "The Truth about Recycling" (June 7). <http://www.economist.com/node/9249262?story_id=9249262>.

______.2007a. "The Price of Virtue" (June 7). <http://www.economist.com/node /9302727?story_id=9302727>.

Environmental Justice Resource Center (EJRC). 2012. "Environmental Justice in the 21st Century." <http://www.ejrc.cau.edu/ejinthe21century.htm>.

Epa.vic.gov.au (EPAV). 2011. "If You've Been Reported—EPA Victoria" (August 24). <http://www.epa.vic.gov.au/litter/if_you_have_been_reported.asp>.

eprworkinggroup.org (EPR). 2003. "Extended Producer Responsibility: A

Prescription for Clean Production, Pollution Prevention and Zero Waste" (July). <http://www.eprworkinggroup.org/>.

Erin, P. 2005. "Environmental Strategies in Industry: Turning Business Incentives into Sustainability." Report 5455. Swedish Environmental Protection Agency.

ESA. 2009. "What are Hypervelocity Impacts?" (February 19). <http://www.esa.int/esaMI/Space_Debris/1112600510257_0.html>.

Ethical Consumer (ETH). 2012. "Why Buy Ethically" (June 26). <http://www.ethicalconsumer.org/shoppingethically/topethicaltips/whybuyethically.aspx>.

Europa (EUR). 2012. "Waste Incineration" (June 21). <http://europa.eu/legislation_summaries/environment/waste_management/l28072_en.htm>.

Everett, Hugh. 1957. "'Relative State' Formulation of Quantum Mechanics" (July). *Reviews of Modern Physics*, vol. 29, no.3. <http://www.univer.omsk.su/omsk/Sci/Everett/paper1957.html>.

Firehammer, Reginald. 2005. "Basic Principles of Ontology" (January 05). <http://usabig.com/autonomist/philosophy/ontology2.html>.

Ford, Matt. 2009. "Orbiting Space Junk Heightens Risk of Satellite Catastrophes" (February 27). *Ars Technica*. <http://arstechnica.com/science/2009/02/orbiting-space-junk-heightens-risk-of-satellite-catastrophes/>.

Foust, Jeff. 2011. "A Sticky Solution for Grabbing Objects in Space." *MIT Technology Review*. <http://www.technologyreview.com/article/425656/a-sticky-solution-for-grabbing-objects-in-space/#.To50S0SzPvk.twitter>.

Gandy, M. 1994. *Recycling and the Politics of Urban Waste*. London: Earthscan Publications.

Gerrard, Michael B. 1993. "The Victims of NIMBY." *Fordham Urban Law Journal*. <http://heinonline.org/HOL/LandingPage?collection=journals&handle=hein.journals/frdurb21&div=25&id=&page=>.

Glasbergen, P. 2007. *Partnerships, Governance, and Sustainable Development: Reflections on Theory and Practice*. Cheltenham: Edward Elgar.

Global Aerospace Corp (GAC). 2010. "Company Floats Giant Balloon Concept as Solution to Space Mess" (August 4). <http://www.world-science.net/othernews/100804_spacejunk>.

GlobalSecurity.org (GLO). 2012. "Chelyabinsk-65/Ozersk" <http://www.globalsecurity.org/wmd/world/russia/chelyabinsk-65_nuc.htm>.

González, Abel J. 1999. "Strengthening the Safety of Radiation Sources & the Security of Radioactive Materials: Timely Action." IAEA Bulletin, 41 (3).

Gonzales, Richard. 2008. "Raids On Recycling Bins Costly To Bay Area" (July 19). <http://www.npr.org/templates/story/story.php?storyId=92705125>.

Gott, J. Richard. 2002. *Time Travel in Einstein's Universe*. Boston : Houghton Mifflin.

Gould, Peter C. 1988. *Early Green Politics: Back to Nature, Back to the Land, and Socialism in Britain, 1880-1900*. New York : St. Martin's Press.

Griffiths, Michael. B.; Flemming Christiansen; & Malcolm Chapman. 2010. "Chinese Consumers: The Romantic Reappraisal." *Ethnography*, 11: 331-57.

Guth, Alan. 1997. *The Inflationary Universe: The Quest for a New Theory Of Cosmic Origins*. New York: Helix Books.

Hanisch, C. 2000. "Is Extended Producer Responsibility Effective?." *Environ Sci Technol*, 34 (7): 170 A-175 A.

Hardy, Dennis. 2000. *Utopian England: Community Experiments 1900–1945*. New York, NY : Routledge.

Harn, Kimberly. 2010. "Just Policy?: The Limits of U.S. Environmental Protection Agency's Environmental Justice Strategy during the Clinton Administration." Rhodes Institute for Regional Studies. <http://www.rhodes.edu/images/content/Academics/Harn_EPA.pdf>.

Harrison, Jeff. 2004. "U.S. Wastes Half of Food Produced" (November 18). <http://uanews.org/node/10448>.

Heartland Institute (HEART). 2002. "Recycling: It's a bad idea in New York" (May 01). <http://news.heartland.org/newspaper-article/2002/05/01/recycling-its-bad-idea-new-york>.

Hechler, M.; & J. C. van der Ha. 1981. "The Collision Probability of Geostationary Satellites." 32nd International Astronautical Congress: 23.

History Magazine (HIST). 2009. "The Plague Of Justinian": 9-12. Moorshead Magazines, Limited.

Hodgkinson, Tom. 2006. *How To Be Free*. London: Hamish Hamilton. <http://books.google.co.uk/books/about/How_to_be_free.html?id=7pZWAAAAYAAJ&redir_esc=y>.

Huesemann, M. H. 2003. "The Limits of Technological Solutions to Sustainable Development." *Clean Techn Environ Policy*, 5: 21–34.

Inglesby, T. V.; D. Dennis; D. Henderson; et al. 2000. "Plague as a Biological Weapon: Medical and Public Health Management. Working Group on Civilian Biodefense" (May). *JAMA*, 283 (17): 2281–90.

International Atomic Energy Agency (IAEA). 1998. "The Radiological Accident in Goiânia." <http://www-pub.iaea.org/MTCD/publications/PubDetAR.asp?pubId=3684>.

Jackson, Emily. 2010. "Vancouver Fireworks a Boon for City" (July 28). *The Vancouver Sun*. <http://www.vancouversun.com/Vancouver+fireworks+boon+city+bottle+collectors/3334521/story.html>.

Jay, Kate. 2008. "First Carbon Neutral Zone Created in the United States" (November 14). *Reuters*. <http://www.reuters.com/article/2008/11/14/idUS164153+14-Nov-2008+PRN20081114>.

Jeffries, Elisabeth. 2011. "What Price Environmental Justice?." <http://www.environmentalistonline.com/article/2011-03-09/what-price-environmental-justice>.

Jenkins, Joseph. 2005. *The Humanure Handbook: A Guide to Composting Human Manure*. PA: Jenkins Publishing.

_____.2005a. *Humanure Compost Toilet System Instruction Manual*. PA: Jenkins Publishing.

Johnson, Nicholas. 2011. "Space Debris Issues." Audio file, @1:03:05-1:06:20. The Space Show. <http://www.thespaceshow.com/detail.asp?q=1666>.

Josyn, Ed. 2006. "About Veganism." <http://www.vegan.org/>.

Kaku, Michio. 1994. *Hyperspace: A Scientific Odyssey Through Parallel Universes, Time Warps, and The Tenth Dimension.* Illustrations by Robert O'Keefe. Oxford: Oxford University Press.

Kessler, Donald. 2009. "The Kessler Syndrome" (March 08). <http://webpages.charter.net/dkessler/files/KesSym.html>.

____.1991. "Collisional Cascading: The Limits of Population Growth in Low Earth Orbit." *Advances in Space Research*, 11 (12): 63-66.

Klinkrad, Heiner. 2006. *Space Debris: Models and Risk Analysis*. NY: Springer-Praxis.

Kurutz, Steven. 2007. "Not Buying It" (June 21). *New York Times*. <http://www.nytimes.com/2007/06/21/garden/21freegan.html?_r=1>.

Landy, M. 2001. *Environmentalism: A New Approach to Policy*. Washington: George Marshall Institute.

Landsburg, Steven A. 1993. *The Armchair Economist*. New York: Free Press.

Lavee. D. 2007. "Is Municipal Solid Waste Recycling Economically Efficient?." *Environmental Management*. <http://www.springerlink.com/content/r4611ju585760316/?MUD=MP>.

League of Women Voters (LEAG). 1993. *The Garbage Primer*. New York: Lyons & Burford.

Leigh, Patricia. 2007. "Highway Debris, Long an Eyesore, Grows as Hazard" (May 11). *New York Times*. <http://query.nytimes.com/gst/fullpage.html?res=9D01E2D81431F932A25756C0A9619C8B63>.

Lewycka, Marina. 2009. "So, I'm a Skip Addict—Avocado Bath Suite, Anyone?" (July 02). *London Evening Standard*. <http://www.standard.co.uk/news/so-im-a-skip-addict--avocado-bath-suite-anyone-6799766.html>.

Lowe, P. 1998. *British Environmental Policy and Europe: Politics and Policy in Transition*. London: Routledge.

Lovgren, Stefan. 2006. "Space Junk Cleanup Needed, NASA Experts Warn" (January 19). *National Geographic News*. <http://news.nationalgeographic.com/news/2006/01/0119_060119_space_junk.html>.

Maccarthy, Fiona. 1981. *The Simple Life: C.R. Ashbee in the Cotswolds*. London: Lund Humphries.

Macfarlane, Christina. 2009. "Golf Balls: 'Humanity's Signature Litter'" (November 10). *CNN*. <http://edition.cnn.com/2009/SPORT/11/04/littering.golf.balls/>.

Mark, Jason. 2010. "How Does Your Garden Grow? Watch and See" (February 26). *Food.change.org*.

Marshall, Peter. 1996. *Nature's Web: Rethinking Our Place on Earth*. Armonk, New York: M.E. Sharpe.

Mazar, Nina; & Chen-Bo Zhong. 2010. "Do Green Products Make Us Better People?" *Psychological Science* (April).

McAleer, Neil. 1987. *The Mind-Boggling Universe*. New York: Doubleday & Co.

McClaughlin, Lisa. 2008. "How to Live with Just 100 Things" (June 05). *Time*. <http://www.time.com/time/magazine/article/0,9171,1812048,00.html>.

McKean, Erin, ed. 2005. *The New Oxford American Dictionary* (second ed.). England: Oxford University Press.

McKie, Robin; & Michael Day. 2008. "Warning of Catastrophe from Mass of 'Space Junk'" (February 24). *The Observer*. <http://www.guardian.co.uk/science/2008/feb/24/spaceexplorationspacejunk>.

Medina, M. 2000. "Scavenger Cooperatives in Asia and Latin America." *Resources,* 31: 51–69.

Mee Jr, Charles L. 2011. "The Black Death, a Bubonic Plague of Great Dimensions–Part 2." <http://wordfocus.com/word-black-death-pt2.html>.

Merkel, Jim. 2003. *Radical Simplicity*. British Columbia: New Society.

Merriam-Webster's Collegiate Dictionary Online (MWD). 2007. "Agency" (March 20). <http://www.m-w.com/dictionary/agency>.

Michaels, Daniel. 2009. "A Cosmic Question: How to Get Rid Of All That Orbiting Space Junk?" (March 11). *Wall Street Journal*. <http://online.wsj.com/article/SB123672891900989069.html>.

Miller, Jr., G. Tyler. 2003. *Environmental Science: Working With the Earth*. Pacific Grove, California: Brooks/Cole.

Monbiot, George. 2010. "It Goes against our Nature; but the Left Has to Start Asserting its Own Values" (October 12). *The Guardian*. <http://www.guardian.co.uk/commentisfree/cif-green/2010/oct/11/left-values-progressive-self-interest>.

Moore, C. J. 2008. "Synthetic Polymers in the Marine Environment: A Rapidly Increasing, Long-Term Threat." *Environmental Research*: 131–139.

Morozov, Evgeny. 2011. *The Net Delusion: The Dark Side of Internet Freedom*. New York: Public Affairs.

Morrison, Denton. 1986. "Environmentalism and Elitism: A Conceptual and Empirical Analysis." *Environmental Management*, 10 (5): 581-589.

<http://www.springerlink.com/content/j052801335x8j717/?MUD=MP>.

msnbc.msn.com (MSNBC). 2012. "What You Need to Know about Bed Bugs" (June 21). <http://www.msnbc.msn.com/id/11915026/ns/dateline_nbc-dont_let_the_bed_bugs_bite/t/what-you-need-know-about-bed-bugs/>.

Nahra, Henry. 1989. "Effect of Micrometeoroid and Space Debris Impacts on the Space Station Freedom Solar Array Surfaces." Presented at the 1989 Spring Meeting of the Materials Research Society (April 24-29). <http://ntrs.nasa.gov/archive/nasa/casi.ntrs.nasa.gov/19890016664_19890166 64.pdf>.

NASA. 2009. "Orbital Debris FAQ: How Much Orbital Debris is Currently in Earth Orbit?" (July). <http://orbitaldebris.jsc.nasa.gov/faqs.html#3>.

NationalArchives.gov.uk (NAT). 1975. "Keep Britain Tidy." <http://www.nationalarchives.gov.uk/films/1964to1979/filmpage_tidy.htm>.

Nearing, Scott. 2006. *Civilization and Beyond.* <http://books.google.co.uk/ books?id=1NeqUvTqGEcC&pg=PA101&dq=#v=onepage&q&f=false>.

Nourish the Spirit (NOUR). 2012. "Annie Leonard Interview & 'Story of Electronics' Release." <http://www.nourishthespirit.com/annie-leonard-interview-story-of-electronics-release/>.

nouse.co.uk (NOUSE). 2008. "Waste Not, Want Not" (November 29). <http://www.nouse.co.uk/2008/11/29/waste-not-want-not-2/>.

Nova Online (NO). 2000. "Time Can Vary." <http://www.pbs.org/ wgbh/nova/time/thinktime.html>.

NRDC. 1996. "Too Good To Throw Away" (June 30). New York Times Magazine. <http://www.nrdc.org/cities/recycling/recyc/appenda.asp>.

NWTRCC. 2007. "Low Income/Simple Living as War Tax Resistance." <http://www.nwtrcc.org/practical5.php>.

Ockerman, Herbert; & Conly Hansen. 2000. *Animal by-Product Processing & Utilization.* Technomic Publishing Company Inc.

Oecd.org (OECD). 2003. "Glossary of Statistical Terms" (March 17). <http://stats.oecd.org/glossary/detail.asp?ID=2896>.

_____.2001. "Extended Producer Responsibility: A Guidance Manual for Governments." Paris, France.

Office of Technology Assessment (OTA). 1990. "Orbiting Debris: A Space Environmental Problem." Background Paper, OTA-BP-ISC-72, U.S. Government Printing Office. <http://www.fas.org/ota/reports/9033.pdf>.

Osborne, Hilary. 2009. "Daniel Suelo: Free spirit or freeloader?" (July 23). *The Guardian* (UK). <http://www.guardian.co.uk/money/blog/2009/jul/23/daniel-suelo-caveman>.

Oskamp, Stuart. 1995. "Resource Conservation and Recycling: Behavior and Policy." *Journal of Social Issues*, 51.4: 157–177.

Palmer, Jason. 2011. "Space Junk Could be Tackled by Housekeeping Spacecraft"

(August 08). *BBC News*. <http://www.bbc.co.uk/news/science-environment-14405118>.

PBS NewsHour (PBS). 2010. "Report on the Zabaleen" (February 16).

Pierce, Linda Breen. 2000. *Choosing Simplicity*. Gallagher Press.

Powys, John Cowper. 1933. *A Philosophy of Solitude*. New York: Simon and Schuster.

Prakash, Bhavani. 2012. "The Light Bulb Conspiracy: The Story of Planned Obsolescence." *Eco Walk the Talk*. <http://www.ecowalkthetalk.com/blog/2011/08/26/the-light-bulb-conspiracy-the-story-of-planned-obsolescence/>.

Pratarelli, M. E. 2010. "Social Pressure and Recycling: A Brief Review, Commentary and Extensions." *S.A.P.I.EN.S.*, 3 (1). <http://sapiens.revues.org/905>.

Princen, T. 1999. "Consumption and Environment: Some Conceptual Issues." *Ecological Economics*, 31: 347-363.

QuotationsBook.com (QUO). 2012. "Unknown" (June 23). <http://quotationsbook.com/quote/26535/>.

Rajvanshi, Anil K. 2012. "How to Live Simply and in a Sustainable Way" (March 24). <http://www.speakingtree.in/spiritual-blogs/masters/self-improvement/how-to-live-simply-and-in-a-sustainable-way>.

Raman, Shankar. 1997. Lectures on Literary Theory. Cambridge, MA: MIT (spring).

Ramboll Consult (RAMB). 2012. "Waste to Energy in Denmark" (June 21). <http://viewer.zmags.com/showmag.php?mid=wsdps>.

Reason Magazine (REAS). 2010. "Peak Everything?" (April 02). <http://reason.com/archives/2010/04/27/peak-everything>.

Renton, Alex. 2007. "Why I Love Bin Diving" (August 17). *Guardian.co.uk*.<http://www.guardian.co.uk/lifeandstyle/wordofmouth/2007/aug/17/wastenotwantnot>.

Rhodes, R. 1997. "The New Governance: Governing without Government." *Public Policy and Administration*, 8 (2): 724.

Rivera, Ray. 2008. "Mayor Calls Electronics Recycling Bill 'Illegal'" (February 15). *The New York Times*. <http://cityroom.blogs.nytimes.com/2008/02/15/mayor-says-hell-ignore-veto-on-electronic-recycling/#comment-39647>.

Roberts, R. Gregory. 1998. "Environmental Justice and Community Empowerment: Learning from the Civil Rights Movement." *American University Law Review*. <http://www.wcl.american.edu/journal/lawrev/48/pdf/roberts.pdf>.

Roeder, Linda. 2004. "Hazardous Waste: Advocacy Group Recommendations Promote Manufacturer Responsibility" (March 16). *DAILY ENV'T REP*.

Royal Commission on Environmental Pollution (RCEP). 2007. "Urban Environment." <http://www.official-documents.gov.uk/document/cm70/7009/

7009.pdf>.

Sachs, Andrea. 2004. "Get Your Mind Into the Gutter" (November 7). *Washington Post*. <http://www.washingtonpost.com/ac2/wp-dyn/A31130-2004Nov6?language=printer>.

Sale, K. 1997. "America's New Luddites" (February). *Le Monde Diplomatique*. <http://mondediplo.com/1997/02/20luddites>.

Salter, Jessica. 2010. "The Man who Lives without Money (August 18). *The Telegraph* (UK). <http://www.telegraph.co.uk/earth/greenerliving/7951968/The-man-who-lives-without-money.html>.

Samuelson, Paul; & William Nordhaus. 2001. *Microeconomics*. New York: The McGraw-Hill Companies.

Saul, John Ralston. 1992. *Voltaire's Bastards: The Dictatorship of Reason in the West*. New York: Free Press.

Schackelford, T. K. 2006. "Recycling, Evolution and the Structure of Human Personality." *Personality and Individual Differences*, 41: 1551–1556.

Schefter, Jim. 1982. "The Growing Peril of Space Debris" (July). *Popular Science*: 48-51. <http://books.google.ee/books?id=q0qVc8dQrpgC&pg=PA48&redir_esc=y#v=onepage&q&f=false>.

Scheffer, Victor. 1991. *The Shaping of Environmentalism in America*. Seattle, WA: University of Washington Press.

Schlosberg, David. 2007. *Defining Environmental Justice: Theories, Movements, and Nature*. England: Oxford University Press.

Scientific American (SCIA). 2009. "Phosphorus Famine: The Threat to Our Food Supply" (June). <http://www.scientificamerican.com/article.cfm?id=phosphorus-a-looming-crisis>.

Sepúlveda, A.; M. Schluep; F. Renaud; M. Streicher; R. Kuehr; C. Hagelüken; et al. 2010. "A Review of the Environmental Fate and Effects of Hazardous Substances Released from Electrical and Electronic Equipments during Recycling: Examples from China and India." *Environmental Impact Assessment Review,* 30: 28–41. <http://ewasteguide.info/files/Sepulveda_2010_EIAR_0_0.pdf>.

Shi, David. 2001. *The Simple Life*. Athens, GA: University of Georgia Press.

Shlain, Leonard. 1991. *Art and Physics: Parallel Visions in Space,Time, and Light*. New York: William Morrow and Co., Inc.

Sierra Club (SIERRA). 2012. "Producer Responsibility Recycling" (June 30). <http://www.sierraclub.org/committees/zerowaste/producerresponsibility/index.asp>.

Slocock, N. 2004. "'Living a Life of Simplicity?' A Response to Francis of Assisi by Adrian House" (May). <http://www.tssf.org.uk/attachments/article/219/Living_a_Life_of_Simplicity.pdf>.

Smith, Julian. 2007. "Space Junk" (August 26). *USA Weekend.*

Smith, M. F. 2001. *Lucretius: On the Nature of Things*. Indianapolis, Indiana: Hackett Pub. Co.

Spacek, Stephen. 2012. "DO MESS WITH IT!: A Sociopolitical Study of Littering and the Role of Southern and Nearby States" (June 24). <https://digital.library.txstate.edu/handle/10877/3763>.

Spaceweather.com (SPACE). 2007. "Rocket Explosion" (February 22). <http://www.spaceweather.com/archive.php?view=1&day=21&month=02&year=2007>.

Spangenberg, J. H.; & S. Lorek. 2002. Environmentally Sustainable Household Consumption: From Aggregate Environmental Pressures to Priority Fields of Action." *Ecological Economics*, 43: 127-140.

Star-tech-inc.com (STAR). 2012. "Space Debris Removal" (June 26). <http://www.star-tech-inc.com/id121.html>.

Stepaniak, Joanne. 2000. *Being Vegan: Living with Conscience, Conviction, and Compassion*. McGraw-Hill Professional.

Stupl, Jan, et al. 2010. "Debris-Debris Collision Avoidance Using Medium Power Ground-Based Lasers. Beijing Orbital Debris Mitigation Workshop, 18–19. <http://swfound.org/media/28344/Stupl-DebrisAvoidanceUsingLasers.pdf>.

Tasaki, T.; T. Kawahata; M. Osako; Y. Matsui; S. Takagishi; A. Morita; S. Akishima. 2006. "A GIS-Based Zoning of Illegal Dumping Potential for Efficient Surveillance." *Waste Manag*, 27 (2): 256-67. <http://www.ncbi.nlm.nih.gov/pubmed/16531036>.

Taylor, K. 2007. "The Year I Saved The World" (August 8). <http://www.nysun.com/arts/year-i-saved-the-world/60056/>.

Than, Ker. 2007. "Rocket Explodes Over Australia, Showers Space with Debris" (February 21). *Space.com*. <http://www.space.com/3493-rocket-explodes-australia-showers-space-debris.html>.

Thinkexist.com (TE). 2012. "William Faulkner" (June 20). <http://thinkexist.com/search/searchquotation.asp?search=trash&page=3>.

_____.2012a. "Isaac Bashevis Singer" (June 20). <http://thinkexist.com/search/searchquotation.asp?search=trash&page=3>.

_____.2012b. "Henny Youngman" (June 29). <http://thinkexist.com/search/searchquotation.asp?search=trash&page=3>.

Thoma, K.; et al. 2006. "New Protection Concepts for Meteoroid/Debris Shields" (April). *Proceedings of the 4th European Conference on Space Debris* (ESA SP-587): 445. <http://www.cosis.net/abstracts/SPACE2005/00155/SPACE2005-A-00155.pdf>.

Tierney, John. 1996. "Recycling Is Garbage" (June 30). *New York Times*. <http://www.nytimes.com/1996/06/30/magazine/recycling-is-garbage.html>.

Trash, Mr. (TRASH). 2012. "Are You Really Interest in Dumpster Diving?" (July 01). *Same Day Dumpsters.*

Union of Concerned Scientists (UCS). 2009. "UCS Satellite Database" (July 16). <http://en.wikipedia.org/wiki/Space_debris#cite_note-60>.

United Nations (UN). 2010. "UN Space Debris Mitigation Guidelines." UN Office for Outer Space Affairs. <http://orbitaldebris.jsc.nasa.gov/library/Space%20 Debris%20Mitigation%20Guidelines_COPUOS.pdf>.

_____.1999. "Technical Report on Space Debris." <http://www.orbitaldebris.jsc.nasa.gov/library/UN_Report_on_Space_Debris 99.pdf>.

_____.1972. "Report of the United Nations Conference on the Human Environment." United Nations Environment Program. <http://en.wikipedia.org/wiki/Simple_living#cite_note-33>.

University of Massachusetts/Lowell (UML). 2012 "Minerals and Forensic Science" (July) Department of Environmental, Earth, & Atmospheric Sciences. <http://faculty.uml.edu/Nelson_Eby/89.215/Assignments/Mineral%20Identifi cation.pdf>.

Unklesbay, Nan. 1992. *World Food and You.* London: Routledge.

U.S. Environmental Protection Agency (EPA). 2010. "Medical Waste Tracking Act of 1988." <http://www.epa.gov/waste/nonhaz/industrial/medical/tracking. htm>.

Vigso, Dorte. 2004. "Deposits on Single Use Containers—A Social Cost-Benefit Analysis of the Danish Deposit System for Single Use Drink Containers. *Waste Management & Research*, 22 (6): 477. <http://wmr.sagepub.com/ content/22/6/477.abstract>.

Walsh, Maureen. 1992. "The Global Trade in Hazardous Wastes: Domestic and International Attempts to Cope with a Growing Crisis in Waste Management." *Catholic University Law Review*, 42: 103–140.

Wang, Jennifer. 2011. "Upcycling Becomes a Treasure Trove for Green Business Ideas" (Marhc 22). <http://www.entrepreneur.com/article/219310>.

Weber, Max. 1930. *The Protestant Ethic and the Spirit of Capitalism.* Talcott Parsons, tr. New York: Scribner.

Weiss, Allie; & Dawn Yukus. 2009. "Freeganism as a Religion." <http://faculty.ncf.edu/hite/Secure_Documents/dawn%20document.pdf>.

Wikipedia (WK). 2012. "Waste" (June 20). <http://en.wikipedia.org/wiki/Waste>.

_____.2012a. "History of Waste Management" (June 20). <http://en.wikipedia.org/ wiki/History_of_waste_management>.

_____.2012b. "Waste Minimization" (June 21). <http://en.wikipedia.org/wiki/ Waste_minimization>.

_____.2012c. "Waste Management" (June 21). <http://en.wikipedia.org/wiki/

Waste_management>.

_____.2012d. "Waste Hierarchy" (June 21). <http://en.wikipedia.org/wiki/Waste_hierarchy>.

_____.2012e. "Hand Me Down" (June 21). <http://en.wikipedia.org/wiki/Hand_me_down>.

_____.2012f. "Reuse" (June 21). <http://en.wikipedia.org/wiki/Reuse>.

_____.2012g. "Recycling" (June 21). <http://en.wikipedia.org/wiki/Recycling>.

_____.2012h. "Waste-to-Energy" (June 21). <http://en.wikipedia.org/wiki/Waste-to-energy>.

_____.2012i. "List of Waste Types" (June 21). <http://en.wikipedia.org/wiki/List_of_waste_types>.

_____.2012j. "List of Waste Management Topics" (June 21). <http://en.wikipedia.org/wiki/List_of_waste_management_topics>.

_____.2012k. "Biowaste" (June 22). <http://en.wikipedia.org/wiki/Biowaste>.

_____.2012l. "Bubonic Plague" (June 22). <http://en.wikipedia.org/wiki/Bubonic_plague>.

_____.2012m. "Extended Producer Responsibility" (June 22). <http://en.wikipedia.org/wiki/Extended_producer_responsibility>.

_____.2012n. "Radioactive Waste" (June 22). <http://en.wikipedia.org/wiki/Radioactive_waste>.

_____.2012o. "Environmental Justice" (June 22). <http://en.wikipedia.org/wiki/Environmental_justice>.

_____.2012p. "Waste Converter" (June 22). <http://en.wikipedia.org/wiki/Waste_converter>.

_____.2012q. "Simple Living" (June 22). <http://en.wikipedia.org/wiki/simple_living>.

_____.2012r. "Freeganism" (June 22). <http://en.wikipedia.org/wiki/Freeganism>.

_____.2012s. "Biodegradable Waste" (June 23). <http://en.wikipedia.org/wiki/Biodegradable_waste>.

_____.2012t. "Clinical Waste" (June 23). <http://en.wikipedia.org/wiki/Clinical_waste>.

_____.2012u. "Medical Waste Tracking Act" (June 23). <http://en.wikipedia.org/wiki/Medical_waste_tracking_act>.

_____.2012v. "Autoclave" (June 24). <http://en.wikipedia.org/wiki/Autoclave>.

_____.2012w. "Litter" (June 24). <http://en.wikipedia.org/wiki/Litter>.

_____.2012x. "Space Debris" (June 25). <http://en.wikipedia.org/wiki/Space_debris>.

_____.2012y. "Basel Convention" (June 25). <http://en.wikipedia.org/wiki/Basel_convention>.

_____.2012z. “Hazardous Waste” (June 25). <http://en.wikipedia.org/wiki/Hazardous_ waste>.

_____.2012aa. “Ethical Consumerism” (June 26). <http://en.wikipedia.org/wiki/Ethical_ consumerism>.

_____.2012bb. “Dumpster Diving” (July 01). <http://en.wikipedia.org/wiki/Dumpster_diving>.

_____.2012cc. “Resource Recovery” (July 02). <http://en.wikipedia.org/wiki/Resource_recovery>.

_____.2012dd. “Thermal Efficiency” (July 03). <http://en.wikipedia.org/wiki/Thermal_efficiency>.

_____.2012ee. “Animal Product” (July 03). <http://en.wikipedia.org/wiki/Animal_product>.

_____.2012ff. “Animal by Products Regulations” (July 03). <http://en.wikipedia.org/wiki/Animal_By-Products_Regulations>.

_____.2012gg. “Rendering (Animals)” (July 03). <http://en.wikipedia.org/wiki/Rendering_(animals)>.

_____.2012hh. “Upcycling” (July 05). <http://en.wikipedia.org/wiki/Upcycling>.

_____.2012ii. “Zero Waste” (July 06). <http://en.wikipedia.org/wiki/Zero_waste>.

_____.2011nn. “Design Elements and Principles” (September 30). <http://en.wikipedia.org/wiki/Design_elements_and_principles>.

_____.2011oo. “Interpretation (Logic)” (April 13). <http://en.wikipedia.org/wiki/Interpretation_(Logic)>.

_____.2011pp. “Unsolved Problems in Linguistics” (February 20). <http://en.wikipedia.org/wiki/Unsolved_problems_in_linguistics>.

_____.2008. “Posthumanism” (October 20). <http://en.wikipedia.org/wiki/Posthumanism>.

_____.2008a. “Transhumanism” (October 20). <http://en.wikipedia.org/wiki/Transhumanism>.

_____.2008b. “Lowest Common Denominator” (March 26). <http://en.wikipedia.org/ wiki/Least_common_denominator>.

_____.2008c. “Infinity” (November 11). <http://en.wikipedia.org/wiki/Infinity>.

_____.2008d. “Transfinite Number” (November 10). <http://en.wikipedia.org/wiki/Transfinite_ number>.

_____.2008e. “Quantum Mechanics” (November 26). <http://en.wikipedia.org/wiki/Quantum_mechanics>.

_____.2008f. “Copenhagen Interpretation” (November 26). <http://en.wikipedia.org/wiki/Copenhagen_interpretation >.

_____.2008g. “Many Worlds Interpretation” (November 26). <http://en.wikipedia.org/wiki/Many_Worlds_Interpretation>.

_____.2008h. "Hyperspace (Science Fiction)" (November 23). <http://en.wikipedia.org/wiki/Hyperspace_(science_fiction)>.

_____.2008i. "Time Travel" (November 26). <http://en.wikipedia.org/wiki/Time_travel#Time_dilation>.

_____.2008j. "Wormhole" (November 25). <http://en.wikipedia.org/wiki/Wormhole>.

_____.2007. "Linguistics" (March 04). <http://en.wikipedia.org/wiki/Linguistics>.

Wikiversity (WKV). 2008. "Why 10 Dimensions" (November 23). <ttp://en.wikiversity.org/wiki/Why_10_dimensions>.

Wilderness Survival Techniques (WST). 2011. "The Do's and Don'ts of Dumpster Diving" (September 22). <http://wildernesssurvivaltechniques.com/the-dos-and-donts-of-dumpster-diving/>.

Willhite, Nikki. 2009. "Dumpster Diving." *All Things Frugal*. <http://www.allthingsfrugal.com/dumpster.htm>.

Wittgenstein, Ludwig.1953. *Philosophical Investigations*. G. E. M. Anscombe and R. Rhees, eds. G.E.M. Anscombe, tr. Oxford: Blackwell.

Wolsink, M. 1994. "Entanglement of Interests and Motives: Assumptions behind the NIMBY-Theory on Facility Siting." *Urban Studies*, 31.6: 851-866.

Worsham, Julia. 2009. "Disparate Impact Lawsuits under title VI, Section 602: Can a Legal Toul Build Environmental Justice?." (September 15).*Boston College Law Review*.

Wright, David, 2997, "Debris in Brief: Space Debris from Anti-Satellite Weapons" (December). Union of Concerned Scientists. <http://www.ucsusa.org/assets/documents/nwgs/debris-in-brief-factsheet.pdf>.

INDEX

•A•

•B•

•C•

•D•

•E•

•F•

•G•

•H•

•I•

•J•

•P•

•Q•

•R•

•S•

•T•

•V•

•W•

•X•

•Y•

•Z•